ECOLOGICAL ISOLATION
IN BIRDS

Ecological Isolation
in Birds

DAVID LACK
F.R.S.

MAIN ILLUSTRATIONS BY
ROBERT GILLMOR

HARVARD UNIVERSITY PRESS

CAMBRIDGE, MASSACHUSETTS

1971

Library of Congress Catalog
Card Number 70–151286

SBN 674 22442 6

First published 1971

Printed in Great Britain

Contents

Text Figures   ix

1 Introduction   1

2 A Complex Example of Coexistence   18
The European species of tits *(Parus)*

3 A World Survey   38
Tits *(Parus)* in Asia, Africa and America

4 Another World Survey   63
Nuthatches *(Sitta)*

5 Coexistence in a Man-modified Habitat   74
The European finches Fringillidae

6 Migratory Birds with Two Homes   93
The European trans-Saharan migrant passerines

7 Review of a Continental Avifauna   112
The other European birds

8 North American Studies   126
Passerines, hawks, peeps, auklets

9 African Studies   138
Usambara, turacos, brood parasites, vultures

10 Other Tropical Studies   155
Fruit-eaters, honey-eaters, suburbs, seabirds

CONTENTS

11 Adaptive Radiations in Archipelagoes   174
    Galapagos finches and Hawaiian sicklebills

12 A Widespread Family on Remote Islands   200
    The White-eyes Zosteropidae

13 A Tropical Archipelago   222
    The West Indies

14 Conclusions   243
    Range, habitat, food, consequences

    Appendixes

    1   Tits *Parus* on Formosa and the Philippines.   271
    2   The species of *Parus* in the mountains of central Asia.   272
    3   Main means of ecological segregation of *Parus* species
        in mountains of central Asia.   273
    4   The *Parus* species of Africa south of the Sahara.   274
    5   Ecological segregation of African species of *Parus*.   276
    6   The North American species of *Parus*.   276
    7   Main means of segregation of North American species
        of *Parus*.   278
    8   Ecological segregation of the European and North
        American species of *Sitta*.   279
    9   Nuthatches *Sitta* of the mountains of central and
        southeast Asia.   280
    10 Probable means of ecological segregation of *Sitta*
        species in the mountains of central and southeast
        Asia.   282
    11 The European finches *Fringillidae*.   283
    12 Main means of ecological segregation in congeneric
        European finches.   285
    13 Summer and winter ecology of congeneric European
        passerine migrants to Africa.   286
    14 Main means of ecological segregation of congeneric
        European transequatorial migrant passerine birds in
        their summer and winter homes.   301
    15 Ecological isolation in the other congeneric European
        passerine birds.   304
    16 Ecological isolation in other (non-passerine) European
        birds.   310
    17 Ecological isolation of congeneric species in selected
        families in Usambara, Tanzania.   322

CONTENTS

18 Habitat restriction of forest and woodland birds in
   selected families in Usambara, Tanzania.   325
19 Habitat restriction in European forest species.   326
20 Geographical and ecological isolation in congeneric
   turacos Musophagidae.   328
21 Geographical and ecological isolation in congeneric
   birds of paradise and bower birds.   329
22 Differences in feeding and morphology in tanagers in
   Trinidad.   332
23 Food species recovered from Ascension Island seabirds.   333
24 Measurements of congeneric species of Hawaiian
   sicklebills Drepanididae.   334
25 Ecological isolation in congeneric Drepanididae in
   Hawaiian Islands.   336
26 Ecological Isolation in White-eyes Zosteropidae.   336
27 Ecological isolation in congeneric West Indian
   passerine birds.   344
28 Ecological isolation in congeneric non-passerine West
   Indian land birds.   349
29 Habitat of forest passerines in three West Indian
   islands.   352
30 Tentative survey of ecological isolation in congeneric
   passerine birds in other tropical archipelagoes.   354

References   357

Index of Birds and other Animals   375

General Index   397

Text Figures

1 Changes in avifauna when Breckland heaths were planted with Scots Pine. 4
2 Morphological differences between warblers in three genera *(Sylvia, Phylloscopus, Hippolais)* and waders in one *(Tringa)*. By Robert Gillmor. 15
3 Feeding stations of Blue Tit, Marsh Tit and Great Tit in oak woodland. By Robert Gillmor. 23
4 Feeding stations of Coal Tit, Crested Tit and Willow Tit in pine forest. By Robert Gillmor. 29
5 Ranges of Blue Tit *P.caeruleus* and Azure Tit *P.cyanus*. 31
6 Typical Blue Tit *P.c.caeruleus* of broadleaved forest and thin-beaked Canarian form *P.c.teneriffae* of pine forest; typical Coal Tit *P.a.ater* of conifer forest and thick-beaked Persian form *P.a.phaeonotus* of oak woods. By Robert Gillmor, adapted from David Snow. 34
7 Range of Great Tit *Parus major* superspecies. 40
8 Three Himalayan tits related to Coal Tit. By Robert Gillmor, adapted from David Snow. 43
9 Ranges of tits in *P.afer* superspecies. 46
10 Ranges of tits in *P.niger* superspecies. 47
11 Ranges of tits in *P.rufiventris* superspecies and *P.funereus*. 48
12 Six North American tits and their European counterparts. By Robert Gillmor. 51
13 European Nuthatch *Sitta europaea* and Corsican Nuthatch *S.whiteheadi*. By Robert Gillmor. 64
14 Ranges and beak-length of Western Rock Nuthatch *Sitta neumayer* and Eastern Rock Nuthatch *S.tephronota*. 67
15 Ranges of 6 species of nuthatches *Sitta* in Himalayas, N. Burma and S.W. China. 69
16 Spread of Serin *S.serinus* from southern Europe. 75
17 British finches feeding in winter. By Robert Gillmor. 79

TEXT FIGURES

18 Northward shift of boundary between northern Brambling *Fringilla montifringilla* and southern Chaffinch *F.coelebs* in Finland. 90

19 The 5 European hirundines at their nests. By Robert Gillmor. 95

20 Ranges of Great Grey Shrike *Lanius excubitor* and Lesser Grey Shrike *L.minor*. 98

21 Summer and winter ranges of Redbacked Shrike *Lanius collurio* and Woodchat *L. senator*. 99

22 Ranges in summer of European *Hippolais* warblers. 101

23 Ranges in winter of European *Hippolais* warblers. 102

24 Summer and winter ranges of the two European nightingales. 107

25 Partly overlapping ranges of the two European treecreepers. 113

26 Three species of *Falco*, a Hobby pursuing martins, a Kestrel hovering above a vole and a distant Peregrine stooping on a pigeon. By Robert Gillmor. 119

27 Ranges of *Alectoris* partridges in western Palaearctic. 120

28 Feeding of dabbling ducks in England in winter. By Robert Gillmor. 123

29 Main feeding zones in spruce of 5 coexisting North American *Dendroica* warblers. 131

30 Ranges of turacos in genus *Tauraco*. 146

31 Black, Griffon and Egyptian Vultures round a carcase. By Robert Gillmor. 153

32 Ranges of birds of paradise in genus *Paradisaea*. 157

33 Heads of *Melithreptus* species in Australia. 159

34 Southern limit of birds of open country spreading south from Burma. 163

35 Beaks of five tropical terns. By Robert Gillmor. 167

36 Map of Galapagos. 175

37 Heads of three monotypic genera of Darwin's finches. 176

38 Heads of ground finches *Geospiza* on central Galapagos islands. 179

39 Variations in beak-depth of small, medium and large ground finches on different islands. 181

40 Resemblance of *G.difficilis* on Tower to *G.fuliginosa*, and on Culpepper to *G.scandens*. 185

41 *G.conirostris* on Tower and Hood compared with *G.scandens* and *G.magnirostris*. 185

42 Heads of insectivorous tree finches *Camarhynchus* in central Galapagos. 187

TEXT FIGURES

43 Heads of insectivorous tree finches on outlying islands.   191
44 Character displacement in *Loxops*.   195
45 Geographical replacement in *Hemignathus*.   196
46 Malay archipelago and adjoining islands with number of species of white-eyes Zosteropidae on each.   204
47 Islands in Western Pacific with number of species of white-eyes Zosteropidae on each.   207
48 The five main islands of the Carolines.   208
49 The white-eyes of the Caroline Islands. By Robert Gillmor.   209
50 Map of the West Indies.   223
51 Ranges of crested flycatchers *Myiarchus* in West Indies.   225
52 The hummingbird genera of the West Indies. By Robert Gillmor, based on plates by D. Eckleberry in Bond 1960.   229
53 Geographical replacement of small hummingbirds in West Indies.   231
54 Ranges of larger hummingbirds in West Indies.   232
55 Summer and winter range of Kirtland's Warbler *Dendroica kirtlandi*.   250
56 Westward limit of breeding range of Waxwing *Bombycilla garrulus* in years of differing abundance.   251
57 Foot of Meadow Pipit and Tree Pipit. By Robert Gillmor.   254
58 Beak difference between male and female Huia *Heteralocha acutirostris*.   262

Chapter 1

Introduction

Two species of animals can coexist in the same area only if they differ in ecology. Such ecological isolation, brought about through competitive exclusion, is of basic importance in the origin of new species, adaptive radiation, species diversity and the composition of faunas, as touched on in my final chapter. The chief aim of this book is to develop the theme of ecological isolation itself, my examples being from birds because, for a critical evaluation, it is essential to 'know one's animal'. Anyway the first, and for a long time the only, field studies on ecological isolation were carried out on birds. But this state of affairs is being improved, so I hope that the ideas put forward here will be of interest not only to ornithologists, but to ecologists working on other kinds of animals. While I have surveyed previous research, I have also analysed various groups of birds not treated in this way before.

It is rather easy to sit in a study devising ecological principles and then searching for proofs of them in books on natural history. This is a modern equivalent of the Devil quoting scripture for his purposes, because nature, like the bible, is so diverse that diligent search will yield examples in support of any principle. The only defence against self-deception is to seek comprehensively, not selectively. Hence while my main interest in writing this book has been in the underlying ideas, I have also aimed to provide a comprehensive survey. Since, however, there are about 8500 species of birds in the world, most of which have not been studied, a review of all birds would be impossible. The groups treated here are simply those few, and all of those, about which enough is known for ecological analysis. Such analysis requires full documentation, and I hope that other ornithologists will be as fascinated as I was in making this search, to see how general trends emerge and, once recognized, how patterns repeat themselves, with revealing variations and exceptions. But though the details will provide the main point of the book for ornithologists, they may be of less interest to workers on other animals, so after some full examples in chapter 2, much of the information is set out in a series of appendices (starting p. 271) and the discussions can be read without them.

The chance is negligible that two species will be equally well adapted, and competitive exclusion follows as a logical consequence. Hence if they coexist, they presumably differ in ecology, at least with respect to those resources of the environment which determine their numbers. Since the numbers of most birds are probably limited by food (Lack 1966), two bird species which live in the same area may be expected to differ either in their feeding or by occupying separate habitats, while if they have similar ecology, they can persist only in separate areas. It accords with this view that closely related species of birds normally differ from each other in range, habitat or feeding, though the situation need not be so simple as this bald summary suggests.

Historical survey

Steere (1894) wrote that 'no two species structurally adapted to the same conditions will occupy the same area' (quoted by Rand and Rabor 1960). No notice was taken by others of this statement, which anyway referred to island birds, most of which would now be treated as subspecies of the same species. Grinnell (1904, 1917a, 1928, 1943) was the first to state explicitly the principle of competitive exclusion (as pointed out by Udvardy, 1959), but he did so incidentally in studies of particular birds, which is presumably why its general importance was not appreciated at the time. The necessity for competitive exclusion was also contained in the mathematical equations of Lotka and Volterra, on the basis of which the Russian worker G. F. Gause (1934) set up model populations in the laboratory to measure the changes in numbers of two species of microorganisms which ate the same limited food supply. He showed that, under these circumstances, one species eliminated the other, and was fully aware of the importance of this principle in nature, illustrating it with examples from birds and mammals. Hence when I concluded that ecological isolation is critical for understanding the distribution of Darwin's finches, the Geospizinae of the Galapagos, I referred to 'Gause's principle' or hypothesis (Lack 1944, 1947), and his name remained attached until it was appreciated that Grinnell had earlier said the same, after which it has been termed 'competitive exclusion' (Hardin 1960).

Grinnell (1917a, 1924, 1928, 1943) also coined the term 'ecological niche', but he again published it with such seeming casualness that others did not notice it. It was independently invented, with a similar but slightly different meaning, by Elton (1927), through whom it came into general use for the particular part of a habitat occupied by a species, especially in relation to where it feeds. The term 'feeding niche' has sometimes been used instead, but there seems no need for it. Although 'ecological niche' is hard to define precisely, and has often been used vaguely, its value is shown when one attempts to find an alternative term or phrase, and I doubt that it would be valuable to try to define it rigorously.

My own interest in the factors limiting habitat distribution in birds was aroused in 1931 when, as an undergraduate at Cambridge University, I was invited by Dr A. S. Watt of the Botany School to join his group studying the changes in plant and animal life due to the afforestation of the

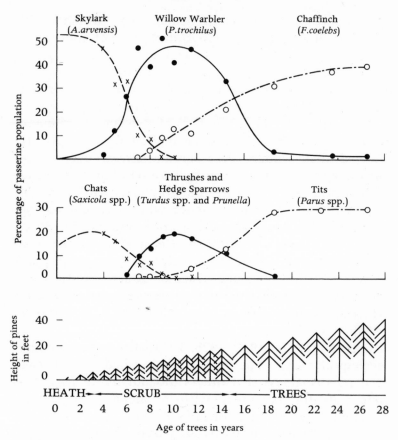

Fig 1. Changes in avifauna when Breckland heaths were planted with Scots Pine (from Lack 1954).

Breckland heaths in eastern England. The changes in bird-life as the pine trees grew up, summarised in fig. 1, were so rapid that I appreciated that they could not be the direct result of the factors at that time thought important in determining the habitat of each species. I therefore postulated that each species selects its habitat by psychological or recognition factors, which are not necessarily those essential to its existence (Lack 1933, 1934; 1937, 1940). This concept of 'habitat selection', which aroused opposition at the time, proved not to be new, being hinted at by Brock (1914) and (once again) Grinnell (1917b), and clearly stated by Sunkel (1928) (see also review by Hilden 1965).

In my paper of 1933, I went on to ask why each species should select a different habitat, and wrongly postulated that differences in habitat selection might themselves lead to the evolution of new species. It will be recalled that, at that time, only the first few pieces of evidence had been collected, by Rensch (1928, 1929, 1933), that new bird species arise from geographically isolated subspecies, and the synthesis by Mayr (1942) was nine years in the future. Nor had Baker (1938) yet clarified biological interpretation by his idea, first applied to avian breeding seasons, that explanation is needed at two levels, that of the 'ultimate factors' concerned with survival value and that of the 'proximate factors' concerned with physiological or psychological means. It is now evident, but was not in 1933, that the psychological factors by which a bird selects its habitat are 'proximate' and leave the evolutionary factors out of account.

Further, it was almost universally believed at that time that the differences between closely related species are non-adaptive, ánd that adaptive differences arise only at the level of the genus, or higher. The idea, in particular, that differences in habitat between congeneric species might be correlated with adaptive differences between them was quite contrary to biological opinion (see, for instance, Robson and Richards 1936). The differences between subspecies of the same species were likewise held to be non-adaptive. The biologists who, in the years just after 1945, attacked the 'Sewall Wright effect' were evidently unaware that Wright put forward his mathematical theory to explain what systematists of the time accepted as fact, that both conspecific subspecies and congeneric species differ from each other in non-adaptive features.

The view that new species of birds arise solely from geographically isolated subspecies was firmly established by

Mayr in his book of 1942 and gradually gained acceptance. In the same year Huxley (1942) postulated that big size differences between congeneric species of birds are a means of ecological isolation. When reconsidering the evolution of Darwin's finches in 1943, I combined these views and postulated that, given two well-marked subspecies with sufficient genetic differences not to interbreed freely, they could persist in the same area only if they also differ sufficiently in ecology for one not to eliminate the other through competition. Their ecological differences might be small when they first meet, but since those individuals with such differences will tend to survive better than those which lack them, they will be intensified by natural selection until the two species no longer compete effectively for essential resources. This explained the otherwise puzzling point that though closely related species of birds arise only through geographical isolation, they often occupy separate habitats. Having formulated this idea in Darwin's finches, I tested it for British passerine birds, and showed that it also holds in this more normal group of birds, as well as in all the closely related pairs of species on islands that had been cited by Mayr (1942) as evidence for the origin of a new species through two successive invasions of the same mainland stock (Lack 1944).*

*The human mind works curiously. I produced this idea slowly, indeed painfully, over a period of more than half a year, thinking late at night, and virtually on my own. Yet in 1939 I had read, and immediately rejected, a single sentence by Gause (1939) on the necessity for ecological isolation, and in 1942 I had read, but completely forgotten, Huxley's view on the significance of size differences in birds. Between them these ideas provided the essence of what I had come to think of as 'my' view. Luckily I recalled Gause's statement before completing my manuscript, while Huxley drew my astonished attention to his view on size differences when I showed him a draft prior to publication. I should add that my book on Darwin's finches of 1947 was written just before my paper on ecological isolation in the *Ibis* for 1944, the delays in war-time book production being responsible for their appearing out of sequence. At the time of writing, the species-pairs cited by Mayr as evidence for the double invasion of islands were unknown to me, and I still recall the

Later (1946), I analysed the recorded prey of European raptors, and found that each species differs from every other in its main prey, except for those which take chiefly lemmings. This led me to the idea that if a food supply is 'temporarily superabundant', so that it is not significantly reduced by the species preying on it, several different species can take it without effectively competing. The lemming predators come in this category, but in the years when lemmings are scarce, each species turns to different types of prey from every other. A superabundant food must, however, be only temporary, since otherwise the numbers of the birds preying on it will tend to rise until there is effective competition for it. Other good examples are seasonal fruits, seed crops, and flying termites, which are preyed upon when in season by a variety of passerine and other birds.

When I put forward 'Gause's hypothesis' at a meeting of the British Ecological Society in 1944, hardly anyone accepted it, except G. C. Varley, with whom I had discussed it during the previous year, and Charles Elton (1946) who, working independently, had found that nearly all the species present in a particular habitat are in different genera. This latter view was challenged by Williams (1947); its validity in fact depends on the definition of a habitat (Bagenal 1951) and on whether a broad or a narrow concept of the genus is used, the latter point being illustrated from passerine birds later in this book.

At that time, the chief criticism raised by ecologists was that, to their knowledge, closely related species often live in the same habitat and have similar ecology. But gradually more and more of these apparent exceptions were shown to be illusory, refined analysis indicating that the species in question differ substantially in some aspect of their ecology.

excitement when, in a hunt through the scattered literature during brief spells of leave at the British Museum (Natural History), pair after pair proved either to occupy separate habitats, or to live in the same habitat but to differ markedly in size.

One such case was that of the Cormorant *Phalacrocorax carbo* and Shag *P.aristotelis*, which fish in the same coastal waters, but hunt below the surface at different levels for mainly different types of prey (Lack 1945a). However, a species whose numbers are not limited by food need not depend on a different food supply from other species. This may be why several species of phytophagous insects may be found on the same leaf; but where this happens, they will doubtless have different species of insect parasites if their numbers are limited by the latter.

Critics then shifted to the view that, though related species often differ in their ecology, this is unimportant and due, not to past competition, but to incidental differences evolved in their period of isolation from each other. For instance, Andrewartha and Birch (1955, pp. 463, 464) wrote: 'the difficulty in this hypothesis (i.e. Lack's) is that, by the very nature of the case, it can hardly be proved or disproved, because we have no evidence that "competition" took place in some past epoch.' Again, 'We are forced to conclude that his interesting results do not in any way demonstrate that "competition" between birds in nature is at all commonplace or usual. On the contrary, his results seem to show that it hardly ever occurs.'

Andrewartha and Birch were right, of course, in thinking that competition between closely related species of birds is uncommon today, but this is not because it is unimportant, but because it is so important that, as already mentioned, the species concerned have evolved differences in range, habitat or feeding habits to mitigate it. If a species will be much more successful in one area or one type of habitat than another, or by taking one type of food than another, there will be strong selection for behaviour by which it favours the one and avoids the other. Nevertheless such behaviour must be reinforced by competition and natural selection at the present day, since otherwise it would disappear. There are sufficient records of a species breeding outside its normal range or habitat, and within that of another species,

8

for one to suppose that this is happening. Moreover, if Andrewartha and Birch had been correct in supposing that differences in habitat are due merely to chance, one would have expected far more overlap between related species than what is found in nature.

Evidence for competitive exclusion

The best circumstantial evidence that ecological differences are due to competition is provided by the instances in which they are restricted to a small area where another species is present. For instance, throughout its extensive range in Europe and North Africa, the Chaffinch *Fringilla coelebs* breeds commonly in both broadleaved and coniferous forest, except on the islands of Gran Canaria and Tenerife, where it breeds in broadleaved but not pine forest, in the latter being replaced by the similar Blue Chaffinch *F.teydea* (Lack 1944). Admittedly, as Andrewartha and Birch said, 'so far as the case is stated, there is no direct evidence that the two species could not live apart if they were put together'. But it would be stretching coincidence too far to suppose that it is just through chance that the only part of its range where *F.coelebs* does not breed in the available pine forests is also the only part where another species of chaffinch is present. Again, two species of nightingales breed in thickets in Europe, *Luscinia megarhynchos* in the west and southwest and *L.luscinia* in the east, and they replace each other geographically except in part of eastern Europe; here, and only here, *L.megarhynchos* is restricted to drier and *L.luscinia* to wetter habitats of the same general type. As already mentioned for Darwin's finches, another means of ecological isolation is by body-size and size of beak. Hence it is significant that the big difference in size between the two rock nuthatches *Sitta tephronota* and *S.neumayer* is restricted to the area in Persia where they coexist, whereas in both the east and west of Persia, where they replace each other geographically, they are similar in size (Vaurie 1951). This last example was termed 'character displacement' by Brown and Wilson (1956).

Finally, instances are known in which one species has replaced another. One must expect there to be few, because if one species can displace another this is likely to have happened already, and in fact, the only recorded instances in birds have occurred in habitats modified in recent times by man. Even here, all that has been observed is 'replacement', and that it is due to 'competitive displacement' is an inference, for direct competition has not been seen. To cite only one instance, rather over a century ago, the Rock Sparrow *P.petronia* bred in both the towns and the countryside of the central and western Canary Islands, but during the last hundred years the Spanish Sparrow *Passer hispaniolensis* has established itself there, after which the Rock Sparrow disappeared from the towns (though not the countryside), presumably because displaced (Cullen et al. 1952).

The principle of competitive exclusion implies that those species which differ primarily in range differ in their adaptations to their respective ranges, that those which differ primarily in habitat differ in their adaptations to their respective habitats, and that those which differ primarily in their foods differ in their feeding adaptations. Earlier, as already mentioned, the differences between closely related species were thought to be non-adaptive, but evidence is now accumulating to the contrary. To take only one example of each type, (i) Brünnich's Guillemot *Uria lomvia* replaces the Common Guillemot *U.aalge* in the far north, and it follows the general (and presumably adaptive) trend in warm-blooded animals for larger size in colder climates; (ii) the Tree Pipit *Anthus trivialis* has a shorter hind claw than the Meadow Pipit *A.pratensis*, in parallel with a general (and presumably adaptive) difference between arboreal and terrestrial birds respectively; and (iii) the smaller Blue Tit *Parus caeruleus* takes smaller prey, and is also more agile in feeding at the tips of twigs, than the larger Great Tit *P.major*, which feeds lower down; and in general, smaller birds feed on thinner twigs and take smaller prey than larger congeners.

The three types of isolation

The three main ways in which birds are segregated from each other, by range, habitat and feeding, are broadly, but not completely, separable. Thus it is normally clear when two species differ in geographical range, but no two areas provide identical habitats or foods, so a difference in range may be associated with differences in ecology, but the latter may be unimportant. Further, a particular habitat, for instance brachystegia woodland in Africa, may be found mainly in one area, so that a species confined to it may inevitably differ in range from a related species which lives in, say, acacia savanna. A borderline case concerns species which differ in altitudinal range. Since the range of a montane species is often discontinuous on isolated mountains, and since a difference in altitude is normally linked with one in habitat, such cases have been grouped under habitat, not range.

Habitats are here considered to be broadly in a horizontal plane, e.g. desert, marshland or forest (and their subdivisions). Where two species differ in their vertical feeding zones, for instance high or low in the same trees, they are here classified as differing in feeding stations, not habitat. The importance of vertical feeding stations in ecological isolation was first pointed out by Colquhoun (1941, and with Morley 1943). Finally, where two species living in the same habitat differ markedly in size, and particularly in size of beak, they have here been assumed to differ in their feeding, as this has been established in many instances, and there are no known exceptions.

Ecological isolation is most clearly seen in primaeval habitats, to which the species in question have become adapted over a long period. Through cultivation, and the associated clearance of forest and drainage of marshes, man has greatly modified the natural scene, long ago in much of Europe and Asia, more recently elsewhere, and most of the new types of habitat are unstable and liable to change with changing agricultural practice. Especially where such changes are recent, birds may not be properly adapted to them. Here, therefore, ecological isolation may be incomplete and feeding adaptations to the primeval habitat may not be used, so that the situation built up through natural selection over a long period may be obscured. At the same time, changes in bird life which have occurred under these conditions provide different problems, and in particular, they help to show the dynamic nature of ecological isolation between species.

Classification

Ecological analysis must rest on sound systematics, and I have not discussed in this book any family or genus in which there are serious doubts about the correct classification. In two of the included groups, nuthatches Sittinae and white-eyes Zosteropidae, there has until recently been much dispute, but I support the findings of the latest revisers, Greenway (1967) and Mees (1957, 1961, 1969) respectively. In both these groups, as also in swifts *Apus* (Lack 1955), confusion has been caused because two subspecies of one species may differ in appearance to a greater extent than certain full species. For a biological interpretation, it does not much matter if subspecies are treated as species, but it matters greatly if they have been allocated to the wrong species, as has sometimes happened in the past, through the difficulty of distinguishing resemblances due to affinity and convergent evolution respectively.

Because competition is likely to be most severe between similar species, this book is concerned primarily with ecological isolation between congeneric species. Difficulty over the delimitation of species arises chiefly with respect to geographically replacing forms. These are normally regarded as subspecies of one species if they freely interbreed, but as full species if they rarely do so (Sibley 1961). Inevitably, however, there are borderline cases, and it is hard to know how to treat well-marked island forms which are more similar to each other than to any other forms, but differ to an extent comparable with that of mainland species, and do not interbreed because they do not meet. Where such forms are distinctive, they are usually classified as full species but are put in one superspecies. If there is any doubt, I have treated them here as species, so as to ensure that all potentially competing forms are considered.

Restriction of the discussion to congeners also presupposes that avian taxonomists have been correct in their use of genera. But while there may be doubts in a few cases, it is safe to say that, in general, bird species which are nowadays put in the same genus are more closely related to each other than to any put in other genera. In addition, though individual ornithologists used formerly to delimit genera rather, and sometimes very, differently from each other, there is much greater agreement now than in even the recent past. Where there is any serious doubt, I have here used broad genera, so as to make sure that all closely related species are included.

At the end of each main section, I have scored the proportion of congeneric species separated respectively by range, habitat or feeding. The figures depend to some extent on the classification used. Thus my tendency, where there is doubt, to treat distinctive geographical forms as species rather than subspecies increases somewhat the proportion of species separated by range, especially in island birds. Similarly my tendency, where there is doubt, to use broad rather than narrow genera increases somewhat the proportion of congeneric species separated by feeding, since the narrower genera of certain other workers are usually based on minor morphological characters adapted to differences in feeding. A further difficulty arises if comparisons are made between different families of birds, because the genus is not an equivalent ecological unit in all groups, and in particular it usually rests on smaller morphological differences in passerine than other birds. This is not because the systematists who have worked on passerine and other birds respectively use different taxonomic standards, but because the criteria for generic separation include not only morphological differences, but the extent to which there are clearcut gaps between groups of species.

For instance, as illustrated in fig. 2, European warblers in the genera *Sylvia, Hippolais* and *Phylloscopus* respectively are more like each other in morphological characters, including body-size and shape of beak, than are the wading birds put in the single genus *Tringa*. The species in each of the three warbler genera are clearly separated from each other by minor differences. But while the smallest and largest *Tringa* species are more different from each other than are any of these warblers, there are species of intermediate appearance at any level at which one might subdivide them into two or more groups, hence they have to be put in one genus. As a result, more species of *Tringa* than of congeneric warblers are segregated by feeding, but if the genera *Sylvia, Hippolais* and *Phylloscopus* had been merged, the proportion of congeneric warblers separated by feeding would have been much increased. Again, one of the few passerine genera in which many species are segregated from each other by feeding is that of the tits *Parus*, but formerly this genus was divided into *Parus (sens. strict.)*, *Cyanistes, Periparus, Lophophanes, Poecile* and *Baeolophus*, and if these had been retained, most instances of congeneric species segregated by feeding would disappear, as will be seen in chapters 2 and 3. Similarly the dabbling ducks in the

Fig 2. Morphological differences between warblers in three genera *(Sylvia, Phylloscopus, Hippolais)* and waders in one *(Tringa)*. By Robert Gillmor.

Willow Warbler
Phylloscopus

Garden Warbler
Sylvia

Icterine Warbler
Hippolais

Spotted Redshank

Greenshank

Redshank

Common Sandpiper

Wood
Sandpiper

genus *Anas*, nearly all of which are separated by feeding, were formerly put in *Anas (sens. strict.)* for the mallards, *Mareca* for the wigeons, *Nettion* for the teals, *Dafila* for the pintails, and *Spatula* for the shovelers, and if these genera were retained, few congeners would be separated by feeding, indeed there would be few congeners. Hence care is needed in interpreting the quantitative differences between different groups of birds with respect to the means of ecological isolation between congeners, and this particularly affects differences between passerine and other birds.

For nomenclature, I have usually followed Vaurie (1959, 1965) for European species, and Peters' *Check-List of Birds of the World* for other groups, but parts of 'Peters' are agreed to be out of date, and others have not yet appeared, so in some chapters I have followed other authorities, as there stated. At its first mention in each chapter, and in most tables, each species is denoted by its vernacular and scientific names, except that only a group vernacular name is used where there is no genuine one for the species. After its first mention, solely the vernacular name is used for European, North American and some other species, and solely the scientific name where the vernacular name is the artificial construction of a museum worker that is rarely used in practice. Although inconsistent, this treatment probably makes the easiest reading for English-speaking ornithologists.

Acknowledgements

This book could not have been written without the field observations of many amateur naturalists, or without the compilers of regional avifaunas who made their observations accessible. In addition, I am extremely grateful to the following specialists who read particular sections of the book in manuscript, and whose criticisms led to much revision: Dr David Snow for chapters 2 and 3, especially on European and Asian tits, and for the section on frugivorous birds in chapter 10; Mrs B. P. Hall for the section on African tits and Dr Keith Dixon for the section on American tits in chapter 3; Dr I. Newton for chapter 5 on finches; R. E. Moreau for chapter 6, especially for the winter ranges in Africa of the European migrants, also for the sections on Usambara birds and turacos in chapter 9; Dr J. Blondel for information on *Sylvia* warblers in Southern France in chapter 6 and Dr O. Hilden for information on Fenno-Scandian passerine birds in chapters 6 and 7; Dr Robert MacArthur for the section on *Dendroica* warblers in chapter 8 and those on species diversity in chapters 9 and 13; Dr Gordon Orians for a discussion of American black-birds, considered in chapter 8; Dr Hans Kruuk and David Houston for the section on vultures in chapter 9; Dr Allen Keast for *Melithreptus*, Dr Peter Ward for Singapore birds, and Dr Philip Ashmole for tropical seabirds, all in chapter 10; Dr G. F. Mees for chapter 12 on white-eyes, Dr and Mrs Cameron Kepler for chapter 13 on the West Indies, Professor R. A. Hinde for the section on how feeding differences might be acquired in chapter 14, and R. E. Moreau, Dr Ian Newton and Dr C. M. Perrins for the whole of the last chapter; finally to Robert Gillmor for his, as always, admirable drawings.*

Various authors and journals kindly gave permission to reproduce published figures, as follows: the Clarendon Press, Oxford for figs. 1 and 18, Dr David Snow for figs. 5–8 inclusive from his D.Phil. thesis, Mrs Hall for figs. 9, 10 and 11, Dr Charles Vaurie for fig. 14, Dr Dillon Ripley for fig. 15, Dr Ernst Mayr for fig. 16, Dr Robert MacArthur and *Ecology* for fig. 29, Dr Allen Keast and *Evolution* for fig. 33, Dr Peter Ward for fig. 34, the California Academy of Sciences for figs. 37, 38 and 40–43 inclusive, *British Birds* magazine for fig. 56, H. F. and G. Witherby and Co. for fig. 57, and the Cambridge University Press for fig. 58.

*While this book was in press, R. E. Moreau died. I owe him an immense debt, because he read in manuscript every book and paper that I wrote over a period of more than twenty years, and he was unexcelled as a critic.

Chapter 2

A Complex Example of Coexistence
The European Species of Tits *(Parus)*

It may be helpful to start this survey by considering in detail a genus which has a wide distribution in the world and has been well studied in parts of its range. For this purpose, the tits of the genus *Parus* are particularly suitable, as they have been the subject of detailed research at the Edward Grey Institute at Oxford since the war. However, they were selected for study owing to their apparent complexity, and so might not be typical of birds as a whole. Tits are small passerine birds which occur mainly in forest of various types, and also in wooded cultivated land, throughout Europe, continental Asia, Africa and North America. They have short beaks and hunt for food chiefly on leaves and twigs, but at times on the ground, eating insects throughout the year, also seeds in winter, and they nest in holes, normally in trees. The mean winglength varies from 59 to 77 mm, so the largest is a little smaller than a House Sparrow *Passer domesticus*. Most of them are brownish grey, some bluish green, above, and white, in some species yellow, below. Many of them have black or other striking markings on the head.

Differences in range and habitat

This chapter is concerned with the nine European species, six of which are widespread. Ranges and habitats (based here and in the rest of the chapter on Snow, 1953) are summarized in table 1. The six widespread species can

Table 1. The European species of *Parus*

Vernacular name	Species	Subgenus	European range and usual habitat
Coal Tit	*P.ater*	*(Periparus)*	Widespread except far north; conifers, especially spruce
Blue Tit	*P.caeruleus*	*(Cyanistes)*	Widespread except far north; broadleaved woods, especially oaks
Siberian Tit	*P.cinctus*	*(Poecile)*	Solely in far north; coniferous taiga
Crested Tit	*P.cristatus*	*(Lophophanes)*	Widespread except far north, most of Britain and Italy; conifers
Azure Tit	*P.cyanus*	*(Cyanistes)*	Eastern part of central European U.S.S.R.; broadleaved woods, especially riverine
Sombre Tit	*P.lugubris*	*(Poecile)*	Solely the Balkans; broadleaved woods
Great Tit	*P.major*	*(Major)*	Widespread except extreme north; especially broadleaved woods, also open pine
Willow Tit	*P.montanus*	*(Poecile)*	Widespread in northern and central, not southern, Europe; in north especially in birch and mixed forest, in centre in broadleaved riverine woods or montane conifers
Marsh Tit	*P.palustris*	*(Poecile)*	Widespread in middle latitudes, absent from most of Fenno-Scandia, much of U.S.S.R., most of Iberia and Mediterranean seaboard; broadleaved woods.

Note: based on Snow (1953)

occasionally be found breeding in the same mixed broad-leaved and coniferous wood, for instance in central Sweden (Snow 1949). Indeed, five of them breed in the Edward Grey Institute's study area in Marley Wood, a pure broadleaved wood near Oxford. But usually only three or four species are found in one wood, because two are normally restricted

Table 2. Typical weights (in grams) and measurements (in mm) of males of European *Parus* species

	Weight	Wing	Culmen from base	Beak-depth	Culmen ÷ wing	Depth ÷ culmen
ENGLAND						
P.ater (Coal)	9·3	60	10·8	3·7	0·18	0·34
P.caeruleus (Blue)	11·4	64	9·3	4·4	0·145	0·47
P.major (Great)	20·0	76	13·0	5·2	0·17	0·40
P.montanus (Willow)	10·1	60	10·6	4·0	0·18	0·38
P.palustris (Marsh)	11·4	63	10·4	4·3	0·165	0·41
SCANDINAVIA						
P.cinctus (Siberian)	13·7	68	11·7	4·0	0·17	0·34
P.cristatus (Crested)	11·0	64	11·5	3·6	0·18	0·31
EUROPEAN USSR						
P.cyanus (Azure)	—	69	10·1	5·2	0·145	0·51
BALKANS						
P.lugubris (Sombre)	(16·7)	76	12·6	c. 5·3	0·165	0·42

Notes: All figures are averages from those of Snow (1953) for the countries stated, except that the beak-depth of *P.montanus* is from the population in north France, which has the same wing-length and culmen as the English one, while the weight of *P.lugubris* includes only one from the Balkans and three more from Persia. All dimensions vary somewhat from country to country, with a tendency for each species to be larger with a proportionately shorter beak in colder regions.

to broadleaved woods and two others to conifers, as can be seen from table 1. Typical measurements for the weight, overall size (indicated by winglength) and the beak of each species have been set out in table 2. The largest species, the Great Tit *P.major,* weighs more than twice as much as the smallest, the Coal Tit *P.ater.* In general, as noted later, the largest species forage on or near the ground and the smallest on the outermost small twigs.

A comparison of the information in tables 1 and 2 shows that the beak is longer relative to winglength, and narrower relative to its own length, in the species which live in conifers (the Coal Tit *P.ater,* Crested Tit *P.cristatus* and Siberian Tit *P.cinctus*) than in those which live in broadleaved woods (the Blue Tit *P.caeruleus,* Marsh Tit *P.palustris,* Azure Tit *P.cyanus* and Sombre Tit *P.lugubris*), while those that live in both habitats (the Great Tit *P.major* and Willow Tit *P.montanus*) are intermediate. As shown later, a similar variation is found in the Asiatic and North American tits, and also within those species in which one subspecies differs from another in habitat. Hence it is safe to regard the beak-differences in question as adaptive.

English broadleaved woods

Three species, the Blue, Great and Marsh Tits, are regularly found together in broadleaved woods, where they are sometimes joined by the Willow Tit and, in England, also by the Coal Tit. To elucidate the extent to which they differ in ecology, their feeding stations were studied in Marley Wood, a wood of mixed native broadleaved trees with a good shrub layer near Oxford (Gibb 1954), while their foods were studied in the more uniform conditions of a mature oak plantation with hardly any shrub layer in the Forest of Dean, also in southern England (Betts 1955). The diet of each species varies greatly throughout the year. All five species depend on leaf-eating caterpillars for their young in early summer, and all except the Willow Tit take

2

Table 3. Feeding stations of *Parus* species in mixed broadleaved wood

| | Percentage of time observed feeding in each station | | | | |
	P.ater (Coal Tit)	*P.caeruleus* (Blue Tit)	*P.major* (Great Tit)	*P.montanus* (Willow Tit)	*P.palustris* (Marsh Tit)
NOVEMBER–APRIL					
ground	17	7	50	0	16
branches	24	8	16	50	30
dead parts	11	16	6	12	7
twigs, buds	24	34	5	27	19
leaves	4	3	4	4	2
elsewhere	20	32	21	7	26
JUNE–AUGUST					
ground	0	0	2	0	5
branches	23	2	4	22	12
dead parts	16	5	1	17	15
twigs, buds	4	2	2	6	7
leaves	56	90	84	50	48
elsewhere	1	1	7	5	12

Notes: Simplified from Gibb (1954, p. 536) for Marley Wood, Oxford. The figures are the means of the mean for each month, except in the Willow Tit, in which they refer to the combined figures for September to April and May to August respectively. In this wood the Blue and Great Tits are abundant, the Marsh Tit fairly common, the Coal and Willow Tits very sparse but regular. The above picture was largely confirmed by Beven (1959) for a dense oakwood in southern England with a larger shrub layer than Marley, except that here the Great Tit fed in the shrub layer to a greater extent than in Marley. Betts (1955, p. 305) also obtained broadly similar results in a pure oak wood in the Forest of Dean, not by field observation but by analysing where the insects live which she obtained from the gizzards of collected tits. The most important differences from Marley (figures for the whole year) are that the Coal Tit fed more on branches and trunks (40%) and less on the ground (2%) and the Marsh Tit more on the twigs (27%) and scarcely at all on the branches; the Willow Tit was absent, and almost half the prey of the Great Tit could not be allocated to particular parts of the trees. These observations refer solely to the animal and not the vegetable foods.

Fig 3. Feeding stations of Blue Tit (top), Marsh Tit (middle) and Great Tit (bottom) in oak woodland. By Robert Gillmor.

much beechmast in the winters when it is plentiful; but these foods are temporarily so abundant that there could hardly be competition for them. Otherwise, each of them feeds in mainly different stations and takes mainly different prey, as summarized in tables 3 and 4.

Table 4. Size of insect prey of *Parus* spp. in English oak wood

Size-range	Percentage of prey of each size			
	P.ater (Coal Tit)	*P.caeruleus* (Blue Tit)	*P.major* (Great Tit)	*P.palustris* (Marsh Tit)
0–2 mm	74	59	27	22
3–4 mm	17	29	20	52
5–6 mm	3	3	22	16
>6 mm	7	10	32	11

Note: from Betts (1955, p. 307), based on stomach analyses in mature oak plantation in Forest of Dean, Gloucestershire.

The Blue Tit feeds mainly on oak trees at all times of year and, correlated with its small size, it feeds high up on the twigs, buds, leaves and galls, where it is more agile than the other species and freely turns upside down when needed, and it also strips bark for insects underneath. Most of the insects which it takes are less than 2 mm long, and it eats hardly any seeds except those of birch, which it takes from the tree itself. In contrast, the large and heavy Great Tit feeds mainly on the ground, especially in winter, and does so to an increasing extent later in the winter when food is scarce; the only time that it is common in the leaf canopy is when taking caterpillars for its young. It has a larger and stronger beak than the other species, most of the insects which it takes are over 6 mm long, and it eats many more acorns, sweet chestnut and wood sorrel *(Oxalis)* seeds than the others, and is the only one which takes the hard nuts of the hazel. The third common species, the Marsh Tit, is intermediate in size and size of beak between the other two. It feeds either in the shrub layer (in Marley especially in elder) or, when it feeds on large trees (chiefly oak), it hunts especially on the twigs and branches below 20 ft. It also feeds in herbage to a greater extent than the other species except the sparse Willow Tit. Most of the insects which it takes are 3–4 mm long, and it also eats various seeds and fruits, including those of burdock, spindle, honeysuckle, violet and wood sorrel. The ecological relationships of these three species are portrayed in fig. 3.

The small Coal Tit is not found in broadleaved woods on the Continent. In England it feeds there chiefly on oak trees and, later in the winter, on ash, especially on the branches. Most of the insects which it takes are less than 2 mm long, on average even smaller than those taken by the Blue Tit (which does not compete with the Coal Tit as it feeds chiefly in the canopy). The Coal Tit takes relatively few seeds in broadleaved woods. Finally, the sparse Willow Tit feeds especially on birch and to a lesser extent elder, its feeding stations being similar to those of the Marsh Tit, like which it often feeds in the herbage, but unlike the Marsh Tit it avoids oak trees, and takes few seeds.

In broadleaved woodland in the Swiss lowlands, Amann (1954) found that the Willow Tit searches mainly amid undergrowth and herbaceous vegetation, and feeds on the ground only in thick cover, while it takes almost entirely insects, and only a few small soft seeds, such as those of hemp-nettle *Galeopsis tetrahit*. The Marsh Tit, on the other hand, readily searches among open bushes and at the wood edge, it also hammers on bark, whereas the Willow Tit merely picks, and takes a variety of fairly hard seeds which the Willow Tit avoids, these latter differences being correlated with the Marsh Tit's stronger beak. (There is also an alpine race of the Willow Tit which feeds predominantly in conifers, so is separated by habitat from the Marsh Tit, as are those individuals in conifer plantations in England.)

This survey shows that the three main European species in broadleaved woods, the Blue, Great and Marsh Tits, are clearly separated from each other at most times of year, by their feeding stations, size of insect prey, and hardness of seeds. The same holds for the Willow Tit where it coexists with them (it is partly separated from the Marsh Tit by habitat) and for the Coal Tit where it joins them in England. The differences in the size and shape of the beaks of these species are adapted both to the size of their insect prey and to the hardness of the seeds which they take, and the differences between them in overall size are adapted to their feeding stations, high or low in the trees.

Differences in nesting sites

Competition for nesting sites is not important in most passerine birds because their generalized requirements can be satisfied in many places in their habitats. The tits, however, nest in holes in trees, of which there may not be enough to go round, so that at times two pairs, of the same or different species, try to nest in the same hole. One might therefore expect competitive exclusion to be evolved, and there is evidence for this. First, the Willow Tit differs from all the rest in that it excavates its own hole, so is able to breed where there are no natural holes, but is restricted to trees with soft wood, especially birch. Secondly, the Coal Tit differs from all the rest in nesting freely in holes in the ground, and in some woods every nest is in the ground, though it uses holes in trees where available.

The problem has been carried further by experiments with nesting boxes in Germany (Löhrl 1966a). With a choice of heights, the Blue Tit selects boxes at least 15 metres off the ground, whereas the Great Tit selects those between 3·5 and 7 metres off the ground. Further, with a choice of differently shaped entrance holes, the Blue and Great Tits prefer a circle and the Coal Tit a slit. These are, however, only preferences since, as is well known, in a wood provided with only one pattern of nesting box at one height, all five species may use them, including the Coal Tit and occasionally even the Willow Tit. There is the further point that, in a direct dispute for the same hole, the larger Great Tit normally ousts the smaller Blue, and either of these normally ousts the Coal Tit, which is the smallest species, but the smaller species can use holes too small for the Great Tit to enter.

Norwegian coniferous forest

The feeding stations of the tits in conifers have been studied in Norway, and the differences between the three common species have been summarized in table 5, and portrayed in fig. 4. The small Coal Tit spends most of its time high up in the trees among the needles, where it is more agile than the two heavier species, and its smaller beak also suggests that it takes smaller insects. In this connection, it is the only one of them to take and store aphids in large quantities. It also stores and eats spruce seed, but hardly ever picks up the seeds of herbaceous plants.

The Willow Tit takes a higher proportion of vegetable food than the other two (38 cf. 20 per cent), especially the seeds of juniper and hemp-nettle *Galeopsis tetrahit*. It is also the only one which feeds to an appreciable extent in the few available broadleaved trees. In conifers, it spends proportionately more time than the Coal Tit on the lower than the upper parts, on the branches than the twigs, and on the areas without than with needles, these differences being most pronounced in winter. The Crested Tit is intermediate between the other two in the extent to which it divides its time between the upper and lower parts of the trees and between areas with and without needles, it does not feed in the herb layer, but spends much more time than either of the others on the ground. Hence it presumably differs from them to a sufficient extent to be isolated, but detailed food studies are needed to prove this. A fourth species is found

27

in open pine woods, namely the Great Tit, but as it feeds mainly on the ground, and has a much larger beak, it almost certainly does not compete with the other three. Probably, therefore, competitive exclusion holds in all six of the common *Parus* species in Europe, a fact that is not at all obvious at first sight. Further, each of these species has evolved adaptations in overall size and beak to its particular habitat, feeding stations, and size of prey.

Table 5. Feeding of *Parus* ssp. in conifer forest in Norway

	P.ater (Coal Tit)	P.cristatus (Crested Tit)	P.montanus (Willow Tit)
	percentage of time spent feeding		
	(a) throughout year		
Spruce *(Picea)* upper half	50	22	14
lower half	25	23	35
Pine *(Pinus)* top quarter	12	5	4
rest	10	20	14
Broadleaved tree or juniper	1	*	5
Herbs *(Galeopsis)*	1	1	22
Ground	2	28	6
	(b) in winter		
Spruce trunk	—	*	8
dead branch	—	11	19
live branch (inner part)	3	30	52
live branch (outer part)	98	58	21
Pine trunk	—	—	7
branches (inner parts and dead)	—	62	74
twigs and needles (outer part)	100	38	19

Notes: compressed from Haftorn (1956, pp. 7, 9, 36 and Pt.2. p. 13 of same work). * = less than 0·5%.

Fig 4. Feeding stations of Coal Tit (top), Crested Tit (middle) and Willow Tit (bottom) in pine forest. By Robert Gillmor.

The three local species

The Siberian or Lapp Tit *P.cinctus* is restricted to the northern part of the coniferous taiga, where the only other tits are the Great and Willow. But the Great Tit is separated in this area because, unlike the other two, it is normally found near human settlements, on which it depends for food. The Willow and Siberian Tits occur together in the taiga from Norway to eastern Siberia. They differ partly in habitat, as only the Willow Tit occurs in the pure birch forest north of and at higher altitudes than the coniferous forest, and it also feeds commonly in the few broadleaved trees in coniferous forest, whereas the Siberian Tit prefers dense coniferous forest where the Willow Tit is sparse, and also barren pines (O. Hilden pers. comm.). As shown in table 2, the Siberian Tit has a thinner beak than the Willow Tit, which supports the view that it feeds to a greater extent among conifer needles. In winter in Lapland, it searches each tree thoroughly before flying to the next, whereas the Willow Tit spends only a short time on each tree and then flies a long way to the next, which suggests that they are looking for different prey (Snow 1952a). Whether they also differ in feeding stations is not known.

The Coal and Crested Tits reach their northern limit near to the southern limit of the Siberian Tit, but the Coal has a much smaller beak so probably differs in diet. The Crested has a beak of similar shape to that of the Siberian Tit though it is a little smaller, so perhaps it is a potential competitor and that is why they replace each other geographically. Fitting with this view, the boundary between them shifted north during the warm period 1930–49 (Merikallio 1951), i.e. the advance of the southern species coincided with the retreat of the northern, which would be unlikely unless they were competitors.

The Azure Tit *P.cyanus* breeds in light riverine broadleaved woodland in the southern part of the taiga and the northern part of the steppe. It is slightly larger than the

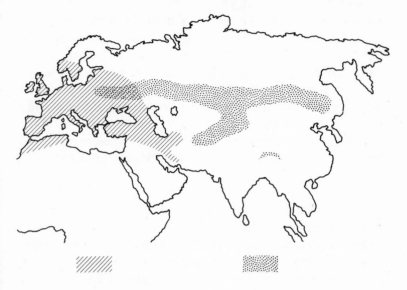

Fig 5. Ranges of Blue Tit *P.caeruleus,* and Azure Tit *P.cyanus* (from Snow 1953).
   Note that westward extension of Azure Tit has varied greatly in recent decades.

Blue Tit, with a similarly shaped beak, and almost certainly fills a similar ecological niche. As shown in fig. 5, these two species replace each other geographically, with an overlap of variable extent in the USSR, where they sometimes interbreed (Pleske 1912). In the 1880s, the Azure Tit spread west into the middle of eastern Europe, but it retreated again at the end of the decade.

   Finally, the Sombre Tit *P.lugubris* is coloured like a Marsh Tit but it is nearly as large as a Great Tit. It breeds solely in the Balkans, Turkey and northern Iran. Published information might suggest that in the north of its range it overlaps extensively with the Marsh Tit, but this is due partly to errors, partly to altitudinal separation, and partly to the Sombre Tit having become scarcer and retracted in range in the last fifty years, while the Marsh Tit has perhaps spread (Reiser 1894, 1905, 1939, cf. Harrison 1933). In the

31

Balkans, the Marsh Tit lives in montane oak and beech woods *(Quercus sessiliflora, Q.robur, Fagus sylvatica)* of the same type as it frequents at lower altitudes further north, whereas the Sombre Tit lives in the more southerly, lower, drier and sunnier submediterranean broadleaved woods of *Quercus pubescens* and *Carpinus orientalis* in western Jugo-slavia and the similar *Quercus cerris* community of Mace-donia (D.Rucner pers. comm., see also Stresemann 1920, Mastrovik 1942, Niethammer 1943, Makatsch 1950, Balat 1962).

In the Balkans the Sombre Tit lives in oak woods with a good shrub layer (Löhrl 1966b), while in Iran it coexists with the Great and Blue Tits in oak woods and is the only species in pistachio scrub (L.Cornwallis pers. comm.), and Paludan (1938) considered it to be more a bush than a tree bird. Hence presumably, though larger, it fills a similar niche to the Marsh Tit, which would explain why it replaces it geographically and why, like the Marsh, it can coexist with the Great and Blue Tits. The two other tits in in the Balkans, the Coal and Crested, are separated from it by habitat, as they are restricted to montane conifer forest.

Variations in habitat

Each species keeps to the habitat set out in table 1 in most of its range, but there are exceptions. Denmark has no natural conifer forest, which is presumably why the Crested and Willow Tits were absent, though the Crested is now established in modern pine plantations (Jespersen 1944). But the Willow Tit is still absent, and this is presumably why the Marsh Tit is common there in pine plantations, a habitat which it does not frequent elsewhere in Europe.

On the Swedish mainland, as elsewhere, the Blue Tit is in broadleaved woods and the Coal Tit in conifers, but on Gotland, 90 km offshore, both species are in both oak and pine woods, which is perhaps linked with the absence of the Willow, Marsh and Crested Tits from the island

(Svärdson 1949). The same three species are absent from Ireland, where the native forest is entirely broadleaved and the Coal Tit has evolved an unusually thick beak. It feeds mainly in the understorey, i.e. the feeding zone of the Marsh Tit in England, but this consists of evergreens in Ireland, instead of deciduous trees as in England.

The Blue Tit is common in the native pine forest of Majorca and also on some of the Aegean islands (Watson 1964) where the Great Tit is the only other species. The Crested Tit breeds in the mixed evergreen and deciduous oak woods in the lowlands of southern Spain, where the Coal, Marsh and Willow Tits do not breed. In both cases the unusual choice of habitat might be linked with the absence of other species. But the presence of the Coal Tit in broadleaved, as well as coniferous, woods in Sardinia (Bezzel 1957), the Appenines, Pyrennes and Cantabrian mountains (in montane beech), and of the Crested in the latter, is not linked with the absence of other species.

The alpine race of the Willow Tit is in conifers, and the English race in broadleaved woods, correlated with which the English race has a broader beak. This difference in habitat might not be hereditary, but due to which kinds of trees are available in the areas concerned, since the Willow Tit breeds in introduced pine plantations in England. Further, the Scandinavian race is regular in both broadleaved and coniferous forest, as already mentioned.

Other variations in habitat are found in countries outside Europe. The Persian Coal Tit occurs in oak woods in both the Elburz and Zagross mountains and, like the Irish Coal Tit, has evolved an unusually thick beak, as portrayed in fig. 6. It occurs here with the Great, Blue and Sombre Tits, but the other European species are absent. In the Caucasus, the Marsh Tit occurs in both broadleaved and coniferous forest; the Willow Tit is absent. In Tunisia, Algeria and Morocco, where the Marsh, Willow and Crested Tits are absent, both Blue and Coal Tits occur in oak, pine and cedar forest, and so, more usually, does the Great Tit. Finally, the

Fig 6. Typical Blue Tit *P.c.caeruleus* of broadleaved forest and thin-beaked Canarian form *P.c.teneriffae* of pine forest; typical Coal Tit *P.a.ater* of conifer forest and thick-beaked Persian form *P.a.phaeonotus* of oak woods. By Robert Gillmor, adapted from David Snow.

Table 6. Beak differences of *Parus* species in atypical habitats in and near Europe.

Country	Habitat	Culmen (mm)	Beak-depth (mm)	Depth ÷ Culmen
		Coal Tit *P.ater*		
	TYPICAL			
C. Sweden	conifer	10·9	3·7	0·34
N. Portugal	conifer	10·2	3·4	0·33
	UNUSUAL			
Ireland	oak	10·7	3·7	0·35
Transcaspia	broadleaf	11·7	c. 4·25	c. 0·36
Elburz Mts.	oak	11·3	c. 4·4	c. 0·39
Tunisia	pine and oak	11·0	3·8	0·35

Table 6 *(continued)*

Country	Habitat	Culmen (mm)	Beak-depth (mm)	Depth ÷ Culmen
		Blue Tit *P.caeruleus*		
	TYPICAL			
S. Scandinavia	broadleaf	9·2	4·8	0·52
N. Portugal	broadleaf	9·1	4·2	0·46
	UNUSUAL			
N. W. Africa	broadleaf			
(4 populations)	and conifers	10·1	4·1	0·41
2 eastern Canary				
Islands	tamarisk	10·4	4·3	0·41
2 central Canary	broadleaf			
Islands	and conifers	10·9	4·1	0·38
2 western Canary				
Islands	mainly pine	10·9	4·0	0·37
		Willow Tit *P.montanus*		
	TYPICAL			
C. Sweden	mixed	11·0	4·1	0·37
	UNUSUAL			
England and				
N. France	broadleaf	10·6	4·0	0·38
Alps	pine	11·5	4·2	0·37

Notes: Measurements from Snow (1953, 1955), who also gave numbers measured and standard deviations. In the righthand column, a higher figure means a thicker beak. The proportions of the beak in the Crested Tit were similar in Spanish oak wood to those in conifers elsewhere.

Blue Tit is the only species of *Parus* in the Canary Islands, and on the central and western islands, where it is common in pines, it has evolved an unusually thin beak (portrayed in fig. 6).

The biggest differences in the shape of the beak correlated with these local variations in habitat have been set out in table 6, preceded by measurements from the typical habitat. Where the Coal Tit is in broadleaved woods it has evolved a broader beak, recalling that of the Blue Tit, and where the Blue Tit is in conifers it has evolved a narrower beak, recalling that of the Coal Tit, and a similar, though smaller, difference is found between those races of the Willow Tit restricted to broadleaved and coniferous woodland respectively.

Conclusion

The European tits were selected for ecological study because of their apparent complexity, and in particular because several species usually coexist in the same habitat. Nevertheless, each species is segregated from every other,

Table 7. Main means of ecological segregation of *Parus* species in Europe

(F by feeding station or type of food, G by geographical range, H by habitat, — no contact)

		1	2	3	4	5	6	7	8	9
1	*P.ater*	1								
2	*P.caeruleus*	H	2							
3	*P.cinctus*	—	—	3						
4	*P.cristatus*	F	H	G?	4					
5	*P.cyanus*	H	G	H	H	5				
6	*P.lugubris*	H	F	—	H	—	6			
7	*P.major*	F	F	H	F	F	F?	7		
8	*P.montanus*	F	HF	F?	F?	F?	—	F	8	
9	*P.palustris*	H	F	—	H	F?	G	F	H(F)	

in a few cases by geographical range, in many by habitat, and in yet more by a difference in food and feeding stations in the same habitat. Differences of this last type are associated with adaptive differences in overall size and in size of beak, the larger species tending to feed lower down, and on larger insects and harder seeds, than the smaller species. Likewise the species which live in coniferous forest have longer and narrower beaks than those in broadleaved woods. Some species have a different, or unusually wide, habitat in a small part of their range, often linked with the absence of another species, and the beak may be appropriately modified. The main way in which each species is isolated ecologically from the rest is summarized in table 7.

Chapter 3

A World Survey
Tits *(Parus)* in Asia, Africa
and America

On the most recent classification (Snow 1967), there are 45 species of tits in the world, 9 in Europe, 23 in Asia (of which 7 also occur in Europe), 10 in Africa south of the Sahara and 10 in North America (of which one also occurs in the Old World). Only in Europe has their ecology been studied in detail, but with this information as a basis, much can be surmised about other regions from the known range, habitat, size and beak-size of each species. Western Asia and North Africa were considered in the previous chapter, and the present chapter is concerned with the rest of Asia, Africa south of the Sahara, and North America.

Semi-species

In all three continents, some widespread species or super-species have given rise to forms which replace each other geographically like subspecies but are so distinctive that some writers, though not others, have treated them as full species. A few of these, which coexist without inter-breeding in a small area of overlap, are clearly good species, but just which of the others should be regarded as such, and which as subspecies, is rather arbitrary. The nomen-clature followed here is, with one exception noted in appendix 6, that of Snow (1967), who in general used a broad species concept. The cases in which treatment has differed in recent publications are included in the foot-notes to appendices 2, 4 and 6.

Asia excluding the central mountains

This and the following section are based on the full review by Snow (1953, 1954). The situation in northern Asia is similar to that in Europe, since the Coal, Siberian, Azure, Great and Willow Tits and (with a gap in the middle) the Marsh Tit, range across the north of the Eurasian continent (except that the Azure is replaced by the Blue Tit in most of Europe, as mentioned in the previous chapter). But the Crested Tit is absent and this is presumably why in Asia, unlike Europe, the Siberian Tit extends to the southern limit of the taiga, as these two species are mutually exclusive (see p. 30).

In general, each of these species lives in the same type of habitat in Asia as in Europe; and, as in Europe, there are some local variations. Thus the Coal Tit, normally in conifers, also occurs in subtropical broadleaved forest in Japan, in mixed larch and broadleaved forest on some of the Kurile Islands, and on one of them, Urup, in pure broadleaved forest. Likewise the Marsh Tit, though often in broadleaved or mixed woodland, occurs in montane pine forest in various parts of Asia, including northern Mongolia, Korea, western Yunnan and Mount Victoria (Burma), as well as the Caucasus mentioned in the previous chapter. As pointed out by Snow (1955), it could hardly be a coincidence that, with one exception, the Willow Tit is absent from these areas, and also Denmark (where the Marsh Tit is likewise in pines, see p. 32). The exception is northern Mongolia (Kozlova 1933). In contrast, the Willow Tit does not appear to have a wider habitat in those areas where the Marsh Tit is absent, but in any case the Willow Tit is at home in both broadleaved and coniferous woods. The Asiatic forms of the Marsh Tit have broader, not narrower, beaks than those in Europe, although several of them occur in conifers.

The only species of tits found in most of northern Asia are the European species already mentioned. But in Korea and Japan, the Azure Tit is replaced by the Varied Tit

P.(Sittiparus) varius, a small agile bird with a relatively long
stout beak recalling that of a Great Tit, but which feeds
primarily in the leaf canopy in broadleaved trees like a
Blue Tit. Although Japan and Britain are some 10,000 km
apart, they have four species of tits in common, while the
Varied replaces the Blue, but the Crested is missing (Austin
and Kuroda 1953).

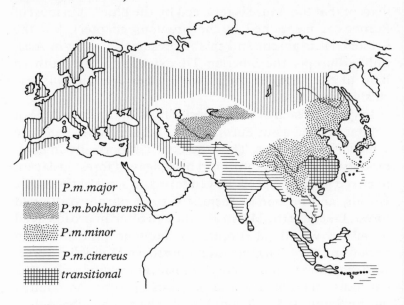

P.m.major
P.m.bokharensis
P.m.minor
P.m.cinereus
transitional

Fig 7. Range of Great Tit *Parus major* superspecies (from Snow 1953).

The European species do not extend to the southern lowlands of Asia, except for the Great Tit, which is represented
by different semispecies in central, eastern and southern
Asia, as shown in fig. 7. In peninsular India it is joined by
two other large species, *P.xanthogenys* and, locally, *P.
nuchalis,* and in parts of China by the large *P.spilonotus* and
the small *P.venustulus.* The ecology of these birds has not
been studied except that, where they meet, *P.major* prefers
drier and more open deciduous woodland and *P.xanthogenys*
evergreen or moist deciduous forest (S. Ali 1953 and pers.

comm.). The latter statement may well apply also to *P. spilonotus*, as it is so similar to *P.xanthogenys* that some workers treat them as conspecific. *P.venustulus* is much smaller with a thinner beak than the other species, so presumably takes different foods. Hence at least most of these species are evidently segregated, and the same holds for the four species on Formosa and the three in the Philippines, set out in appendix 1 (p. 271).

The mountains of central Asia

Whereas in most of Asia the number of *Parus* species in any one area is similar to that in Europe, as many as 14 species are found in the high mountains of central Asia, in the region bounded in the north by the Tian Shan, Kunlun and Hanshan and in the south by the Himalayas, northern Burma and Yunnan. Admittedly this is a large area, and some of these species do not meet each other. Indeed a few of them are very local, notably *P.davidi* and *P.superciliosus,* both solely in part of western China and eastern Tibet, and *P. melanolophus* solely in the western Himalayas. But most of the others have a wide range, and Schäfer (1938) met ten of them in his journey through eastern Tibet and the adjoining Haifan mountains of China. The habitats and measurements of the species concerned have been summarized in table 8 and their ranges, English names and subgenera in Appendix 2 (p. 272).

Many of these species, especially those in the same subgenus, differ from each other in habitat. The Great Tit, for instance, occurs in open forest, and the similar but rather smaller Green-backed Tit *P.monticolus* in true (but still not dense) forest. Again, four species in the subgenus *Poecile* occur in the mountains of southwestern China, *P.davidi* in bamboo jungle, the Marsh Tit *P.palustris* in dry valleys with xerophytic vegetation, the Willow Tit *P.montanus* in montane conifer forest and *P.superciliosus* in the alpine bush zone of rhododendrons and dwarf willows above the

Table 8. Habitats and measurements of *Parus* species in mountains of central Asia (for subgenera see Appendix 2)

Species (including subspecies if very different)	Main habitat	Average measurement in mm of		
		wing	culmen	beak-depth
P.ater				
P.a.aemodius (east)	conifers	59	9·6	2·9
P.a.rufipectus (west)	conifers	63	10·9	3·4
P.cyanus	riverine broadleaf	68	10·5	4·7
P.davidi	broadleaf	66	10·5	4·5
P.dichrous	conifers	70	10·9	3·8
P.major	open broadleaf (or pine)	75	—	—
P.melanolophus	conifers	64	10·7	3·3
P.montanus				
P.m.affinis (east)	mainly broadleaf, also mixed	67	10·7	—
P.m.songarus (west)	spruce	70	13·0	4·5
P.monticolus	broadleaf (or mixed)	68	11·6	4·3
P.palustris	xerophytic vegetation, conifers and mixed	66	10·3	—
P.rubidiventris				
P.r.rubidiventris (east)	conifers	70	10·6	4·0
P.r.rufonuchalis (west)	conifers	77	12·6	4·4
P.superciliosus	high alpine bush zone	66	10·7	4·5
P.spilonotus	mixed, subtropical	c. 75	c. 10·5	—
P.venustulus	mixed, subtropical	c. 66	c. 9·3	—
P.xanthogenys	mixed, subtropical	c. 71	c. 10·0	—

Notes: habitats from Schäfer (1938) and Snow (1953, 1954), measurements from Snow except for *P.spilonotus* and *P.xanthogenys* from Baker (1922) and *P.venustulus* from La Touche (1925–30). Because *P.major* is in very open and *P.monticolus* in less open forest, *P.monticolus* is usually at a higher altitude than *P.major*, but *P.major* occurs above *P.monticolus* on some mountains where there is light open woodland above the true forest.

Fig 8. Three Himalayan tits related to Coal Tit. By Robert Gillmor, adapted from David Snow.

forest (Schäfer 1938). Hence their differences in habitat are linked with differences in altitude.

Certain other species coexist in the same habitat. Thus two species in the subgenus *Periparus* coexist in coniferous forest in most of the Himalays, but *P.rubidiventris* is much larger than either the Coal Tit *P.ater*, or its geographical replacement in the west *P.melanolophus* (as portrayed in fig.8), hence it is reasonably certain that they do not compete for food. These species are smaller in the east than the west of the Himalayas, as shown in table 8, and their parallel variation suggests character displacement, i.e. that they avoid competition through differing in size (see p. 10). However *P.rubidiventris* remains equally large in Baluchistan, and equally small in Nepal and Yunnan, although the smaller species is absent there (Snow 1953, 1954). *P.dichrous* (in a different subgenus) coexists with them and has, like them, the narrow beak typical of the tits in conifers. It is similar in size to *P.rubidiventris,* and how it might be separated is not known. There are a few further instances in which two species in different subgenera coexist in the same habitat, but nearly all of them differ from each other in size of beak, so presumably take mainly different foods.

The probable way in which each of the 14 species is isolated ecologically from the rest has been summarized in appendix 3 (p. 273). This is more doubtful than the corresponding table for the European species because the Asiatic birds are less well known. Even allowing for possible errors, however, it is clear that most species in the Asian mountains are isolated from each other by habitat (including altitude). Hence the unusually large number of species there is attributable to the unusual diversity of forest types, which is correlated with the great altitudinal range, and is not due to more species coexisting in the same habitat than in Europe.

A few of these species differ in habitat in different parts of their range. Thus both the Great Tit *P.major* and the related *P.monticolus* occur mainly in broadleaved forest, but in pine forest in northwestern Yunnan. Since, however, the Great Tit is regularly in open pine forest in Europe, this may not be a true exception. The occurrence of the Marsh Tit *P.palustris* in pine forest in western Yunnan and on Mount Victoria in Burma has already been mentioned (p. 39). *P.melanolophus* and *P.rubidiventris* live, as already mentioned, in coniferous forest, but in the northwestern Himalayas they occur in montane oak woods at a similar altitude to the coniferous forest which they frequent elsewhere.

There are also a few instances in which two subspecies of the same species differ markedly in size of beak, the most striking being included in table 8. It is not known whether these differences are related to differences in feeding ecology, but they are apparently not related to differences in habitat. While, also, *P.rubidiventris* varies in parallel with *P.ater* as already noted, it is not known why both of them should be larger in the west than the east. The small east Himalayan form of the Coal Tit *P.ater aemodius* is the smallest of all tits. It is portrayed with two other subspecies of Coal Tits in fig. 8.

Africa south of the Sahara

The ranges and habitats of the 10 African species south of the Sahara have been set out in appendix 4 (from Hall and Moreau *in preparation*). These species are more like each other than they are like any other tits, so presumably one palaearctic form, in the *Major* subgenus, became established in Africa and there evolved into different species. Nearly all of them live in open deciduous woodland or thorn scrub, but one in lowland evergreen forest.

The ranges of the species in the *P.afer* superspecies are set out in fig. 9. *P.a.afer*, the Grey Tit of South Africa, lives in dry bush with scattered trees, another subspecies *cinerascens* replaces it immediately to the north in acacia steppe, and another subspecies *thruppi* occurs in acacia steppe much further north in Somalia. *P.griseiventris* lives in the rather richer habitat of brachystegia woodland in central and east Africa, and hence between *cinerascens* and *thruppi*, but though it replaces them geographically, this is because of the distribution of their respective habitats; where *P.griseiventris* overlaps with *P.a.cinerascens* in part of Rhodesia, each keeps to its particular habitat. The third species in this superspecies, *P.fasciiventer,* lives in montane gallery forest so differs in habitat, and therefore also in range, from the others.

The ranges of the species in the *P.niger* superspecies are shown in fig. 10. *P.niger*, the Black Tit of South Africa, frequents mopane and other dry woodland. Two other species, *P.leucomelas* and *P.albiventris*, live further north in the drier habitat of acacia savanna, also the wood edge. *P.leucomelas*

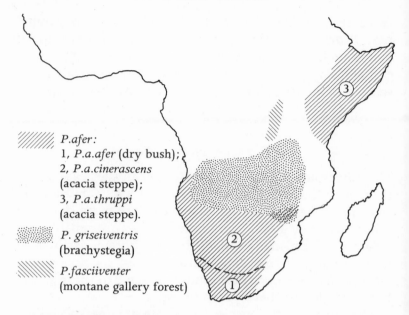

P.afer:
1, *P.a.afer* (dry bush);
2, *P.a.cinerascens*
(acacia steppe);
3, *P.a.thruppi*
(acacia steppe).

P. griseiventris
(brachystegia)

P.fasciiventer
(montane gallery forest)

Fig 9. Ranges of tits in *P.afer* superspecies (from Hall and Moreau *in preparation*).
Note small area of overlap between *P.a.cinerascens* and *P. griseiventris* where they are separated by habitat.

is separated from *P.niger* largely by range, but by the habitat difference just noted in a small area of overlap in Malawi and Zambia. *P.albiventris* is likewise separated by range from *P.leucomelas*, but with a small area of overlap on both sides of the continent, where it is at higher and *P.leucomelas* at lower altitudes, though elsewhere neither is restricted in altitude in this way. The fourth species, *P.leuconotus*, is confined to woodland above 2000 m in Abyssinia and Eritrea.

A species in the *P.afer* superspecies and another in the *P.niger* superspecies are often found in the same area, but are segregated by habitat. In South Africa, *P.afer* is in acacia steppe and *P.niger* in the richer habitat of mopane woodland. Further north, the habitat preferences of the two superspecies are reversed, for *P.griseiventris* (in *afer*) lives in

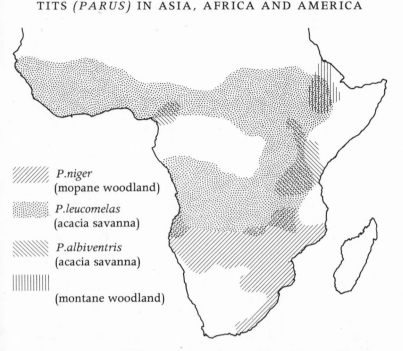

P.niger
(mopane woodland)

P.leucomelas
(acacia savanna)

P.albiventris
(acacia savanna)

(montane woodland)

Fig 10. Ranges of tits in *P.niger* superspecies (from Hall and Moreau *in preparation*).
Note that *P.leucomelas* overlaps with *P.niger* in part of South Africa, with *P.albiventris* in parts of east and west Africa, and with *P.leuconotus* in Abyssinia, where it is separated by habitat (from *P.albiventris* by altitude).

brachystegia woodland and *P.leucomelas* (in *niger*) in the drier habitat of acacia savanna. *P.fasciiventer* (in *afer*) in montane gallery forest is likewise separated by habitat, including altitude, from *P.leucomelas* (in *niger*).

The *P.rufiventris* superspecies consists of two smaller species than the others. *P.rufiventris* and the more northern *P.fringillinus* replace each other geographically, as shown in fig. 11. *P.rufiventris* coexists in brachystegia woodland with *P.griseiventris* (in *afer*), but it feeds mainly in the canopy, and *P.griseiventris* mainly on the lower branches and dead boughs (C. W. Benson and M. P. S. Irwin pers. comm., Praed and Grant 1955). Further north, *P.fringillinus*

47

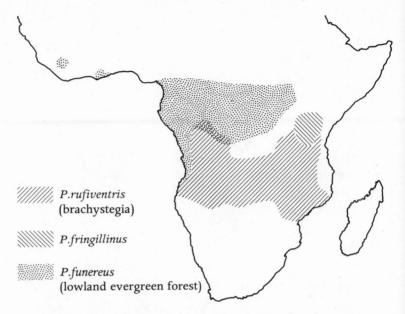

P.rufiventris
(brachystegia)

P.fringillinus

P.funereus
(lowland evergreen forest)

Fig 11. Ranges of tits in *P.rufiventris* superspecies and *P.funereus* (from Hall and Moreau *in preparation*).

meets *P.albiventris* (in *niger*), but they are partly separated by habitat, *P.albiventris* at the wood edge and *P.fringillinus* in drier country, and where they overlap in Kenya, Angwin (1968) found *P.albiventris* feeding on larger branches, especially of *Acacia xanthophloea* (which has a peeling bark) and *P.fringillinus* on the leaves, twigs and bark, especially of *Acacia kirkii* (which has smooth bark with narrow cracks). Finally *P.funereus* is the only species in lowland evergreen forest, which also means that it differs in range from the rest, as shown in fig. 11.

The probable way in which each of the African species of *Parus* is isolated is summarized in appendix 5. Each of their main habitats, brachystegia woodland, acacia steppe, evergreen forest, and so on, occurs over huge stretches of country, so it is hard to know how to classify those species which occupy separate ranges owing to the ranges of their respective habitats. But even if some pairs had been classified differently, the situation differs greatly from that in Europe or central Asia, owing to the large number of species separated by range, and also because only two pairs of species coexist in the same habitat with a difference in feeding. These points suggest that the evolution of *Parus* in Africa is at a relatively early stage. All the species appear to be derived from a single form, and they have not diverged ecologically to anything like the same extent as the palaearctic tits. This may be because they arrived in Africa in relatively recent times, when many ecological niches which might otherwise have been suitable for them were already occupied by other kinds of birds. Moreover only one species has become established in lowland evergreen forest, a habitat which the palaearctic migrants to Africa have also found hard to enter (Moreau 1966). In saying this, I do not mean that an African species has filled the particular niche suitable for a tit, but rather that the available foods are shared out in a different way, so that a tit would have partly to displace several species to become established.

North America

The situation in North America is rather like that in Africa, but a little more complex, as two subgenera are involved, the tufted tits *Baeolophus,* a group peculiar to America, and the chickadees *Poecile,* a group also widespread in the palaearctic. The following account is based on Dixon (1961). The species are set out in appendix 6 and portrayed in fig. 12, while measurements are shown in table 9.

The tufted tits (*subgenus* Baeolophus)

Three species in the same superspecies replace each other geographically, the Tufted Tit *P.bicolor* in eastern USA, the Black-crested Tit *P.atricristatus* in Texas and northern Mexico, and the Plain Tit *P.inornatus* in western USA. They are found mainly in broadleaved woods, the Plain Tit also

Table 9. Habitat and dimensions of North American species of *Parus*

Species	Main habitat	Mean weight (grams)	Mean wing (mm)	Mean culmen (mm)	Beak-length and depth	European equivalent
BAEOLOPHUS (tufted tits)						
P.atricristatus (Black-crested Tit)	broadleaf	16–17	71, 77	9·4, 10·7	long and broad	
P.bicolor (Tufted Tit)	broadleaf	22	80	12·1	long and broad	*P.major* Great Tit
P.inornatus (Plain Tit)	broadleaf (pine)	16–17	69, 72	11·2, 12·9	long and broad	
P.wollweberi (Bridled Tit)	broadleaf (pine)	10	65	9·0	short and broad	*P.caeruleus* Blue Tit
ATRICAPILLUS (black-capped chickadees)						
P.atricapillus (Black-capped Chickadee)	broadleaf and open conifers	12	62, 70	8·9, 9·7	medium and medium	*P.montanus* Willow Tit
P.carolinensis (Carolina Chickadee)	broadleaf	10	62, 63	8·4, 8·8	medium and broad	*P.palustris* Marsh Tit
P.gambeli (Mountain Chickadee)	montane pine	11–12	70	10·4	long and thin	*P.cristatus* Crested Tit
P.sclateri (Mexican Chickadee)	montane pine	—	69	9·2	long and thin	
CINCTUS (brown-capped chickadees)						
P.cinctus (Siberian Tit)	edges of taiga	—	70	9·2	long and thin	*P.cinctus* Siberian Tit
P.hudsonicus (Boreal Chickadee)	heavy taiga	—	64, 66	8·6, 9·6	long and thin	
P.rufescens (Chestnut-backed Chickadee)	spruce	10	60, 61	9·1, 9·2	long and thin	*P.ater* Coal Tit

Notes: habitat and mean weights from Dixon (1961), except for weight of *P.atricapillus* from Lawrence (1958), wing length and culmen from Ridgway (1904). Where two figures are given, they refer to the means of the smallest and largest subspecies of the species concerned.

Fig 12. Six North American tits (left) and their European counterparts (right, cf. Table 9. By Robert Gillmor.

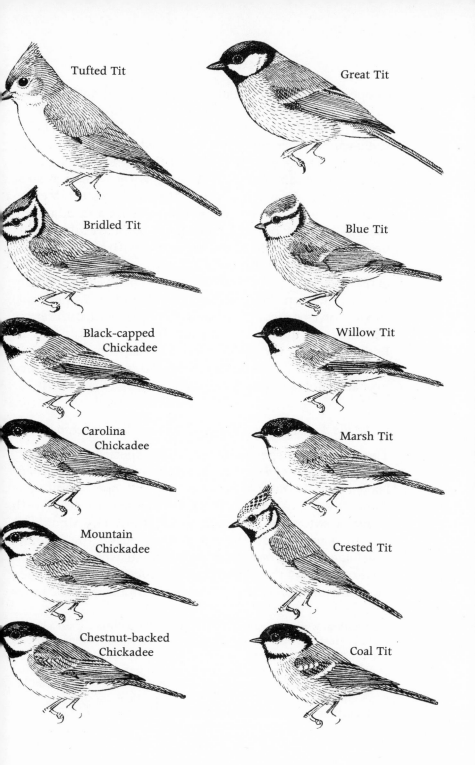

Tufted Tit

Great Tit

Bridled Tit

Blue Tit

Black-capped
Chickadee

Willow Tit

Carolina
Chickadee

Marsh Tit

Mountain
Chickadee

Crested Tit

Chestnut-backed
Chickadee

Coal Tit

in open pine, and are relatively large heavy birds, similar in size, shape of beak and feeding habits to a Great Tit.

The Bridled Tit *P.wollweberi* is much smaller, being similar in size, shape of beak and feeding habits to a Blue Tit. Its habitat is similar to that of the tufted tits, which it replaces geographically in Mexico, with a small area of overlap with the Black-crested Tit in northeastern Mexico and with the Plain Tit in eastern Arizona and adjoining Mexico. As can be seen from table 9, the Bridled Tit is nearly half the weight of the tufted tits, with a shorter beak. Where it overlaps with the Black-crested Tit, it feeds more in the foliage (51 per cent cf. 27 per cent of observed feeding stations), less on the branches (17 per cent cf. 34 per cent) and less on tree seeds (1 per cent cf. 10 per cent), differences which recall those between the Blue and Great Tits in England (p. 22), though they are less marked. As the Plain Tit is similar in size to the Black-crested, a similar difference probably holds where it overlaps with the Bridled Tit.*

*The black-capped chickadees (*P*.atricapillus *superspecies)*

There are four species in this superspecies. The Black-capped Chickadee *P.atricapillus* occurs both in pure broad-leaved and in mixed broadleaved and open coniferous woods over most of northern, eastern and central North America, extending west to the area of the Rocky Mountains and in places yet further. It is replaced by the Carolina Chickadee *P.carolinensis* in broadleaved woods in south-eastern USA, by the Mountain Chickadee *P.gambeli* in open montane pine forest in the Rocky Mountains and the inner

---

*In addition, where they meet, the Bridled Tit usually occurs at a higher altitude than the Black-crested, with an overlap between 1000 and 1400 m, and similarly where it meets the Plain Tit, it is said to occur more in oak and less in pine than the Plain Tit, though they occupy similar types of woodland outside their area of overlap. These partial differences in habitat seem much less important than the differences in feeding stations, but the point needs checking.

coast ranges of the west, and by the Mexican Chickadee *P.sclateri* in montane pine forest in Mexico and a small part of south-central USA. In agreement with the variations in beak in the same subgenus in Europe, the Carolina Chickadee (in broadleaved woods) has a slightly shorter and broader beak than the Mountain and Mexican Chickadees (in conifers), while the Black-capped Chickadee (in mixed woods) is intermediate. Hence the Carolina Chickadee is the ecological equivalent of the Marsh Tit, the Mountain and Mexican Chickadees of the Crested Tit and the Black-capped Chickadee of the Willow Tit. The adaptive differences in beak in the subgenus *Poecile* are set out in table 10.

Table 10. Adaptations of beak-shape to habitat in the subgenus *Poecile*

Species	Habitat	Beak-length relative to wing-length	Beak-depth relative to beak-length
EUROPE			
P.cinctus	conifers	0·17	0·34
P.montanus	mainly conifers	0·17	0·37
P.montanus	mainly broadleaf	0·18	0·38
P.palustris	broadleaf	0·16	0·41
P.lugubris	broadleaf	0·16	0·45
ASIATIC MOUNTAINS			
P.superciliosus	broadleaf	0·16	0·42
P.davidi	broadleaf	0·16	0·43
NORTH AMERICA			
P.gambeli	conifers	0·17	0·29
P.rufescens	conifers	0·17	0·35
P.hudsonicus	conifers	0·16	0·36
P.sclateri	conifers	0·15	0·36
P.atricapillus	mixed, mainly broadleaf	0·14	?
P.carolinensis	broadleaf	0·15	0·42

Note: from Snow (1954), but he there merged *P.montanus* in *P.atricapillus*, as was then the custom, so separate figures have been obtained here from earlier tables.

Where the Black-capped and Mountain Chickadees meet inland in British Columbia and northwestern USA, they are separated by habitat, the Mountain chickadee in montane pines and the Black-capped in broadleaved riverine woodland lower down, and there is a subsidiary difference in foraging (Godfrey 1966). Where the Black-capped and Carolina Chickadees meet in southeastern USA, both occur in broadleaved woods and have similar ecology. Near the borders of Virginia, Kentucky, Tennessee and North Carolina, the woods at a lower altitude are occupied solely by the Carolina Chickadee, while only the Black-capped is found on nearly all the mountains above about 1000 m in Western Virginia and rather higher further south. In both areas, the Black-capped is absent from a few of the higher mountains and here the Carolina Chickadee is found (Tanner 1952), suggesting competitive exclusion. The boundary between the two species has gradually shifted northward, by about 50 km in 50 years in part of Illinois (Brewer 1963), presumably because it is determined by their respective adaptations to climatic factors, which have somewhat changed.*

Members of the black-capped chickadee superspecies regularly coexist with tufted tits. In eastern USA, either the Black-capped or Carolina Chickadee is in broadleaved woods with either the Tufted or Black-crested Tit, but the latter are twice as heavy, with proportionately heavier beaks, they forage to a much greater extent on the ground and in the lower storey of the forest, and also take more mast, hence their diets are mainly different. The same presumably applies where the Plain Tit coexists with the Black-capped Chickadee in part of Oregon and the Great Basin and with the Mountain Chickadee in parts of California (but in addition they are partly separated by habitat and altitude,

*Apparently in some areas the ranges of the Black-capped and Carolina Chickadees almost adjoin but with a gap where neither occurs, possibly because hybrid young would be disadvantageous (Brewer 1963), but in Kansas their ranges adjoin and they frequently interbreed (Rising 1968).

Dixon 1961). The Mexican and Bridled Tits are much more similar in size. Where they overlap in Mexico, the Mexican Chickadee is mainly in conifers above 2000 m and the Bridled Tit mainly in oak woods from 600 m to about 2500 m, and though they overlap in mixed woods at 2000–2500 m, the Mexican Chickadee becomes commoner as the pines increase higher up, and the Bridled Tit as the oaks increase lower down, so they are largely separated by habitat (Marshall 1957, Dixon pers. comm.). Further, the Mexican Chickadee has a narrow and the Bridled Tit a broad beak, presumably adapted to feeding in coniferous and broadleaved trees respectively.

The brown-capped chickadees

The Boreal (or Hudsonian) Chickadee *P.hudsonicus* breeds from eastern Canada and the northern USA almost across the Continent. It is replaced geographically in the coast ranges of the west by the Chestnut-backed Chickadee *P.rufescens,* which extends from southern Alaska south to the middle of California. Where they adjoin in southern British Columbia and Washington State, the Boreal Chickadee breeds inland at higher altitudes in the boreal and subalpine forest, the Chestnut-backed lower down in the spruce, fir and Douglas fir of the wet coastal forest and also inland in the Columbian forest (Munro and Cowan 1947). Hence they have separate habitats. The Chestnut-backed is smaller than the Boreal Chickadee, with a finer beak. It is an ecological equivalent of the European Coal Tit *P.ater,* and like the latter feeds largely in spruce needles. The third American brown-capped species is the Siberian Tit *P. cinctus,* which breeds in the interior of northern Alaska and northern Yukon, where it overlaps with the similar Boreal Chickadee, but is at least partly separated by habitat, as it occurs chiefly at the forest edge, notably along rivers, and the Boreal in heavy, dense and shady coniferous forest.

The only contact between a brown-capped chickadee and a tufted tit is that between the Chestnut-backed and

Plain Tit in the coastal areas of central California, where the Chestnut-backed is mainly in coniferous forest and the Plain Tit in oaks, but they meet in broadleaved riparian woodland, where the Chestnut-backed hunts primarily in the terminal twigs of the canopy and the much heavier and heavier-beaked Plain Tit mainly below the canopy; the Plain Tit also takes seeds (Root 1964).

The Siberian Tit does not come in contact with any of the American black-capped Chickadees. The Chestnut-backed meets the Mountain Chickadee on the western slopes of the Rockies, and the Black-capped Chickadee in Washington State and southern British Columbia, but is separated from both by its habitat, dense wet spruce and fir at lower altitudes (Jewett et al. 1953, Munrow and Cowan 1947). However, it sometimes coexists with the Black-capped Chickadee in mixed broadleaved and coniferous woodland in winter, where the Chestnut-backed feeds almost the whole time in conifers, especially at 14–15 m above the ground, and the Black-capped feeds most of the time in broadleaved trees, usually within 2 m of the ground (Smith 1967). In one area, the San Juan Islands, the Chestnut-backed is in broadleaved trees, but here the Black-capped Chickadee is absent (Sturman 1968). There is also a small overlap between the Boreal and Mountain Chickadees in part of British Columbia, but they are separated by their habitats, already noted (Munro and Cowan 1947).

The most extensive overlap between brown-capped and black-capped chickadees is between the Boreal and Black-capped, as the Boreal breeds in the Hudsonian and Canadian zones, the Black-capped in the Canadian and Transition zones, so both occur in the wide band of the Canadian Zone across North America. Here they are at least partly separated by habitat, the Boreal in heavy dark coniferous forest and the Black-capped in lighter and more open mixed broadleaved and coniferous woods; a difference similar to that between their palaearctic counterparts, the Siberian and Willow Tits, though the situation is not identical.

Comparison between North America and Europe

From museum skins, one would assume that the ecology of *Parus* tits was similar in North America and Europe, for as already mentioned, each North American species has a European counterpart, and lives in the same type of habitat and feeds in the same parts of the trees as its counterpart. The close resemblance in size and beak between each of these pairs can reasonably be attributed to convergent adaptation. Whereas, however, in Europe up to six species breed in one area, and three, sometimes four or five, in the same type of wood, in North America only two normally breed in the same area (except in a few small areas of overlap between two species which elsewhere replace each other geographically). At first, I wondered whether this might be due to a difference in the distribution of forest, American tits being much more restricted in range than the European tits because the type of tree to which each species is adapted does not occur together with the types of trees to which other species are adapted. For instance, the spruces and firs preferred by the Chestnut-backed Chickadee occur in the west coast ranges, and the open pines preferred by the Mountain Chickadee higher up inland. Hence if the two species are restricted to trees of these types, they will inevitably occupy different areas. In Europe, on the other hand, spruce and pine often occur in the same forest, so their European counterparts, the spruce-loving Coal Tit and pine-loving Crested Tit, can occur alongside each other. But this is not a sufficient explanation, for though the Coal Tit prefers spruce it often feeds in pine, and the Crested Tit likewise feeds in spruce as well as pine. Each of them, however, selects mainly different parts of the tree in which to feed, and so they coexist in both pure pine woods and pure spruce woods without effective competition. Similarly the Willow Tit coexists with the Coal and Crested Tits in most European coniferous woods, whereas its American counterpart, the Black-capped Chickadee, is normally separated by range and, where it meets them, by habitat and altitude, from the Chestnut-backed and Mountain Chickadees. Similar differences hold in broadleaved woods. The Blue and Great Tits occur together over most of Europe, but the Bridled and the Black-crested Tits in the main replace each other geographically. This is the more striking since, in their small area of overlap, the Bridled differs in feeding stations from the Black-crested in the same kind of ways as does the Blue from the Great Tit, though the differences are less clear-cut. Similarly the Black-capped and Carolina Chickadees do not overlap at all in range, whereas their European counterparts, the Willow and Marsh Tits, regularly breed in many of the same woods. Widespread co-existence in the same habitat is found in America only in representatives of the tufted tit and black-capped chickadee superspecies in broadleaved woods in eastern USA, and between the Black-capped and Hudsonian Chickadees in the Canadian zone.

The ways in which the North American species are isolated from each other have been summarized in Appendix 7, which may be compared with table 7 (p. 36) for the European species. Europe is, of course, much smaller than North America, but even if the European table were extended to the whole palaearctic, the proportion of species segregated by range would still have been much less than in North America. On the other hand, the situation in North America is rather similar to that in Africa south of the Sahara (p. 49). The contrast between the four main areas analysed here is summarized in table 11.

Since six species of *Parus* coexist over much of Europe, but only two do so over much of North America, one might have expected that each American species would be more abundant than its European counterpart, but this is not so. To cite only one example, in 32·5 ha of central hardwood forest in the District of Columbia counted in six of the years 1948–54, there were on average a little over 5 pairs of

Table 11. Ecological segregation of *Parus* species in different regions

	Europe	Asiatic mountains	Africa south of Sahara	North America
Number of species	9	14	10	10
Number segregated primarily by				
range (G)	3	2	7	13
habitat (H)	$11\frac{1}{2}$	51	13	5
feeding station (F)	$15\frac{1}{2}$	16	2	2
no contact (—)	6	22	23	25
Proportion separated primarily by				
habitat	32%	56%	30%	11%
feeding station	43%	18%	4%	4%
range (incl. no contact)	25%	26%	67%	84%

Note: derived from table 7 and appendices 3, 5 and 7. A single symbol in one square has been scored as 1, two symbols (without brackets) in the same square as $\frac{1}{2}$ each, two symbols, the second being in brackets, as 1 for the first symbol, and any symbol followed by a ? as if the ? were not there. In Asia, Africa and America, it was not always easy to determine how to allocate the segregation of two species which differ in range and also in habitat, where the habitat restricts the range.

Tufted Tits and just under 5 pairs of Carolina Chickadees breeding each year, or roughly 31 pairs of tits per square kilometre (Stewart and Robbins 1958). In the 22·5 ha of Marley Wood, near Oxford, studied by workers at the Edward Grey Institute between 1948 and 1967 inclusive, there were on average 38 pairs of Great Tits, 28 of Blue Tits in nesting boxes and perhaps 10 more in natural holes, about 7 of Marsh Tits and one each of Willow and Coal Tits, making in all about 380 pairs of tits per sq km, or 12 times the American figure. Indeed the Great Tit or the Blue Tit alone is much commoner than both the American species combined. There is no reason to think these censuses atypical.

It has been suggested that, in North America, certain other families, such as the warblers Parulidae in coniferous forest and the vireos Vireonidae in broadleaved trees, occupy some of the feeding niches in woods which in Europe are occupied by tits (Mayr in Dixon 1961). If so, this would help to explain the paucity both of coexisting species and of overall numbers of tits in American woods. The American tits are derived from only two subgenera, the European from five, which suggests that tits may have existed for longer in the Old than the New World. If they reached North America in relatively recent times, they might perhaps have found parts of the niches which they occupy in the palaearctic already filled by species in indigenous American families.

The origin of coexistence

The differences in beak, body-size and feeding behaviour of the American tits are adaptations to different habitats, not to coexistence in the same habitat. Perhaps coexistence with segregation by feeding takes a long time to be evolved. The distribution of the American tits suggests that it is most likely to be evolved where two species, which in the main replace each other geographically, meet along a common boundary. Here it can be evolved gradually, as the two species are in contact the whole time, in contrast to a remote island, where any new arrival must compete immediately with the established species and if it fails, dies out. The evidence also suggests that, even after coexistence has been evolved along a common boundary, it may take a long time to spread further. Otherwise, one might have expected that the pairs which coexist in a small area, such as the Bridled and Black-crested Tits, would do so over a much wider area. This pair differ less in size than their European counterparts, the Blue and Great Tits, which might provide one reason for their inability to coexist over a wide area, but the size differences between other American tits are similar to those between their European counterparts, as can be seen by comparing tables 9 and 2, so this does not provide a general explanation. Perhaps differences in feeding behaviour take a long time to evolve. Another possibility is that there are only a few special areas in North America where the ecological resources are sufficiently diverse to permit different species of tits to coexist, but it is not obvious why such areas should be so much more plentiful in Europe, though possibly the presence of vireos and parulid warblers in North America is critical.

Since this comparison has been based on European and North American tits, it may be recalled that the African tits present an even greater contrast with those of Europe, as they belong to only one subgenus and there is usually only one species in any one place, except for the two in brachystegia woodland. However, the African species do not show the morphological diversity of the American ones. The special feature of the American tits, as compared with the European, is that their parallel morphological differences are associated with different means of ecological isolation.

Conclusion

The European species of *Parus* were selected for review owing to their ecological complexity, but those in the other continents merely because they belong to the same genus. The situation in the other continents is very different from that in Europe, and in particular many fewer species coexist in the same habitat and are separated by feeding. As many as 14 species reside in the mountains of central Asia, of which most are separated by habitat, associated with a big altitudinal range and a diversity of forest types. In most of Africa south of the Sahara and most of western North America, there is only one species in any one habitat (though two in brachystegia woodland in Africa and two in broadleaved woodland in the eastern USA). Perhaps this is because the African species, derived from a single subgenus, and the North American, derived from two, are at an earlier stage in their evolution than the European. In addition, the presence of passerine species in other families with similar feeding requirements perhaps restricts the niches available for tits. Each North American species has a European counterpart, but the adaptations of the American species are for different habitats, not for coexistence in the same habitat. Probably, coexistence is most easily evolved along a common boundary, but even so perhaps takes a long time to spread, since several North American species coexist only in small areas where their ranges adjoin. These conclusions are uncertain because past history is involved, but though the latter is unobservable, it requires discussion if, as seems likely, it has been important.

Chapter 4

Another World Survey
Nuthatches *(Sitta)*

One further continental genus, that of the nuthatches *Sitta*, has been selected for treatment on a world basis, initially because it was in this genus that 'character displacement' (see p. 10) was discovered (Vaurie 1951, Brown and Wilson 1956). There are 21 species of nuthatches (Greenway 1967a), distributed in Europe, Asia and North America. They are adapted for climbing on trunks and branches, eat both insects and tree seeds, and have short tails and legs but long beaks, the beak being thinner in the species of coniferous than of broadleaved forest. The upper parts are bluish grey, bright blue in a few tropical species, and the underparts are reddish in tropical species, pinkish buff in broadleaved temperate woodland and usually white in conifers. Various white-breasted, thin-beaked nuthatches of conifer forest in the Old and New Worlds were formerly treated as conspecific, but it is now appreciated that their similarities are due to convergent evolution (Vaurie 1957, Löhrl 1960–1). Species typical of broadleaved and coniferous forest respectively are shown in fig. 13.

Europe

The comparatively large and thick-beaked European Nuthatch *S.europaea* is widespread in Europe in broadleaved woods (and occasionally in conifers, e.g. in Switzerland, Glutz 1962). The other three European species are very local. The Western Rock Nuthatch *S.neumayer*, which is

slightly larger with a proportionately longer beak, is found in rocky areas with scrub, especially on limestone, in southeastern Europe and feeds almost entirely on the ground. The Corsican Nuthatch *S.whiteheadi* (shown in fig. 13), is white-breasted, thin-beaked and much smaller than *S. europaea*, which it replaces in Corsica, where it lives in pine forest; it seems best regarded as separated by habitat rather than range. Another white-breasted but larger species in conifers, *S.krüperi*, just reaches Europe on Mytilene (Löhrl 1965). It replaces *S.whiteheadi* geographically in Turkey, where *S.europaea* and *S.neumayer* also occur, but differs from them in habitat (Danford 1878). The means of segregation of the four European species are summarized in appendix 8A. There is no small nuthatch in the northern conifer forest of Europe corresponding to *S.canadensis* in the New World.

North America

Four species breed in North America. The small Redbreasted Nuthatch *S.canadensis* is characteristic of the coniferous forest of the Canadian Zone, so is found in Canada, a small part of northeastern USA and high up in the western mountains from British Columbia south to California. It is separated by habitat, including altitude, and in large part also by range, from the other three American species, which are typical of the Transition Zone, which is found south of, and in the mountains below, the Canadian Zone.

The much larger Whitebreasted Nuthatch *S.carolinensis* breeds primarily in the Transition Zone, mainly in broadleaved woodland, but in pines in southeastern USA and the west (e.g. Howell 1924, 1932, Grinnell and Miller 1944). It breeds in most of the United States and small parts of Canada and Mexico. Its range is mainly south of, and in the

Fig 13. European Nuthatch *Sitta europaea* (top) and Corsican Nuthatch *S.whiteheadi* (bottom). By Robert Gillmor.

western mountains lower than, that of the Redbreasted Nuthatch.

The Brown-headed Nuthatch *S.pusilla* is found in open pine forest in southeastern USA, where its much smaller size probably segregates it by feeding from the White-breasted Nuthatch, and it also differs in that it does not frequent broadleaved woodland or store the nuts of broadleaved trees (Howell 1924, 1932). The similar-sized Pygmy Nuthatch *S.pygmaea* lives in the western mountains and part of the Pacific coast, from British Columbia to Mexico, mainly in open forest of Yellow Pine. It likewise coexists with the White-breasted Nuthatch, but is much smaller, and it usually feeds high in the trees, searching the needles and cones on the topmost outer branches and not, like the White-breasted Nuthatch, the trunks or larger limbs (Grinnell and Miller 1944, Norris 1958). The main ways in which the four American species are isolated from each other are summarized in Appendix 8B, with their measurements in a footnote.

Rock Nuthatches

The two rock nuthatches, the Western *S.neumayer* and Eastern *S.tephronota*, differ in habitat from all the rest since, as already noted, they frequent rocky hill slopes and feed on rocks and the ground. *S.neumayer* occurs in the southern half of the Balkans through Asia Minor to the Caucasus and Iran, *S.tephronota* from the Caucasus and Iran through Afghanistan and Baluchistan to the Tian Shan. Where only one of them is present they are closely similar in size and appearance, but where they coexist in Iran and the Caucasus, they are completely separated, *S.neumayer* being smaller (male wing 73–81 mm, culmen 21–25 mm) and *S.tephronota* larger (male wing 87–100 mm, culmen 27–31 mm) (Vaurie 1950, 1951, and fig. 14). In Iran they live in the same habitat (L. Cornwallis pers. comm.) and are presumably separated by feeding.

Their distinctiveness in the area of overlap was termed 'character displacement' by Brown and Wilson (1956), and rightly excited attention, but the equally remarkable question has hitherto been overlooked of why, since they

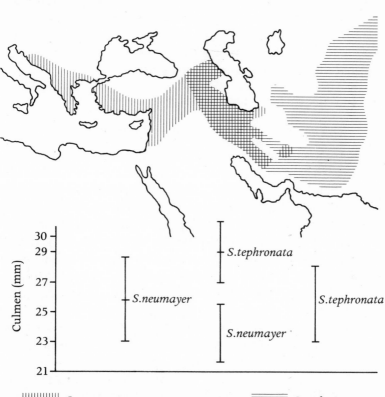

Fig 14. Ranges (above) and beak-length (below) of Western Rock Nuthatch *Sitta neumayer* and Eastern Rock Nuthatch *S.tephronota* (from Vaurie 1951).

Note that they differ completely in size where they overlap in Persia, and are of similar size elsewhere.

have evolved coexistence in one area, they have not done so elsewhere. There seem two possible explanations, one that the food resources are unusually diverse in Iran, so that only here can their niche be subdivided, the other that they came in contact in relatively recent times and that, after ecological segregation had been evolved at their boundary, it is taking a long time to spread further. For the first alternative there are probable parallels among Darwin's finches and the island white-eyes, considered in later chapters, and for the second perhaps among the North American tits discussed in the previous chapter. On the face of things, the second explanation seems the more likely for the rock nuthatches, but the point needs study.

The mountains of central and southeast Asia

The greatest number of species of nuthatches in any one region is 12, in the mountains of east-central Asia (as set out in Appendix 9), but many of them have restricted ranges, and the most in any one area is 5, in northern Burma. There has been much difference of opinion concerning taxonomy in this area (see note to Appendix 9) and Voous and Van Marle (1953) and Ripley (1959) have drawn attention in different ways to the unusual ecological situation. Ripley's map of the ranges of the critical species is reproduced as fig. 15.

Two species, the Chestnut-bellied *S.castanea* and the Velvet-fronted Nuthatch *S.frontalis*, occur in tropical and subtropical forest in the lowlands and foothills up to about 1300 m in India, Burma and Thailand. They differ in habitat and feeding, *S.castanea* living in drier and more open deciduous woodland, often feeding on the ground, and *S.frontalis* living in evergreen and moist deciduous woodland, and feeding chiefly on the ends of branches, not on the trunks; its bright colouring also suggests that it is a bird of the canopy (S. Ali 1953, amplified pers. comm.,

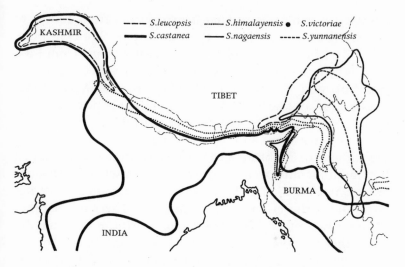

Fig 15. Ranges of 6 species of nuthatches *Sitta* in Himalayas, N. Burma and S.W. China (from Ripley 1959).

Delacour and Jabouille 1931, de Schauensee 1934, Smythies 1953, Whistler 1941).

These two species are normally replaced above 1200–1400 m in temperate montane broadleaved forest (mainly oaks, also open pine) by one of two small thick-billed species, and in a limited area by both. The White-tailed Nuthatch *S.himalayensis* is found from the Punjab across northern India to N. Burma and part of W. Yunnan, and south through eastern Burma to northern Laos and Tonkin. The Naga Hills Nuthatch *S.nagaensis* is found mainly to the northeast of *S.himalayensis*, in S. Kanzu, W. Szechwan, S.E. Tibet and Yunnan, but there are also partly isolated populations in northern and central Burma (south to Mount Victoria), northern Thailand, S. Annam and Fokien. For the most part, these two species live in the same altitudinal zone in similar types of forest and replace each other geographically, with some interdigitation in Indochina. But they overlap in N. Burma (including a little of eastern India

69

and W. Yunnan), where *S.nagaensis* occurs from 1200m to 2100 m (but usually lower than this upper limit) in pine, oak, elder, and larch, while *S.himalayensis* occurs from 1400 to 2700 m, nearly always in heavy broadleaved forest (Stanford and Ticehurst 1938, Stanford and Mayr 1941, Voous and van Marle 1953). Hence in their zone of overlap, they are segregated by habitat and altitude.*

At the western end of the Himalayas, in Kashmir, *S.himalayensis* is absent and is replaced in temperate forest by a montane subspecies of the normally lowland *S.castanea (S.c.cashmirensis)*. The latter has varied in colour, but not beak, in the direction of *S.himalayensis*. There is one other montane race of *S.castanea (S.c.tonkinensis)* on Doi Hua Mot in northern Thailand and some other mountains in Tonkin and Laos, where *S.nagaensis* is absent (Deignan 1945). Here, however, *S.himalayensis* is present but is largely separated by altitude, *S.c.tonkinensis* occurring from 1200 to 2200 m and *S.himalayensis* above 2000 m (Voous and Van Marle 1953, Delacour and Jabouille 1931). These two exceptions strongly suggest that, everywhere else in its range, *S. castanea* is held below about 1300 m through competitive

---

* *S.nagaensis* is a little larger with a slightly longer (but not thicker) beak than *S.himalayensis*, but the difference is too small to suggest that they differ to any important extent in feeding. Ripley (1959) stated that *S.nagaensis* has a thinner beak where it overlaps with *S.himalayensis* than elsewhere, which he attributed to character displacement, but though some populations of *S.nagaensis* differ a little in their dimensions from others, I could not find a consistent difference in the direction postulated by Ripley, and, if it exists, it is much too small to attribute to character displacement. Ripley similarly claimed that *S.himalayensis* has a broader beak where it overlaps with *S.nagaensis* than elsewhere, which he again attributed to character displacement, but though *S.himalayensis* has a proportionately rather longer beak in the Himalayas, where *S.nagaensis* is absent, than in N. Burma where they coexist, it has a yet shorter and proportionately broader beak in Tonkin and Laos, where *S.nagaensis* is again absent. Hence the variation is not correlated with the presence or absence of *S.nagaensis*. Conceivably its longer beak in the Himalayas is correlated with it there occurring in oaks and pines (Ali 1949), but elsewhere only in broadleaved forest.

exclusion, but why it should displace *S.nagaensis* in Tonkin and *S.himalayensis* in Kashmir is not clear.

In the mountains of S.E. Tibet, Szechwan and Yunnan, a further small species, the Yunnan Nuthatch *S.yunnanensis*, occurs in coniferous forest in the altitudinal zone frequented by *S.nagaensis* (Schäfer and de Schauensee 1938). Elsewhere, as just noted, *S.nagaensis* occurs in both broadleaved trees and open pine, but in this area it is restricted to broadleaved trees (see also Ludlow 1944). This restriction is presumably due to competitive exclusion by *S.yunnanensis*. Hence while in N. Burma *S.nagaensis* coexists with *S.himalayensis* and often occurs in coniferous woods (though also in broadleaved trees), in S.E. Tibet it coexists with *S.yunnanensis* and occurs solely in broadleaved woods, an illuminating example of competitive exclusion in relation to the other species present.

The situation is different again on Mount Victoria in central Burma (Stresemann and Heinrich 1940). Here *S.castanea* and *S.frontalis* in the foothills are replaced as usual at about 1400 m by *S.nagaensis*, while the latter is replaced at 2600 m in humid evergreen broadleaved forest rich in epiphytes by the endemic *S.victoriae*. *S.victoriae* is closely related to *S.himalayensis* but has a much thinner beak. The difference in altitude between *S.nagaensis* and *S.victoriae* is in the same direction as that between *S. nagaensis* and *S.himalayensis* in north Burma, but on Mount Victoria there in no overlap.

One further thin-beaked and white-breasted nuthatch, the White-cheeked *S.leucopsis,* occurs in the alpine spruce and cedar forests of the Himalayas, one race in Kashmir and the other far removed in S.E. Tibet and adjoining China (Bates and Lowther 1952, Schäfer and de Schauensee 1938, Schäfer 1938, Löhrl and Thielke 1969). In both areas, the type of forest which it frequents is at a higher altitude than that frequented by any of the other resident nuthatches. It feeds mainly on the undersides of the upper branches of the trees. The beak is of very different length in the two subspecies, but they are not found alongside any other nut-hatches, so this is not due to character displacement (cf. Ripley 1959).

Two other species are found in montane forest, usually above 1400 m, and coexist with *S.himalayensis* or *S. nagaensis*, but they are much larger than the two latter, and obviously do not compete with them for food. The Beautiful Nuthatch *S.formosa*, recorded in Sikkim, northern Burma, Assam, northern Tonkin and Laos, is everywhere rare. Its glossy blue plumage, which recalls that of *S.frontalis*, suggests that it may live in the treetops. The even larger, but dully coloured, Giant Nuthatch *S.magna* has a proportionately huge thick beak. It is found in east central Burma, western Yunnan and northern Thailand, mainly in dense evergreen forest (de Schauensee 1934, Riley 1938, Deignan 1938, 1945) but also in open pine forest (Smythies 1953). Its range is mainly east and south of that of *S.formosa*, and apparently these two species do not anywhere coexist. In any case, their colouring suggests that they have different feeding stations.

S.europaea (in the sense used by Greenway 1967a) occurs only in the extreme north of the mountains of central Asia, in Kansu, part of Szechwan, and the Tian Shan. In Kansu and the Kokonor, there is also the thin-beaked, white-breasted Chinese Nuthatch *S.villosa*, which extends from there to northeast China. It lives in coniferous forest and is separated by habitat and altitude from *S.europaea*, which in most of this area is subtropical. It is separated by range from *S.leucopsis* and *S.yunnanensis*, which live in coniferous forest further south.

Finally, the rock nuthatch *S.tephronota* occurs in the Tian Shan, alongside *S.europaea*, and in the western Himalayas with various other species, from all of which it is separated by its habitat preference for rocky hill slopes.

The summary in Appendix 10 shows that most species of *Sitta* in central and southeast Asia are separated from each other by habitat, including altitude, or have such restricted ranges that they do not come in contact with each other. Few species coexist in the same habitat, the two main examples being (i) *S.castanea* with *S.frontalis* in the lowlands, (ii) either *S.formosa* or *S.magna* with either *S.himalayensis* or *S.nagaensis* in montane forest.

Other species in southeastern Asia

Only two more species of *Sitta* remain to be discussed, both in southeast Asia. Both are brightly coloured, so presumably feed in the canopy. The Lilac Nuthatch *S.solangiae,* found in part of Tonkin and Annam, is apparently a well-marked form of *S.frontalis,* which it replaces geographically. The Blue Nuthatch *A.azurea* of Malaya, Sumatra and Java, occurs in Malaya with *S.frontalis,* but is found above 900 m and usually above 1200 m, while *S.frontalis* occurs below it (Glenister 1951).

Conclusion

Nuthatches have nowhere been studied in the field with respect to competitive exclusion, yet as this survey shows, there is circumstantial evidence that each species is isolated ecologically from every other, and also as to the way in which each is isolated. Evidently competitive exclusion is normal, and the means by which it is achieved are usually simple. Of the 21 species of *Sitta,* 12 occur in the mountains of central and southest Asia, where nearly all of them are separated from each other by range or habitat (forest type). Although the classic case of character displacement occurs in this genus, the latter provides only a few examples of species living in the same habitat and differentiated by size and presumably feeding.

Chapter 5

Coexistence in a Man-modified Habitat
The European Finches *(Fringillidae)*

The European finches have been selected for detailed treatment next because they introduce the reader to the problem of birds in a man-modified environment. Many of the species concerned live in natural forest or scrub, but many others are common in cultivated land and some are, in Europe, confined to it, which helps one to see what happens when a natural ecological balance has been disturbed. Their main diet of seeds is studied with relative ease, and their adaptations in beak and legs for feeding are, with those of Darwin's finches discussed later, better known than those of any other group of closely related birds. The British species have been studied intensively by I. Newton at the Edward Grey Institute, and the present chapter is based on his findings (1964, 1967abc, amplified pers. comm.).

The colonization of man-made habitats

The habitat where a species now lives successfully in farmland need not bear any close resemblance to its natural habitat. This is best illustrated by the (Lesser) Redpoll *Carduelis flammea,* which has spread over the cultivated land of southern England only in the present century. Here it breeds mainly in young conifers planted by man and feeds chiefly on seeds of farmland weeds, while in the last few years it has taken to eating fruit buds in commercial orchards in winter. Anyone who knew the bird only in this new environment could not guess that its main natural

74

habitat in Europe is the boreal birch forest, where it depends for food primarily on birch seed (Newton 1967a). Curiously, it started taking fruit buds earlier in its introduced home in New Zealand than in England (Stenhouse 1962, Newton 1967b).

A more notorious species which invades English orchards in the late winter and early spring to eat fruit buds is the Bullfinch *P.pyrrhula*, but though this habit was recorded in England in the sixteenth century, the bird has become a pest only in recent times. Moreover, whereas formerly it returned to woods to breed, it nowadays nests freely in hedgerows and gardens and has increased enormously in numbers (Newton 1964, 1967c). In eastern Europe, it feeds in villages and towns in winter but still returns to woods for breeding.

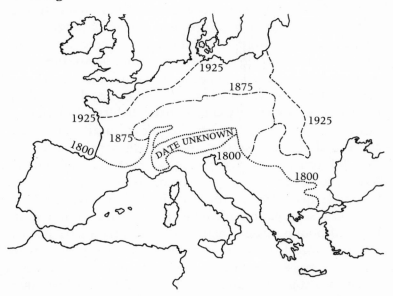

Fig 16. Spread of Serin *S.serinus* from southern Europe (from Mayr 1926).

Note: Since Mayr wrote, the Serin has reached the southern side of the Baltic, also Denmark, south Sweden and (in 1967) England.

A change in habitat or feeding habits may lead not only to an increase in numbers, but also to a big extension in range. The best documented case is that of the Serin *S.serinus,* which occurs naturally in open woodland, especially conifers, in southern Europe, but which, more than a century ago, found suitable breeding places, similar in appearance to open forest, in city parks and gardens with scattered trees, where it feeds on a variety of seeds and buds. Its success in this new habitat enabled it to spread north into much of central Europe in the course of the nineteenth century, as shown in fig. 16 (Mayr 1926). This spread has continued, as the species became established in Holland, north Germany, Denmark and Sweden in the present century (Voous 1960) and first bred in Britain in 1967 (Ferguson-Lees 1968).

In view of these known changes, it is reasonable to suppose that various other European finches changed their feeding habits, range or habitat before there were ornithologists to record them. Hence present distributions may be very different from what they were a thousand years ago. This raises two difficulties. for the ecologist. First, farmland is not merely an artificial habitat, but has repeatedly changed as one type of agricultural practice has succeeded another, so that there could hardly have been time for birds to evolve ecological isolation in farmland. Indeed some of them, as just noted, are still changing their habits. Hence whereas the British finches show clear-cut ecological segregation in woods, they do not do so in farmland. Although modern managed woods are a pale shadow of virgin forest, they are evidently like enough for many woodland passerine birds to live there under fairly natural conditions; but farmland is extremely different from their natural haunts. Secondly, the size and shape of their beaks, legs and other structures were evolved in their natural habitats, and have not had time to become modified appreciably in relation to their changed feeding habits in farmland, where their adaptive significance may therefore be hard to interpret. Indeed, some of their adaptations might not be used at all. On the other hand, the fact that ecological isolation is absent, or at best partial, in farmland provides valuable evidence that the balance found in natural habitats may have taken a long time to evolve. This balance is, of course, dynamic, and natural habitats also change; but such changes are usually slow, or simply involve the replacement of one natural habitat by another in existence elsewhere, whereas changes in farmland may be rapid and may result in habitats unlike any found before.

Winter foods in southern England

Ten species of finches spend the winter round Oxford, in southern England. They depend almost entirely on seeds formed in summer and autumn, hence their foods are sparsest at the end of the winter, and it is then that one would expect potential competition between them to be most severe, and that competitive exclusion, if it applies, would apply most strongly. The diet of each species in late winter has been summarized in Table 12, and the seven woodland species are portrayed in fig. 17.

In the nearly natural habitat of English woodland, six of the seven resident species depend basically on different species of seeds in winter, so do not compete. However, both the Chaffinch *Fringilla coelebs* and Brambling *F.montifringilla* eat much beechmast, but they do so mainly in those winters, often alternate ones, when beechmast is abundant, when there may be no effective competition for it, and in other winters many Chaffinch move to arable land to eat grain and other seeds, and though some Bramblings do likewise, most of them move abroad in search of other places where beechmast is abundant. Hence all seven species are separated ecologically, as portrayed in fig. 19. But in autumn, when seeds are more plentiful, some of them take the same kinds. In particular, the Redpoll and Bullfinch then eat many seeds of meadowsweet *Filipendula ulmaria*,

Table 12. Main seeds taken by finches Fringillidae round Oxford in late winter

A. WOODLAND SPECIES	Hornbeam Carpinus	Blackthorn Prunus	Alder Alnus	Birch Betula	Ash Fraxinus	Bramble Rubus	Spruce Picea	Larch Larix	Beech Fagus
Hawfinch C.coccothraustes	+	+							
Siskin C.spinus			+	(+)					
Redpoll C.flammea				+					
Bullfinch P.pyrrhula				(+)	+	+			
Crossbill L.curvirostra							+	+	
Chaffinch F.coelebs									+
Brambling F.montifringilla									+

B. FARMLAND SPECIES	Cabbage, charlock Brassica Sinapsis	Goosefoot Chenopodium	Bramble Rubus	Persicaria, knotgrass Polygonum	Dock Rumex	Nettle Urtica	Teasel Dipsacus	Burdock Arctium	Thistles Cardium Cirsium	Cultivated cereals
Greenfinch C.chloris	+			+				+		+
Goldfinch C.carduelis							+	+	+	
Linnet C.cannabina	+	+		+		+				
Bullfinch P.pyrrhula		+	+		+					
Chaffinch F.coelebs	+	+		+						+

Notes: based on Newton (1967a), who gave a table for the whole year. Note that beechmast was a subsidiary food of the Hawfinch and the main food, in good seasons, of both Chaffinch and Brambling, but in the other years most Chaffinches moved to farmland and most Bramblings went elsewhere in Europe. Bullfinches ate many tree buds in late winter, especially in those years when the ash seed failed, and this held in both woods and farmland. Especially with snow on the ground, Greenfinches depended greatly on foods put out by householders in gardens (not shown in the table). Many Linnets and Goldfinches migrated to the Continent for the winter.

Fig 17. British finches feeding in winter. By Robert Gillmor.

Hawfinch on hornbeam

Siskin on alder

Redpoll on birch

Bullfinch on ash

Crossbill on spruce

Chaffinch and Brambling feeding on beechmast

and both they and the Siskin *Carduelis spinus* take much birch seed, but after the early winter, both Siskin and Bullfinch normally turn to their preferred seeds listed in table 12, and only the Redpoll stays on the birch. As mentioned in chapter 1, it is a general principle that several species of birds may utilize the same main source of food at times when it is particularly abundant and when competition for it is unimportant. But competitive exclusion may apply so soon as food becomes scarce, which in these finches is in the late winter.

The situation is different in farmland where, as shown in the lower half of table 12, only two of the five resident finch species are segregated by diet in late winter. These two are the Goldfinch *C.carduelis,* which specialises on the seeds of Compositae, notably thistles and burdock, and the Bullfinch, which takes mainly the seeds of brambles, docks and nettles, together with the buds of various trees, which the other three species rarely take. These latter three overlap in their foods to a considerable extent, though less than table 12 suggests, because though on English farmland they take mainly the same kinds of seeds in winter, many of the Chaffinches are then feeding in woods on beechmast, many Greenfinches *C.chloris* feed to a large extent round houses, both on weed seeds and on foods, including peanuts, put out by householders, and most Linnets *C.cannabina* have migrated to continental Europe.

Associated adaptations

As shown in table 13, each of these finches has a beak of different size and shape, ranging between the small Redpoll and the large Hawfinch. The latter has a huge conical beak adapted for cracking hard seeds, which is helped by two finely ridged knobs on the palate and another pair in the lower jaw, between which large seeds such as cherry stones are held firmly; although the bird weighs only $1\frac{1}{2}$ oz, it can exert a pressure of up to 159 lbs per square inch (Sims 1955).

Table 13. Size and shape of beak, body-weight and relative length of leg in English finches

Species	length in mm	Beak depth in mm	length/ depth	Mean body-weight (in Nov. in grams)	Ratio of leg-length to body-weight
Hawfinch C.coccothraustes	20·5	17·7	1·2	55	1·4
Greenfinch C.chloris	13·1	11·5	1·2	29	2·2
Goldfinch C.carduelis	12·4	7·5	1·7	16	3·3
Siskin C.spinus	10·4	7·5	1·5	12	3·9
Linnet C.cannabina	9·6	7·6	1·3	19	3·0
Redpoll C.flammea	8·4	6·6	1·2	12	4·1
Bullfinch P.pyrrhula	10·5	10·0	1·1	24	2·6
Crossbill L.curvirostra				40	1·7
Chaffinch F.coelebs	13·0	8·0	1·6	24	2·7
Brambling F.montifringilla	13·0	9·0	1·4	25	2·7

Notes: from Newton (1967a tables 11 and 14). The beak measurements of Hawfinch and Goldfinch are solely of males, those of the females being about 1 mm. shorter and less deep in proportion.

The beak is proportionately long in the Chaffinch and Brambling, this perhaps being an adaptation to their taking many more insects, chiefly in summer, than the other species. It is relatively short and broad in the Linnet, Greenfinch and Redpoll, which feed more than the others on exposed seeds, such as those of grasses, which they enclose in the beak. Not only does each of these three have a beak of different size but, as shown in table 14, each prefers seeds

Table 14. Relation of depth of bill to commonest size of seeds
eaten by four cardueline finches in southern England

| | Mean depth of beak in mm | Percentage composition by weight of seeds taken | | | |
		up to 0·5 mg	0·5–1·0 mg	1–10 mg	10–100 mg	over 100 mg
Redpoll C.flammea	6·6	80	18	2	—	—
Linnet C.cannabina	7·6	15	31	52	2	—
Greenfinch C.chloris	11·5	8	9	27	54	—
Hawfinch C.coccothraustes	17·7	—	<1	1	30	68

Note: from Newton (1967a)

of a different range of sizes. The short and rounded beak of the Bullfinch is adapted for a different purpose, that of tearing into buds and fleshy fruits. The long and narrow beak of the Goldfinch and Siskin is different again, and they use it like a pair of tweezers, to extract the seeds of thistle heads and alder cones respectively. The cock Goldfinch has a rather longer beak than the hen, which enables it to reach teasel seeds, which lie at the bottom of long tubular structures, and which the hen rarely tries to take. Finally the Crossbill *Loxia curvirostra* has a compressed and decurved beak in which the upper mandible crosses over the lower. It inserts the tip of its beak sideways between two bracts of a spruce or pine cone, then raises the bract by means of its upper mandible, sometimes by rotating or opening its beak, and the seed is taken on the protrusible tongue.

Observations on six of these species in captivity showed that each of them prefers a different range of seeds, which is broadly related to the size of its beak, and that each tackles different kinds of seed-heads, which is broadly related to the shape of its beak.

As shown in table 13, the various species also differ in

weight. The Goldfinch, Redpoll and Siskin, which are the three smallest, are the most agile in feeding at the ends of branches, where they often hang upside down (like the small but not the large species of tits discussed in chapter 2). The Crossbill, though heavy, can also hang upside down, being helped in this by its unusually short legs. In contrast, the Chaffinch and Brambling feed more on the ground than the other species and have relatively long legs.

The Crossbill often holds down a cone by its foot when feeding, while the Goldfinch, Redpoll and Siskin, and at times the Crossbill, pull up catkins on long stalks by using their beak and foot in combination, first pulling in part of the stalk and holding this part in the foot, then reaching for the next part and so on. This habit was formerly exploited by birdfanciers, who were able to induce a caged Goldfinch to pull up a miniature bucket of water on a string—hence its old name of 'draw-water'.

The differences in diet of the various English finches are therefore linked with a series of adaptations in structure and behaviour. The differences are greatest between species in different genera, but also hold between congeneric species, notably in *Carduelis*.

Diet in southern England in summer

All the species in the subfamily Carduelinae feed their young on seeds alone, or on seeds and insects, but the Fringillinae bring almost entirely insects. The four species of finches which breed in woods in southern England bring mainly different food to their young, the Hawfinch caterpillars and elm seeds, the Bullfinch chiefly the seeds of various woodland herbs such as Dog's Mercury *Mercurius perennis,* but also elm seeds, the Crossbill conifer seeds, and the Chaffinch almost entirely insects. Hence they are largely segregated from each other.

In contrast, of the five species which breed in farmland near Oxford, only the Chaffinch differs markedly from the rest, because it brings primarily insects, as already mentioned. The commonest seeds brought to their young by the other species are those of the dandelion *Taraxacum officinale* by all four, groundsel *Senecio vulgaris* by the Goldfinch, Greenfinch and Bullfinch, wych elm *Ulmus glabra* by the Linnet, Goldfinch and Greenfinch, brassicas by the Linnet, Greenfinch and Bullfinch, and chickweed *Stellaria media* by the Linnet and Bullfinch. However, certain other seeds are taken frequently by only one of them, those of thistles *Cirsium* spp. by the Goldfinch, and to some extent the Linnet, garlic mustard *Allaria petiola* and sorrel *Rumex acetosa* by the Linnet, and shepherd's purse *Capsella bursa-pastoris* and sowthistle *Sonchus oleraceus* by the Bullfinch.

Hence in summer, as in winter, each of the finch species resident in English woods takes different main foods from the rest, whereas those in farmland, except for the Chaffinch, overlap considerably. In general, the degree of overlap in diet is greater in summer than winter. Since woodland and farmland support mainly different finch species, the most relevant comparison here is between the Bullfinch and Chaffinch, which live in both habitats. In woods, they show virtually no overlap in diet at any time of year, but in farmland the main foods of both include the seeds of *Chenopodium* between October and March, those of *Stellaria media* in April and May and those of brassicas in August, and they have various other minor foods in common. Both are also more clearly separated from the other species of finches in woods than in farmland.

Some of the seeds taken by the farmland species are temporarily so abundant that there may be no effective competition for them. But probably a more fundamental reason for their big overlap in diet is that they have not yet had time to evolve the adaptions necessary for segregation. Ecological isolation has presumably been evolved in birds because those individuals survive best which search for different foods from those taken by other species, but food selection of this type and the associated morphological differences can hardly be evolved quickly, and evidently there has not been time for it in the changing environment of farmland.

A European survey

This concludes my discussion of the main point considered in the present chapter, but in the following chapters ecological isolation is considered in all congenic European birds, so it is convenient to cover all the finches here; the others have not been studied in detail and what follows is based on general works, notably Voous (1960), unless otherwise stated. The ranges of all 18 European species have been summarized in Appendix 11 (p. 283). The species in different genera have such different beaks that it is reasonable to suppose that they do not effectively compete with each other for food (except that some *Carduelis* species look rather like *Serinus* or *Carpodacus*), and attention is concentrated on species in the same genus.

First, however, the four genera may be mentioned each of which has only one European species. Two of these, the Hawfinch (in *Coccothraustes*) and the Bullfinch (in *Pyrrhula*) have already been considered. The Hawfinch, found in broadleaved forest and city parks, depends on such hard seeds as those of hornbeam, yew *Taxus baccata,* blackthorn *Prunus spinosa* and other species of *Prunus*, which separates it from all the rest. In most of western Europe the Bullfinch is resident in broadleaved woods throughout the year, but in the north and in mountain areas in central Europe, it breeds mainly in spruce and fir forest, where its diet includes conifer seeds, and these populations move south or lower down in winter, chiefly to broadleaved woods. The main foods in broadleaved woods have already been described.

The Pine Grosbeak *Pinicola enucleator* breeds in light coniferous and birch forest in the taiga, where it eats buds and shoots, and its beak is a larger version of that of the bud-eating Bullfinch. In winter, it tends to move somewhat further south, particularly in occasional invasion years. Its winter foods include spruce seeds and rowan berries, which are eaten by various other passerine species, and also birch buds. Its invasions, which are presumably correlated with food shortage, occur at longer intervals than, and are not coincident with, those of any other 'invasion bird', such as Crossbills *Loxia* spp., the Waxwing *Bombycilla garrulus* or the Nutcracker *Nucifraga caryocatactes* (Lack 1954), which presumably means that it does not depend on the same combination of foods as any of the others. In overall size and beak it is so much larger than the Bullfinch that it obviously takes mainly different foods, and its beak is so different from that of any other European finch that it seems safe to say that it takes mainly different foods from them also. Anyway it lives so far north that it does not come into contact with most of them.

The fourth species, the Scarlet Rosefinch (or Grosbeak) *Carpodacus erythrina,* is much smaller, about the size of a Linnet. It breeds chiefly in riverine thickets of willow and alder, at times also in gardens. Its diet has not been studied sufficiently to say how it differs from other finches of similar size, but it is separated from some of them by habitat.

Crossbills Loxia

The three European crossbills provide a clear case of ecological segregation by diet with associated adaptations in size of beak. The smallest, with the smallest beak, is the Two-barred or White-winged Crossbill *Loxia leucoptera,* which depends mainly on the seeds of larch *Larix,* the cones of which are softer than those tackled by the other two species. The medium-sized Common Crossbill *L.curvirostra* depends mainly on the seeds of spruce *Picea*, and the larger and large-beaked Parrot Crossbill *L.pityopsittaca* on the seeds of Scots pine *Pinus sylvestris,* which has harder cones

than either larch or spruce. Because of their food preferences, each of these species is normally found where
there are large areas of larch, spruce and Scots pine respectively, hence they are partly separated by habitat.

In confirmation of the correlation between the size of
beak and the preferred type of cone, the Common Crossbill
has evolved a number of isolated subspecies which feed
primarily on pine seeds and have unusually large beaks.
One of these occurs in the Balearic islands and others outside Europe, in northwest Africa, Cyprus, Indochina, the
Philippine Islands and central America. There is also an
unusually slender-beaked subspecies in the Himalayas
which feeds chiefly on larch. Again, the normally thin-
beaked White-winged Crossbill has evolved an unusually
large-beaked form which eats pine seeds on the island of
Hispaniola in the West Indies.

Carduelis

Five of the seven European species of *Carduelis* have
already been discussed, but the Siskin only in winter.
In summer, the Siskin breeds in coniferous forest and feeds
its young mainly on conifer seeds and insects, though it
may turn to other foods in years when conifer seeds fail.
The Redpoll, as already mentioned, breeds in birch forest
and feeds its young mainly on birch seeds, and the very
similar Arctic Redpoll *C.hornemanni* differs in habitat, as it
breeds above the birch forest in bushy tundra with arctic
willows and birches.

The Linnet, discussed earlier, breeds in low and often
thorny scrub, including gorse-clad heaths, hawthorn scrub,
maquis, thorn bushes near the shore, and in southern
Europe high up in the mountains, also occasionally in very
open woodland and at the wood edge. In addition, it is
common where there are bushes near cultivated fields. The
very similar Twite *C.flavirostris* breeds mainly north and
west of the Linnet in Norway and Britain and, where their

ranges overlap, at higher altitudes. It frequents generally similar habitats, but favours treeless and more barren moorland, often near the Atlantic coast or in mountains, and also bare cultivated land in similar places. These species seem best regarded as separated by range. Like various other northern species the Twite has disappeared in recent decades from the southern parts of its range in Britain (Parslow 1967–8), presumably owing to the recent amelioration of the European climate, but whether the Linnet has taken its place is not recorded.*

The natural habitat of the Goldfinch and Greenfinch was probably very open woodland and the wood edge, also scattered clumps of trees in steppe country, but nowadays they are found almost entirely in cultivated land. The Goldfinch sometimes occurs alongside the Linnet, but nests in trees not bushes. More important, all three species differ markedly in beak and in their preferred types of seeds, as already discussed.

In the breeding season, therefore, 5 of the 7 European species of *Carduelis* occur in different natural habitats, though in the Siskin and Redpoll this is determined by their choice of seeds, and in the Twite and Linnet by their geographical range. In winter, the Siskin on alder, the Redpoll on birch and the Goldfinch on Compositae are separated by their diets, and it may be inferred from the size and shape of their beaks that the same held in natural habitats for the Linnet and Greenfinch (though they occur together on farmland). In winter in Britain, the Twite is separated from the Linnet partly by range and partly by habitat, as it is

*The Linnet and Twite, like other geographically-replacing species co-exist in a small area. They do so in the Orkney Islands, for instance, where the Linnet is scarce and almost restricted to light woods and bushes alongside cultivated land, whereas the Twite is common, occurring on moors without trees or bushes, and also in cultivated land, most of which has no bushes, though on a few islands the Linnet also occurs in cultivated land without bushes, hence they differ partly in habitat (Lack 1942–3).

found chiefly on the sea coast from the Wash northward, whereas the Linnet winters chiefly in cultivated land, and on the coast chiefly from the Wash southward. The Arctic Redpoll stays mainly north of *C.flammea* in winter.

Serinus

Under natural conditions in southern Europe, the Serin *S.serinus* breeds at the wood edge and in open sunny woodland, often coniferous. Its main habitat at the present day, however, is in city parks, suburban gardens and cultivated land with trees, with a preference for conifers. The Citril *S.citrinella* breeds primarily in open montane conifer forest in the Alps and adjoining mountains, northern Spain, Sardinia and Corsica, in the last two of which it also breeds in maquis without any trees, down to sea level, but it is much more common in the mountains (Jourdain 1911–12). Hence these two species are largely separated by habitat, including altitude, and probably also by diet, as the Serin eats the seeds of grasses and various herbs, whereas the Citril specialises on the seeds of conifers and Compositae.

The ecological relationship of the two species of *Serinus* to those or *Carduelis* of similar size which occur in similar habitats needs further study, but probably they are segregated. For instance the Citril breeds mainly south of the Siskin, prefers the forest edge and often feeds on or near the ground in clearings and above the treeline, whereas the Siskin prefers true forest and feeds in the trees. The Serin likewise is sometimes seen with various *Carduelis* species, but chiefly in cultivated land, rather than in open woodland in southern Europe.

Fringilla

Finally, of the two species of *Fringilla*, the Chaffinch breeds in all types of woods and in cultivated land with trees, while the Brambling replaces it geographically in the boreal

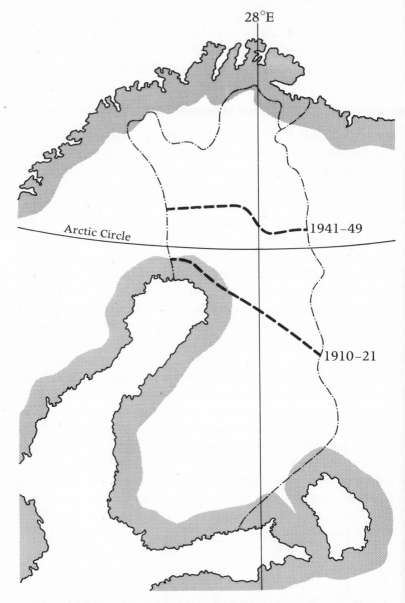

Fig 18. Northward shift of boundary between northern Brambling *Fringilla montifringilla* and southern Chaffinch *F.coelebs* in Finland (from Merikallio 1951).

birch forest of the far north and the Scandinavian mountains, with a narrow zone of overlap where the Chaffinch is chiefly in cultivated land. Both species feed their young on insects. As shown in fig. 18, the boundary between them has shifted gradually north during the last few decades, correlated with the gradual amelioration of the summer climate in Fenno-Scandia (Merikallio 1951). Various other southern species have extended north in Fenno-Scandia in the same period, but since the spread of the Chaffinch has been accompanied by a corresponding retreat of the Brambling, it is reasonable to suppose that the area where they meet is determined by competition, the Brambling being more efficient than the Chaffinch solely in the far north. In this same period, as already mentioned, the Twite has been retreating and the Linnet perhaps advancing in Britain, probably for the same basic reason. Whether the same holds for the replacement of the (Mealy) Redpoll by the Arctic Redpoll is not known.

Conclusion

Six of the seven species of finches present in winter in woods round Oxford take mainly different kinds of seeds (and the seventh, the Brambling, is present only in those winters when beechmast is abundant). Likewise the four species which breed in these woods bring mainly different types of food to their young. In contrast, the five species found in winter in farmland in this area overlap considerably in their diet, though two of them are mainly segregated in late winter; and in the breeding season only one of the five brings mainly different sorts of food to its young, though there are partial differences between the other four. This contrast between a nearly natural habitat and an unstable one greatly modified by man suggests that ecological isolation may take a long time to evolve in birds. Each of the species concerned is adapted in body-size, size and shape of beak, and relative length of leg, to the ways in which it feeds and to the types of food which it takes. A survey of all the European Fringillidae shows that, in natural habitats, most congeneric species are isolated from each other primarily by the types of food which they take, though this means that some of them also differ in habitat; the situation is summarized in appendix 12. During the last hundred years, some of them have greatly increased in numbers or spread in range through colonizing farmland, orchards or suburbs, and others probably did the same at an earlier time.

Chapter 6

Migratory Birds with two homes

The European Trans-Saharan
Passerine Migrants

Although a few of the birds discussed in previous chapters are partial migrants, they usually stay within the breeding range of their species. A more complex problem is presented by species which are wholly migratory, with separate summer and winter ranges. The winter environment may provide very different climatic conditions from the breeding area, especially for species which breed in the arctic and winter in the tropics, though this difference is often smaller than that between the summer and winter weather experienced by northern birds which stay on their breeding grounds. Birds can survive a wide range of physical conditions, so if they live in separate geographical areas, this is not likely to be due to the physical conditions as such, but to the effects of the latter on their food supplies, or on the results of competition with related species of birds.

The two main questions which come to mind for migrants are, first, whether it is necessary for them to be isolated from their congeners in both summer and winter, and secondly whether the means of isolation, when present, are the same at the two seasons. These questions will be considered here for all the European passerine genera with at least one trans-Saharan migrant species. None of the birds in the families concerned have been studied as intensively in the breeding season as the tits or finches discussed earlier, and the information for this time of year is derived from Voous (1960) unless otherwise stated. Ignorance about the situation in winter in Africa is far greater, but fortunately, when I started this analysis, I was able to interest R. E. Moreau in the problem, who devoted himself, aided by many correspondents, to surveying all that could be discovered about the palaearctic migrants to Africa south of the Sahara, and my summaries here are based entirely on his (Moreau *in prep.*). The main information and the references (where not to Voous and Moreau) have been set out in appendix 13, in which the families follow the same sequence as here in the text. While my first aim is to compare the situation in summer and winter, the breeding survey may also be compared with those for the tits, nuthatches and finches discussed earlier.

Swallows Hirundinidae

The three European swallows in the genus *Hirundo* differ in habitat and in nesting sites, and the two European martins in other genera also differ in their nesting sites, as portrayed in fig. 19. This clearcut separation for nesting suggests competitive exclusion, but since three of the species concerned nest chiefly on buildings and another in sand quarries, one cannot now know whether their sites were so sparse in primaeval times that they competed for them. Most other passerine birds have simple nesting sites, on the ground, in bushes, or in trees, and there are far more than enough to go round, while congeners often use similar sites. But hole-nesting species, notably the tits, may sometimes go short, and as already mentioned (p. 26), there is competition for, and partial separation by, nesting sites in this genus. In Africa in winter, the three European swallows have separate ranges.

Fig 19. The 5 European hirundines at their nests. By Robert Gillmor. Left, top to bottom: Swallow on ledge in cave, Red-rumped Swallow on roof of cave and Sand Martin in hole in sand cliff. Right: top, Crag Martin in niche on rock face, and bottom, House Martin under overhang.

Pipits Anthus

Pipits feed on insects in ground vegetation in open country, caught mainly by running along the ground. Five species breed in Europe, but the alpine and coastal races of *A.spinoletta* are so distinctive ecologically that it is simplest to treat them separately. At least five of these six forms occupy separate habitats, and this probably applies also to the Redthroated *A.cervinus* in relation to the Meadow Pipit *A.pratensis* in the arctic. Local variations suggest that at least some of these habitat preferences are due to competitive exclusion. Thus the Rock Pipit *A.spinoletta petrosus*, normally restricted to the sea shore, breeds up to a kilometre inland on spray-blown grassland with rocks, from which the Meadow Pipit is absent, on certain northern British islands (e.g. Lack 1942–43). Conversely the Meadow Pipit, not the Rock, breeds on rocky islets on the inner (but not the outermost and most salty) islands of the Finnish Baltic (P. Palmgren pers. comm.). Further, in the absence of the Alpine Pipit *A.s.spinoletta*, the Meadow Pipit breeds in the alpine meadowland of the Vosges and Abruzzi, while in the absence of the Tree Pipit, *A. trivialis*, it breeds in open birch forest in Iceland (Huxley 1950).

In winter, three of these forms remain in Europe, in similar habitats to those of the summer, so are separated by range from the three in Africa, which there differ from each other in habitat in much the same way as in Europe in summer.

Wagtails Motacilla

Wagtails feed in a similar way to pipits, but more often near water. The three European species occupy different habitats in summer, and keep to these same habitats in winter when in Africa. The Grey Wagtail *M.cinerea* frequents rocky streams, but is absent from northern Europe, where the White Wagtail *M.alba* is often on rocky streams, especially in Iceland.

Shrikes *Lanius*

The predatory, hook-beaked, shrikes usually hunt by flying out from a perch on a bush to pick up large insects or small vertebrates from the ground. The two largest species, the Great Grey *L.excubitor* and the Lesser Grey *L.minor*, are so much larger than the others that it may be presumed that they take mainly larger prey than the others. In particular, the Great Grey takes more vertebrate prey, including small birds, than the rest. How the Great and Lesser Grey are separated is not known, but it could be by size of prey. The Redbacked *L.collurio* and Woodchat *L.senator* differ from each other in habitat where they breed in the same area, but either species may occupy a wider habitat in the absence of the other (Durango 1954). The Masked *L.nubicus* differs from them partly in habitat and partly in feeding habits.

Hence in summer most of the shrikes are separated from each other by feeding, including size of prey. In winter, however, the Great and Lesser Grey are widely separated by range from each other, and so are the Redbacked and Woodchat. This contrast between their ranges at the two seasons is shown in figs. 20 and 21.

Flycatchers *Muscicapa (sens. lat.)*

Three species of pied flycatchers (two of them merged into one by some authors) are separated from each other by range in Europe in summer, and this evidently holds for at least two of them, and maybe the third, in Africa in winter. These three feed especially on caterpillars and other insects on the leaves and by dropping from a perch to take insects off the ground, whereas the other two species fly from a perch to catch flying insects, the Spotted Flycatcher *M.striata* in the open and below the canopy and the Redbreasted Flycatcher *M.parva* above it, and the latter, being smaller, doubtless takes mainly smaller insects.

97

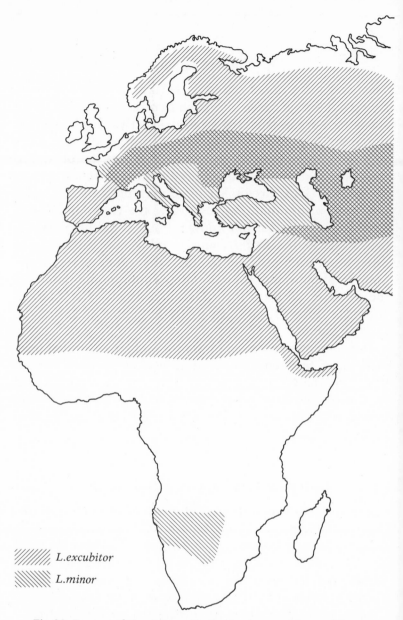

Fig 20. Ranges of Great Grey Shrike *Lanius excubitor* and Lesser Grey Shrike *L.minor* (from Voous 1960, Stresemann *et al.* 1960, 1967, Moreau *in preparation*).

Note that much, though not all, of summer range of Lesser Grey Shrike is within that of Great Grey, but that winter range is entirely separate.

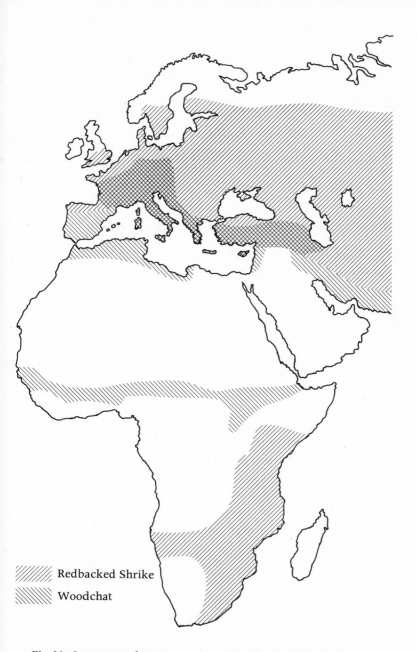

Fig 21. Summer and winter ranges of Redbacked Shrike *Lanius collurio* and Woodchat *L.senator* (from Voous 1960, Stresemann *et al.* 1960, 1967, and Moreau *in preparation*).
Note that summer ranges overlap, winter ranges are separate.

99

These feeding differences are presumably maintained in winter, but the Spotted Flycatcher is mainly south of the three pied species in Africa, and in deciduous not evergreen woodland, and the Redbreasted is in India, so here is another instance of species separated by feeding in summer but in part by range in winter.

Reed and marsh warblers *Acrocephalus*

The reed and marsh warblers skulk in thick herbaceous vegetation, usually over or beside water. Two species occupy similar habitats and are separated by range, two others live in the same habitat and are separated by size, and presumably feeding. The rest are separated from each other by their breeding habitats, as shown for the Neusiedlersee in Austria, where six of them breed, by Koenig (1952). In other parts of Europe, there are fewer species in any one area, and under these conditions, the Sedge Warbler *A.schoenobaenus* may expand into the habitats occupied elsewhere by the Marsh *A.palustris* and Reed *A.scirpaceus*, and probably other species.

In winter, one species in India and another north of the Sahara are separated by range from each other and the other four, which live south of the Sahara. The latter overlap in range, and how they might be separated has not been studied, but one is much larger than the rest. Whereas in Europe the Reed Warbler usually breeds in reeds over water and the Sedge Warbler in thick vegetation near water, in western Uganda the Reed winters in dry *Euphorbia-Capparis* scrub and the Sedge in reeds by or over water (M.Fogden *pers.comm.*). I know no other instance in which the winter habitats are thus transposed, but the difference does not hold in all of the winter range, and while the nesting habitats are as described, in summer the Reed Warbler often feeds on dry land and the Sedge Warbler over water (C.K.Catchpole *pers.comm.*).

100

Hippolais warblers

Of the arboreal and bush-frequenting warblers, those in the genus *Hippolais* are intermediate in size between those of *Phylloscopus* and *Sylvia*, with relatively thick beaks and

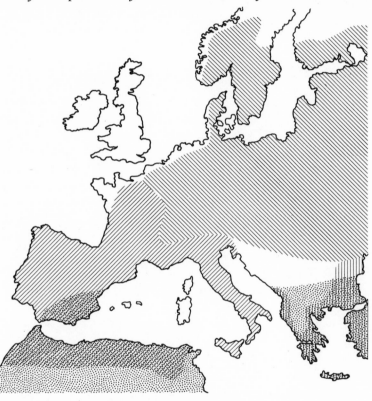

Icterine Warbler	Olive-tree Warbler		
H.icterina	*H.olivetorum*		
Melodious Warbler	Olivaceous Warbler		
H.polyglotta	*H.pallida*		

Fig 22. Ranges in summer of European *Hippolais* warblers (from Stresemann *et al.* 1960,1967).

Note that Icterine, Melodious and Olive-tree Warblers adjoin but do not overlap, while Olivaceous overlaps with Melodious in S.E. Spain and N. Africa, and with Olive-tree in S.E. Europe.

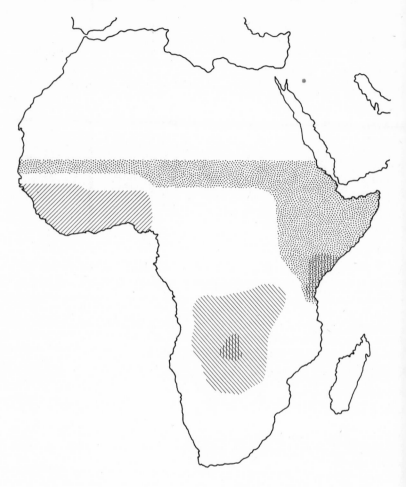

Icterine Warbler	‖‖‖‖‖ Olive-tree Warbler
H.icterina	*H.olivetorum*
▨ Melodious Warbler	░ Olivaceous Warbler
H.polyglotta	*H.pallida*

Fig 23. Ranges in winter of European *Hippolais* warblers (from Moreau *in preparation*).

Note that ranges are mainly separate, except that the sparse Olive-tree Warbler overlaps with the Olivaceous in part of East Africa and with the Icterine in part of southern Africa.

short tails. As shown in fig. 22, three of the four European species replace each other geographically; however, they are not exact ecological equivalents, as the Olive-tree Warbler *H.olivetorum* is decidedly larger than the others, and while it and the Icterine *H.icterina* frequent the canopy in open woodland, often with few or no bushes, the Melodious *H.polyglotta* frequents thick bushes among well-spaced trees. Where the Melodious meets the Icterine in France, there is a narrow zone of overlap in which each male keeps other males of both species out of its territory (Ferry and Deschaintre 1966). The Olivaceous Warbler *H.pallida* overlaps in range with the other three in southern Europe and probably differs from them in habitat. As shown in fig. 23, three of the four species are also separated by range from each other in winter, but at this season it is the sparse Olive-tree Warbler which overlaps with the others, and it is not known how it might be separated (unless by size).

Grasshopper warblers Locustella

The grasshopper warblers skulk to an even greater extent than the reed warblers. The three European species are separated by habitat in summer, and the same holds for at least two of them in winter; the wintering grounds of the Grasshopper Warbler *L.naevia* appear to be unknown, unless the records on passage in northern and western tropical Africa refer also to the winter; if they do, it differs in habitat from the other two (as it does in Europe), and also in range from the River Warbler *L.fluviatilis*.

Leaf warblers Phylloscopus

The small leaf-warblers in the genus *Phylloscopus* live in forest and bushes. Each of the four in northern Fenno-Scandia has a different habitat, but at middle latitudes in Europe two of them, the Chiffchaff *P.collybita* and Willow Warbler *P.trochilus,* occur in the same habitat and differ in feeding higher and lower respectively in the vegetation (but they do not differ appreciably in size). Most of the other species in middle Europe differ in habitat, but the Chiffchaff and Bonelli's Warbler *P.bonelli* coexist in southern Europe and presumably differ in feeding stations like the Chiffchaff and Willow Warbler further north, as Bonelli's is very much a bird of bushes. Yet further south, Bonelli's is the sole species in nearly all the Maghreb (Snow 1952) and the Chiffchaff is the sole species in the Canary Islands (Lack and Southern 1949), and here each of them occupies a wider range of habitats than elsewhere, presumably linked with the absence of congeners.

Two of the six species winter in Asia, largely apart from each other, and the other four in Africa, where Bonelli's winters further north than the Willow and Wood Warbler *P.sibilatrix*, and the Chiffchaff is largely north of the Willow. Some of these species also differ in habitat in winter, but in general, the leaf warblers are separated by habitat in summer and by range in winter. Remarkably, the Chiffchaff and Willow Warbler are separated by habitat in northern Europe, by feeding stations in mid-Europe, and (largely) by range in winter.

Sylvia warblers

The relatively large *Sylvia* warblers include 12 European species, 6 of which are restricted to southern Europe, while a seventh is much commoner there than further north. Most of those at middle latitudes in Europe are isolated from each other by habitat, but the Blackcap *S.atricapilla* and Garden Warbler *S.borin* occur in the same woods and feed higher and lower respectively in the vegetation; like the Chiffchaff and Willow Warbler, they do not differ appreciably in size. The Lesser Whitethroat *S.curruca* in western Europe is in tall deciduous scrub in the lowlands, but in central Europe is in the highest zone of montane coniferous forest; as noted later, the Dunnock *Prunella modularis* is similar.

Some of the Mediterranean species are separated from each other by habitat and a few by range, but there is much apparent overlap between the Subalpine Warbler *S.cantillans*, Sardinian Warbler *S.melanocephala* and Dartford Warbler *S.undata* in dry scrub (macchia). Partial differences in habitat in some areas, for instance the preference of the Sardinian for rather tall scrub, do not hold in others. The Dartford is more resistant than the Sardinian to cold winters, and this is probably why it can occur further north in western Europe, and at higher altitudes in Spain, than the Sardinian; and the Subalpine escapes the winter by migrating, which may be why it can occur at higher altitudes than the Sardinian in the south of France. But elsewhere, the two latter species are at least partly separated by habitat, and where they meet in the south of France, the Subalpine hunts higher and the Sardinian very low (without an appreciable size-difference). When all these factors are taken into account, the three species seem to be more or less separated, but the situation is unusually complex. Perhaps their segregation was much more clearcut before man removed the primaeval forests of the Mediterranean region. Another unusual situation involves Marmora's Warbler *S.sarda*, which looks extremely like the Dartford, and occupies similar habitats. It might have been thought that it was a wellmarked form of the Dartford on the islands of the western Mediterranean, but the Dartford breeds sparsely in most of its range.

Four of the southern species stay on their breeding grounds in winter. The Blackcap and Garden Warbler are separated from the others in Africa by their preference for evergreen montane vegetation (where they might presumably be separated from each other by feeding stations, as in summer). The other six winter in dry country with acacias of one type or another and apparently overlap extensively, especially in northeast Africa, but at least some of them differ in habitat.

Chats Turdinae (part).

As shown in fig. 24, the two European nightingales in the genus *Luscinia*, which live in thick bushy areas, replace each other geographically in both summer and winter, except that their breeding ranges overlap in part of eastern Europe, where the one is in drier and the other in wetter habitats of the same general type. The third species in this genus, the Bluethroat *L.svecica*, occupies a remarkable diversity of breeding habitats, including arctic willow tundra, lowlying lake shores in mid-Europe, coastal salt-marshes in France and the arid scrub on Spanish mountain tops. It is separated from the nightingales in summer by habitat and in winter by range (further north in Africa).

In most of Europe where they are found, the Rock Thrush *Monticola saxatilis* lives on rocky slopes in the mountains and the Blue Rock Thrush *M.solitarius* in rocky places lower down, but the Rock Thrush is absent from Crete (also from the Mahgreb and some Asian mountains), and here the Blue Rock Thrush extends high up. In winter, the two species occupy largely separate ranges.

Two of the European wheatears *Oenanthe* are separated by breeding range, another by habitat. The Common Wheatear *Oe.oenanthe* is mainly north of the Black-eared *Oe.hispanica*, but they are separated by altitude in Southern Europe, and in a complex way on the Aegean islands, summarized in appendix 13 (p. 299). In winter in Africa, only one of these species, the Common Wheatear, is numerous south of the Equator, but several of them occur north of the Equator up to the southern border of the Sahara, together with resident African species, and they coexist to an unusual extent for congeneric species. Some of them prefer drier and more barren country than others, and possibly the extent of their overlap would be found to be less than now thought if critical study had been made of their natural habitats (unnatural habitats such as golf-courses are irrelevant here). Even so, several species are commonly seen side by side. Further, among the various wheatears wintering in North Africa, each male holds an individual feeding territory from which it excludes other

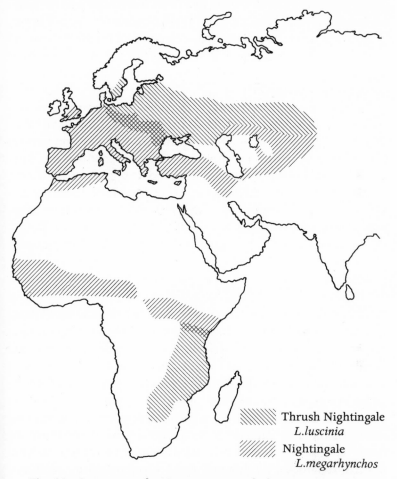

▨ Thrush Nightingale
 L.luscinia

▧ Nightingale
 L.megarhynchos

Fig 24. Summer and winter ranges of the two European nightingales (from Voous 1960 and Moreau *in preparation*).

Note that the area of overlap in eastern Europe is not well documented, that the eastern limits of the breeding range of both species are uncertain, and that the small area of overlap in winter in East Africa involves Asiatic, not European, populations of *L.megarhynchos*.

species of *Oenanthe* as well as its own (Hartley 1949). The occurrence of such behaviour suggests that it might be normal for several species to winter side by side. Hence in the species of *Oenanthe* there is possibly ecological overlap in winter. But further study is needed to check this, especially since the existence of separate breeding habitats in the 7 Moroccan species of *Oenanthe* has only recently been recognised (Brosset 1961) and I would guess that the same holds in winter.

The two European redstarts *Phoenicurus* are separated by habitat in summer, and by both range and habitat in winter. The two European chats in the genus *Saxicola* are separated by feeding stations in summer, the one feeding on herbaceous vegetation and the other on the ground (without an associated size-difference); their feeding preferences also mean that they partly differ in habitat. In winter, they are separated by range.

Discussion of breeding ecology

The way in which each of the species discussed in this chapter is separated from its congeners, in both summer and winter, has been summarized in Appendix 14, and the results are set out in table 15. So far as the breeding season is concerned, three-fifths of the species are separated from their congeners by habitat, a higher proportion than in European tits and finches, which confirms the suggestion made earlier that the latter birds are unusual in having so many species separated by feeding. Only 10 per cent of the species discussed in this chapter are separated by feeding, in most cases, notably in warblers, through a difference in vertical feeding stations, and unlike what is found in tits under these circumstances, the species concerned do not differ appreciably in size. A big size-difference is found only in various shrikes *Lanius*, and in one case in flycatchers. The shrikes are miniature raptors, a group in

Table 15. Main types of ecological isolation in congeneric European transequatorial migrant passerines

Segregated by	Breeding season number of instances	proportion	Winter home number of instances	proportion
Range (G)	14 ⎫	26%	⎫ 104 ⎫	64%
No contact (—)	29 ⎭		⎭ ⎭	
Habitat (H)	100½	62%	38	23%
Feeding (F)	16½	10%	4	2%
Unknown (?)	3	2%	17	10%

Notes: based on appendix 14, ignoring subsidiary means of segregation (shown by a letter in brackets), and doubts (shown by a ? after the letter), but scoring as ½ each case in which two letters have been shown (corresponding to differerent parts of the range). Segregation by range (G) has been scored as such in the breeding season solely where two species have adjoining ranges, and otherwise—for no contact has been used; but the distinction between these two categories is hard to maintain for the winter ranges. Where the means of segregation are unknown (indicated by ?) in the breeding season, it is almost certainly by feeding, but this does not necessarily apply on the wintering grounds, about which little is yet known. Differences in feeding are associated with a clear difference in size solely in certain shrikes *Lanius* and in one instance in flycatchers *Muscicapa*.

which size differences are frequent, as discussed in the next chapter.

A few points are of special interest. First, the swallows and martins are separated not only by habitat but also by nesting sites, being the only European passerine birds to which this applies except, partially, the hole-nesting tits. Secondly, where the Rock or the Alpine or the Tree Pipit is locally absent, the Meadow Pipit may extend its breeding habitat. Thirdly, the Chiffchaff and Willow Warbler are separated by habitat in northern Europe and by feeding stations in mid-Europe. Fourthly, the Chiffchaff and Bonelli's Warbler have unusually broad habitats in the Canary Islands and the Maghreb respectively, where each is the only breeding species of *Phylloscopus*. Fifthly, the two species of nightingales differ primarily in range, but by habitat in a small area of overlap. Sixthly, the Blue Rock Thrush extends to high ground where its montane congener, the Rock Thrush, is absent. Finally, the complex differences and seeming overlap between three Mediterranean warblers (perhaps accentuated by forest clearance) need further analysis.

Isolation in winter

The broad comparison of the means of isolation in summer and winter in table 15 shows some important differences. In particular, many more species are separated by range in winter. This is unquestionable, but the associated reduction in the number of species separated by habitat in winter is partly misleading, because a difference in range has been given priority over a difference in habitat in the analysis. In fact, some of the species separated by range in winter also differ in habitat, though this is irrelevant to their isolation if, anyway, they do not live in the same area. A further difficulty, already mentioned for tits in chapter 3, is that many habitats in Africa occupy huge areas, hence two species in different habitats may inevitably differ in range as well. But even if allowance is made for these points, the number of species separated by habitat is smaller in winter than summer.

The proportion of species separated by feeding is also smaller in winter than summer, though this is probably due, in part, to lack of observations on feeding stations in Africa. Where two species are separated primarily by feeding in summer and coexist in Africa in winter, it may be presumed that they are also separated by feeding in Africa. There are, however, a number of instances in shrikes, flycatchers, *Saxicola* chats and warblers, in which species separated by feeding in summer occupy separate ranges in winter. This suggests that the diversity of their ecological resources is reduced in winter quarters, either by the nature of the habitat, which for many of them is arid, or by the presence of many indigenous African insectivorous species.

Since isolation by feeding is the hardest type to observe, the few species in which the means of separation are unknown in summer are presumably separated by feeding. This might also apply to some of the much greater number of species in which the means of separation in winter in Africa are unknown, but various others could well be cases of separation by habitat, which has been little studied in Africa. In one genus, the wheatears *Oenanthe*, there is suggestive evidence for overlap in winter, but this needs critical study and I myself greatly doubt it.

This chapter has been concerned solely with the European breeding species in winter. Various of them also meet Asian or African congeners in their winter quarters, and these will have to be included for a full assessment of ecological isolation in winter, but this is not attempted here (see Moreau *in prep.*).

Conclusion

In summer in Europe, virtually all the trans-Saharan migrant passerine species are isolated ecologically from their congeners, over three-fifths of them by differences in habitat. Only one-tenth of them are separated by feeding, most of them without an associated difference in size or size of beak. There are several good examples of species which have broader habitats in those parts of their range where congeners present with them elsewhere are absent, there are also examples of species separated partly or mainly by range but which differ in habitat in an area of overlap, and there is one example of two species separated by habitat in one part of their range and by feeding in another part. At least nine-tenths of the species concerned are also isolated from each other in winter, most of them by range, including many of the species which are separated from each other primarily by habitat or by feeding in summer.

Chapter 7

The Review of a
Continental Avifauna
The other European birds

In the course of the preceding chapters, ecological isolation
has been analysed in many families of European passerine
birds. It therefore seems worth completing this type of
analysis for all European congeneric species, but in less
detail. The information on the remaining passerine species
has been set out in appendix 15, which shows that the
means of ecological isolation in the breeding season is
known, at least with strong probability, in almost every
case. It therefore seems reasonable to assume that, except
where there is positive evidence to the contrary, all
European birds are ecologically isolated. Hence for the
other (non-passerine) birds, a much briefer analysis has
been set out (in appendix 16), restricted to those factors
thought to be critical in isolation, and omitting further
information on range or ecology where this does not
appear to be important in keeping congeneric species apart.

Comments on the other passerines

The summary at the end of appendix 15 shows that, in
most other passerine families analysed here, segregation by
habitat is much commoner than by any other means, being
found in 62 per cent of the possible instances. The segregation of similar species by range is rare (but many less similar
congeners occupy separate, and often widely separated,
ranges, so have no contact with each other). Segregation by

feeding holds in 15 per cent of possible instances, nearly all of these being associated with a big difference in size or size of beak. But all these latter examples are found in the seed-eating buntings or in crows, which include many seeds in their diet (though they also take a variety of other foods). Hence the association of size-differences with a seed diet found in finches and to some extent in tits, holds good also in other seed-eating families.

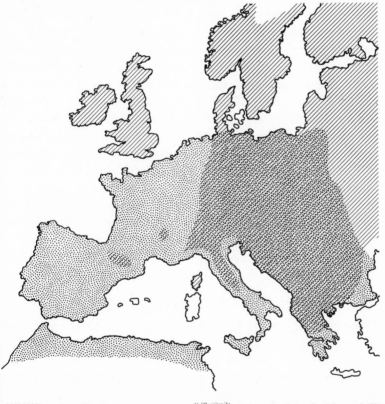

///// *C.familiaris*　　　　　:::::: *C.brachydactyla* (short-bill)

Fig 25. Partly overlapping ranges of the two European tree-creepers (from Voous 1960, Stresemann *et al.* 1960, 1967).
Note that *C.familiaris* is primarily montane in southern Europe.

A few points of special interest in the families considered here may be noted. First, in each of two genera, the goldcrests *Regulus* and tree creepers *Certhia*, (see fig. 25), two species occupy mainly separate ranges but coexist in a relatively large area in central Europe. In both pairs, the two species differ in habitat in the area of overlap, while each occupies a wider range of habitats where its congener is absent. Secondly, the Ring Ousel *Turdus torquatus*, like the Rock-Alpine Pipit *Anthus spinoletta* discussed in the previous chapter, provides a good example of a species with a markedly different habitat in northern and central Europe, in the alpine zone and coniferous forest respectively. Thirdly, the Song Thrush *Turdus philomelos* and Blackbird *T.merula* are separated by habitat in northern Europe, but by feeding stations in part of western Europe. Fourthly, the Cirl Bunting *Emberiza cirlus* has a much more restricted habitat in the extreme north of its range, in England, than where it is abundant further south. Finally, the relationship of the Spanish Sparrow *Passer hispaniolensis* to the House Sparrow *P.domesticus* is exceptionally complex, as in a few places they replace each other geographically in similar habitats, in others they live in the same area but are segregated in different habitats, and in yet others they live together and interbreed.

Discussion of all passerines

If the totals given at the end of Appendix 15 are added to those for the European species discussed in earlier chapters, there are, in all, 325 possible contacts between congeners. Of these, 7 per cent are separated by geographical range, another 18 per cent have separate ranges and no contact, while 55 per cent are separated primarily by habitat, and 19 per cent by feeding. The means of segregation are unknown in only 2 (less than 1 per cent), and there is coexistence with interbreeding in the sparrows just mentioned, scored as $\frac{1}{2}$ because elsewhere they are segregated.

114

The number of instances of separation by range partly depends on the classification adopted, since the Carrion and Hooded Crows *Corvus corone* and *C.cornix* are here treated as full species separated by range, as is the flycatcher *Muscicapa semitorquata,* discussed in Appendix 13, p. 289 but various other workers have treated them as subspecies. It is also hard to know, in some instances, whether to classify two species as differing in range or as having no contact. Normally, species are considered here to be segregated by range only if they occupy similar habitats and their ranges adjoin. But Marmora's Warbler *Sylvia sarda* occupies scrub on the islands of the western Mediterranean and Rüppell's Warbler *S.rüppelli* occupies scrub on the islands of the eastern Mediterranean, they are not closely related, and their habitats, though similar, are not identical, so it is hard to know in which of these two categories to put them. There are also instances, like that of the Common and Black-eared Wheatears *Oe.oenanthe* and *Oe.hispanica,* discussed in the previous chapter, in which a difference in range is reinforced by a minor difference in habitat, especially where they meet, and this probably holds in other instances for, as mentioned earlier, no two geographical areas provide identical habitats. The best evidence for strict geographical replacement is found when the boundary between the species concerned is shifting, as in the Chaffinch *Fringilla coelebs* and Brambling *F.montifringilla* in Fenno-Scandia, discussed in chapter 5.

The number of instances of separation by feeding also depends partly on the classification adopted, this time at the generic level. At one time the European tits now placed in *Parus* were divided among five genera and the finches in *Carduelis* among three, while the flycatchers grouped here in *Muscicapa* are usually put in two genera; and in each of these instances, the finer generic divisions correspond with differences in feeding. Hence on a narrower concept of the genus, the number of instances of separation by feeding would have been much reduced.

As already noted, separation by feeding is often associated with a big difference in the size of beak, notably in the seed-eaters, or with a big difference in body-size, notably in shrikes, while when two species are separated vertically by their feeding stations, usually in forest, the smaller species often feeds higher in the trees than the larger. There are, however, a number of other instances, notably in *Sylvia* and *Phylloscopus* warblers, discussed in the previous chapter, in which a difference in feeding stations is found between species of similar size. Sometimes, a difference in food or feeding station is so closely linked with a difference in habitat that it is hard to know which is primary, and for this and other reasons already discussed, the totals just mentioned provide only an approximate summary of a complex situation.

The most frequent means of ecological isolation in European passerine congeners is by a difference in habitat, and it is striking how often this is clear-cut and easily recognized. For instance, the difference is between broad-leaved and coniferous forest in treecreepers, tits and others, between the presence or absence of scattered trees on plains in certain larks and pipits, and between the presence or absence of standing water below marsh vegetation in various acrocephaline warblers. Only in a few instances, such as the Redbacked and Woodchat Shrikes *Lanius collurio* and *L.senator* and the Mediterranean *Sylvia* warblers, are the differences in habitat not immediately obvious.

Of particular interest are the species which replace each other geographically in part of their range but in habitat where they overlap. At one extreme, the Common and Spotless Starlings *Sturnus vulgaris* and *S.unicolor* replace each other apparently without overlap, and though the Carrion and Hooded Crows *C.corone* and *C.cornix* slightly overlap, they do not there differ in ecology but interbreed. In the two nightingales *L.luscinia* and *L.megarhynchos*, discussed in the previous chapter, the zone of overlap is rather larger, and within it they differ in habitat, and it is yet larger in the two treecreepers and the two goldcrests, though it is still reasonable to describe these pairs of species as separated partly by range and partly by habitat. These examples lead on to others in which two species differ primarily in habitat, though there are some parts of Europe in which only one or the other occurs.

Other land birds

Based on appendix 16a, the probable means of ecological isolation in the other (non-passerine) European land birds have been summarized in table 16. It should be noted that many large edible birds and many raptors have been greatly reduced in numbers by shooting, and that the removal of primaeval forests and swamps has had an even more destructive effect on their numbers. If, for instance, the birds in the remaining primaeval forests of Bialowies in Poland or Plitvicke in Jugoslavia (Rucner 1956) are compared with those in modern managed woodland consisting of the same kinds of trees, the passerine species are very similar but many more species of woodpeckers and owls survive in the primaeval than the managed woods. Hence the ranges and habitats of many non-passerine land birds are much more restricted than formerly, and it may be hard today to be sure what factors limited their distribution under natural conditions. Hence the figures in table 16 are less certain than those given earlier for passerine birds.

In 11 per cent of instances in these other land birds, the means of isolation are not known, but there is no reason to think that the species concerned are not segregated from each other, since these other groups have been much less well studied than passerines, and differences in feeding, which are common, are much less easily observed than differences in habitat. In all, isolation by feeding is known in 44 per cent of instances, and if the unknown instances are included, they comprise over half the total. It should be

Table 16. Means of ecological isolation of other European land birds

	Number of instances of separation by				
	range	habitat	feeding	no contact	unknown
Hawks, Eagles, etc. *Accipitridae*	2	10	5	0	3
Falcons *Falconidae*	4	1	25	14	1
Gamebirds *Galliformes*	4	2	0	0	0
Bustards *Otididae*	0	0	1	0	0
Sandgrouse *Pteroclididae*	0	0	0	0	1
Pigeons *Columbidae*	0	$2\frac{1}{2}$	$1\frac{1}{2}$	0	0
Owls *Strigidae*	0	1	1	0	1
Nightjars *Caprimulgidae*	0	0	0	0	1
Swifts *Apodidae*	0	0	3	0	3
Woodpeckers *Picidae*	1	3	6	0	1
TOTAL	11	$19\frac{1}{2}$	$42\frac{1}{2}$	14	11

Note: based on appendix 16a

added that if two species live in the same habitat but differ greatly in size, it has here been assumed that they differ in feeding, but though there is direct evidence for this in some cases, in others it is as yet lacking. Differences in feeding are particularly common in falcons, and the marked difference in the hunting methods of three of them are portrayed in fig. 26. In the Goshawk *Accipiter gentilis*, the

Fig 26. Three species of *Falco*: a Hobby pursuing martins, a Kestrel hovering over a vole and a distant Peregrine stooping on a pigeon. By Robert Gillmor.

larger female takes mainly different prey from the smaller male (Höglund 1964). While in various raptors a difference in body-size is associated with a difference in size of prey, in some woodpeckers it is associated with a difference in the size of the branches on which they forage. Thus the Lesser Spotted Woodpecker *Dendrocopos minor* forages high in the trees on small branches and twigs and the Great Spotted *D.major* on large branches and trunks lower down.

Only a few of the non-passerine land birds are separated by range, the most striking example being the four species of *Alectoris* partridges, as shown in fig. 27. Where the Rock Partridge *A. graeca* meets the Redlegged *A.rufa* in the

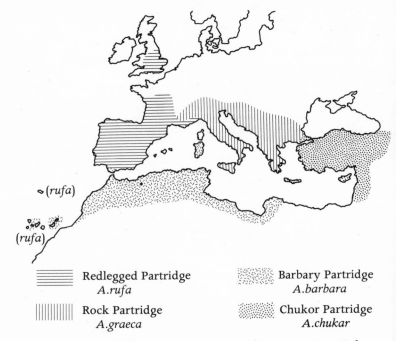

☰	Redlegged Partridge *A.rufa*	⠿	Barbary Partridge *A.barbara*
⦀	Rock Partridge *A.graeca*	░	Chukor Partridge *A.chukar*

Fig 27. Ranges of the 4 *Alectoris* partridges in western Palaearctic (from Watson 1962a, b).

Note that *A.barbara* was introduced to Gibraltar and perhaps to Sardinia, and that *A.rufa* was introduced to Gran Canaria, Madeira and England. Where *A.graeca* meets *A.rufa* in west and *A.chukar* in east, it is at higher altitudes.

west and the Chukor *A.chukar* in the east, it is at a higher altitude than they are, but outside these areas, the other two species may also occur high up. Fewer species are separated primarily by habitat in non-passerine than passerine birds, perhaps partly due to their respective size. The difference, for instance, between broadleaved and coniferous trees is relatively greater for a passerine bird which takes its food off the leaves than for a raptor which hunts between the trees, or even for a woodpecker which climbs the trunks. In swifts *Apus*, as in hirundines (see p. 95), each European species has a different type of nesting site, and the same holds for the three European pigeons in the genus *Columba*.

Wading, freshwater and marine birds

The situation in European wading birds has been set out in appendix 16b, in freshwater swimming birds in 16c and in seabirds in 16d, and the probable means of ecological isolation have been summarized in table 17. In all three groups, more species are separated by feeding than by any other means, and if most of the instances in which the

Table 17. Possible means of ecological isolation in summer of European congeneric wading and water birds

Separated by	Waders	Freshwater swimmers	Seabirds	Total
Range	$4\frac{1}{2}$	$9\frac{1}{2}$	$2\frac{1}{2}$	$16\frac{1}{2}$
No contact	7	8	22	37
Habitat	$12\frac{1}{2}$	11	10	$33\frac{1}{2}$
Feeding	23	$40\frac{1}{2}$	$27\frac{1}{2}$	91
Unknown	3	7	3	13

Notes: based on appendix 16. This table is much less reliable than the comparable tables for land birds, as so much less work has been done on wading and water birds. Among marine birds, a species which feeds further out to sea than another has been classified as differing in habitat, but perhaps it should have been classified as differing in feeding station.

means of separation are unknown are really due to feeding, then the latter accounts for more than half the total, as in the non-passerine land birds.

A few points from these summaries may be singled out as of special interest. First, the southern Curlew *Numenius arquata* and northern Whimbrel *N.phaeopus* replace each other geographically, and as in several other instances of this sort in Fenno-Scandia, the boundary between the two species has been shifting northward with the current amelioration of the climate (cf. p. 90). In the present example, the process has been assisted by man, since in its small area of overlap with the Whimbrel, the Curlew is restricted to cultivated land in the valleys, and cultivation has been extended northward. Secondly, big differences in the size and shape of the beak and the length of the leg in sandpipers in the genera *Tringa* and *Calidris* are presumably correlated with differences in their feeding, and differences in food have in fact been established for three shore-feeding species of *Calidris,* but not yet for the others.

Big differences in size are also found in various freshwater birds, such as grebes *Podiceps* and sawbill ducks *Mergus,* and at least the largest and smallest species in these two genera take different types of prey. Differences in feeding between the various dabbling ducks *Anas* are more complex and have not yet been fully analysed; they are associated with differences in body-size, length of neck, the structure of the beak, and feeding methods, as set out in appendix 16c and portrayed in fig. 28.

Among seabirds, there is a good example of separation by range in the southern Common and northern Brünnich's Guillemots *Uria aalge* and *U.lomvia.* In their area of overlap, but not outside it, they differ in their nesting sites, which is indicative of competitive exclusion. Each of the other European auks, the Razorbill *Alca torda,* Puffin *Fratercula*

Fig 28. Feeding methods of dabbling ducks in England in winter: Wigeon grazing, Shoveller filtering, Pintail up-ending, Mallard (large) and Teal (small) varied. By Robert Gillmor.

Wigeon

Shoveller

Pintail

Mallard

Teal

arctica, Black Guillemot *Cepphus grylle* and Little Auk *Plautus alle,* also differs in nesting site from every other, recalling the situation in hirundines (p. 95) and swifts (p. 313). There is a big overlap in the prey species brought to their young by most of these species (Kartaschew 1960), but the differences in overall size and size of beak between the Guillemot, Razorbill and Puffin are linked with differences in the size of their prey (Fisher and Lockley 1954, Harris 1963 and *in prep.*). Four species of terns *Sterna* and the Kittiwake *Rissa tridactyla* on the Farne Islands in N.E. England likewise bring the same species of prey to their young (Pearson 1968). As discussed in chapter 9, however, tropical terns take the same main prey species but differ in where they obtain it, and in the size of their prey, and the same might well apply to the European seabirds. Two others, the Cormorant *Phalacrocorax-carbo* and the very similar Shag *P.aristotelis,* are segregated by taking their prey on the sea bottom and in the water above it, respectively, and they also differ in their nesting sites (Lack 1945a, Pearson 1968).

The comments in this section apply mainly to the breeding season. In winter, various freshwater wading birds move to the seashore and many travel to the tropics, while various freshwater and coastal aquatic birds move out to sea. How they might be separated, if they are, at this season of the year, has not, in most cases, been studied.

Overall comparison

The frequency of each type of ecological isolation in European congeneric species has been summarized for passerine birds, other land birds and water birds respectively in table 18. The other land birds agree with the water birds in that roughly half of them (if the unknown are included) are isolated from each other by feeding, often associated with a difference in size. But this holds for only about one-fifth of the passerine birds, and nearly all of those in which it holds

Table 18. The proportion of each type of ecological isolation in different groups of birds in Europe

Separated by	Passerines	Other land birds	Wading and water birds
Range	7%	11%	9%
No contact	18%	14%	19%
Habitat	55%	20%	18%
Feeding	19%	43%	48%
Unknown	1%	11%	7%

are seed-eaters for at least part of the year, or are miniature raptors (the shrikes). On the other hand, separation by habitat is much more frequent in passerine than other species, and if the finches, tits, crows and shrikes are omitted, it applies to about three-quarters of them.

Conclusions for Europe

Virtually all congeneric European passerine birds, and at least nine-tenths of those in other orders, are known to be isolated ecologically from each other in the breeding season, and there is no good reason to think that any congeneric species are not so isolated. (The big apparent overlap in the foods of the seabirds might be misleading, as discussed in chapter 9.) In the breeding season, only a few species are separated by range, most passerine species are isolated by habitat, and most other species, together with the seed-eating passerine and the shrikes, by feeding, often but not always with an associated difference in size. Congeneric species which nest in holes or on cliffs, but no others, tend also to be separated by their nesting sites. In winter, most congeneric passerine migrants have separate ranges; how most migratory wading and swimming birds might be isolated at this season has not been studied.

Chapter 8

North American Studies
Passerines, Hawks, Peeps, Auklets

Apart from the tits and nuthatches discussed in earlier chapters, few studies have yet been carried out on ecological isolation in North American birds. The most important are reviewed in this chapter.

Hybrid zones after recent contact

Sibley (1961) has studied several cases in which two forms, previously isolated geographically, have recently come in contact with each other and now interbreed. These forms are treated as well-marked subspecies if the genes peculiar to each population are spreading rapidly through the other, and as full species if they are not doing so, even though there is some interbreeding. As Sibley pointed out, hybrid zones are usually of brief duration, since if the hybrids are at no appreciable disadvantage the two forms soon merge, while if they are at a disadvantage, barriers to prevent breeding are evolved. In most cases, the two forms have met owing to human activities. Thus the planting of trees in the Great Plains has allowed various arboreal forms from either side to meet; elsewhere the clearing of forest has allowed various birds of open country or scrub to meet.

For instance, the Red-eyed Towhee *Pipilo erythrophthalmus* of North America breeds mainly north of the Mexican Collared Towhee *P.ocai*, but the two meet in parts of Mexico (Sibley 1950, 1954, Sibley and Sibley 1964). They frequent bushy undergrowth, and where they meet under natural conditions in Oaxaca in southern Mexico, they are separated by habitat, the Red-eyed in oak woods and the Collared in conifers, and do not interbreed. Presumably they have had time here to evolve both genetic and ecological isolation. But other populations of the two species in Mexico have been brought into contact through clearance of forest by man and its replacement by brush, and in these areas they freely interbreed. Similar examples are found in flickers *Colaptes*, indigo buntings *Passerina* (Sibley and Short 1959), the oriole *Icterus galbula* (Sibley and Short 1964) and two Colombian tanagers in the genus *Ramphocelus* (Sibley 1958), probably also in two Mexican subspecies of the Redwinged Blackbird *Agelaius phoeniceus* (Hardy and Dickerman 1965, Hardy 1967). These populations are on the borderline between subspecies and species.

Interspecific territorial exclusion

Two other species, once geographically isolated, which have now met through the replacement of forest by grassland, are the Eastern and Western Meadowlarks *Sturnella magna* and *S.neglecta* (Lanyon 1956ab, 1957). Where they meet, the Eastern prefers damper and the Western Meadowlark drier meadows, but in part they coexist and here each male excludes other males of both species from its breeding territory. Likewise in western North America, the Dusky and Grey Flycatchers *Empidonax oberholseri* and *E.wrightii* are separated primarily by habitat, in woods with and without a brush understorey respectively, but they occur side by side on the western edge of the Great Basin where forest has been replaced by brush through fire and lumbering, and here there is interspecific territorial exclusion (Johnson 1966).[*]

[*]Johnson seemed puzzled that these two flycatchers do not show character displacement (p. 10) where they overlap, but one would not expect it if they differ primarily in habitat (see also Ashmole 1968a). A third similar flycatcher *E.hammondi,* differs in habitat from the other two (Johnson 1963).

Interspecific territorial exclusion has also been reported between some of the North American tits where their ranges adjoin (Dixon 1961), and in the Red-bellied and Golden-fronted Woodpeckers *Melanerpes (Centurus) carolinus* and *M.(C.)aurifrons* (Selander and Giller 1959). The Red-bellied breeds in much of eastern USA and the Golden-fronted mainly in Mexico, but both occur in central Texas, the Red-bellied in open woodland with large trees in lowlying swampy ground, and the Golden-fronted in lighter and much drier mesquite woodland. But while they differ here in habitat, they have recently met and coexist in the town of Austin, Texas, where they defend mutually exclusive territories. Again, the Great-tailed Grackle *Cassidix mexicanus* occupies a mainly different range from the Boat-tailed Grackle *C.major,* and in most of their area of overlap on part of the coast of the Gulf of Mexico, the Great-tailed inhabits cultivated land and the Boat-tailed coastal marshes, but in places they breed in mixed colonies, and hold mutually exclusive nesting territories (Selander and Giller 1961).

Interspecific exclusion of a different type is found where the Redwinged Blackbird *Agelaius phoeniceus* and the much larger Yellowheaded Blackbird *X.xanthocephalus* breed alongside each other in western USA (Orians and Willson 1964, Miller 1968). The Redwings arrive before the Yellowheads and take up territories in reeds over water and in similar tall vegetation on damp ground. When the Yellowheads arrive, they displace many Redwings from the territories over water on more open lakes of higher productivity, but not elsewhere; the Redwings occupy a much wider range of habitats. Both species nest in their territories, where the Redwings also obtain part, and the Yellowheads much, of their food, but they differ considerably in diet and in where they feed, also in size, and hence presumably in size of prey. This case, therefore, differs in two respects from those mentioned earlier. First, interspecific territorial exclusion is not confined to a narrow zone of contact between two species normally separated by either range or habitat, and secondly the larger species always ousts the smaller. In these respects they seem to be unique so far as competition for habitats is concerned, but there are close parallels with competition for sites in hole-nesting species, which may likewise occur throughout the common range of two species, and while the larger ousts the smaller from holes suitable for either, other holes suit only the smaller species. Miller (1968) argued that such active displacement is the most efficient and most frequent means of ecological exclusion in general, but I support Orians and Willson (1964) that natural selection will usually act against it, as it wastes time and energy. Moreover the evidence in this book indicates that, normally, species with adjoining ranges or habitats do not actively displace each other, but instead each selects its own type of area. Hence such species meet only in border zones where, in a few of them, each male excludes the males of both species from its territory, but this latter behaviour does not help to separate their ranges or habitats. The displacement of the Redwinged Blackbird from its habitat by the Yellowhead is highly abnormal, and its points of resemblance with competition for nesting holes make me wonder whether it might be a special form of nest-site competition.

Warblers and vireos

Five similar species of North American parulid warblers breed in boreal forest in northeastern USA and feed on the same species of spruce tree; these are the Bay-breasted *Dendroica castanea*, Myrtle *D.coronata*, Blackburnian *D. fusca*, Cape May *D.tigrina* and Black-throated Green *D. virens* (MacArthur 1958). Their minor differences in size and size of beak are set out in table 19. Repeated observations on their feeding stations and feeding methods show that, despite a superficial appearance of overlap, they differ greatly; the parts of the tree in which each feeds most

Table 19. Beak differences between five coexisting species of
North American warblers *Dendroica*

	culmen	Mean (in mm) of		male wing
		height at nares	width at nares	
Bay-breasted *D.castanea*	13·0	3·7	3·6	73
Myrtle *D.coronata*	12·5	3·3	2·1	74
Blackburnian *D.fusca*	13·0	3·2	3·4	68
Cape May *D.tigrina*	12·8	2·9	2·9	66
Black-throated green *D.virens*	12·6	3·4	3·2	64

Note: from MacArthur (1958), except for winglengths from Ridgway (1902).

frequently are shown in fig. 29. The Bay-breasted feeds mainly in the shady interior of the tree, working radially out from the base, searching slowly; it is clumsy in the tips of the branches and rarely takes flying insects on the wing. The Myrtle is the most varied in feeding, and takes frequent flights from tree to tree; it feeds mainly on the bottom part of the tree, moving radially, tangentially and vertically, and often takes flying insects on the wing. The Blackburnian is intermediate in hunting methods between the Bay-breasted and Black-throated Green and feeds mainly in the uppermost parts of the tree, usually moving radially and working from the base to the tip of a branch, with much rapid peering. The Cape May also feeds mainly at the top of the tree, but moves vertically instead of radially, feeding especially in the outer shell of the tree and catching flying insects on the wing to a much greater extent than the Blackburnian. The Black-throated Green feeds mainly in the dense parts of the branches and on the new buds at middle elevations, where it usually moves tangentially, and

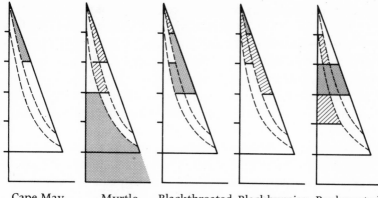

Cape May   Myrtle   Blackthroated Blackburnian Baybreasted
D.tigrina   *D.coronata*   Green   *D.fusca*   *D.castanea*
                         D.virens

Fig 29. Main feeding zones in spruce of 5 coexisting North American *Dendroica* warblers (from MacArthur 1958).

Each diagram represents one half of a spruce tree, subdivided every 10 feet from the ground, and also into an outermost zone of new needles and buds, a middle zone of old needles, and an innermost bare or lichen-covered base. Dotted shading shows those parts of the tree in which the species concerned was recorded both (a) on half of all observations, and (b) in half of the total time observed, while striping shows those parts of the tree for which either (a) or (b) applied, but not both.

it often peers up (whereas the Bay-breasted and Black-burnian peer down) and takes insects from the underside of branches by hovering in the air, though it rarely takes flying insects on the wing.

MacArthur concluded that though each of these species overlaps in its feeding stations and foods with others, each differs to a sufficient extent to coexist. He added that the Cape May, and to some extent the Bay-breasted, are fugitive species, moving to breed in areas where spruce budworm larvae *Choristoneura fumiferana* are particularly abundant. Recently Morse (1968) has shown that in at least 3 of these 4 species, and one other *D.magnolia*, the males hunt higher in the trees than the females. Hence these warblers present a more complex case of ecological interdigitation than any found in European passerine species. Further their differences in feeding stations are associated with much less conspicuous morphological differences than are found between most other coexisting species. All of them migrate south for the winter, but they have not been studied in winter quarters. Nor have critical studies been made of the other North American species in the genus.

In contrast to the *Dendroica* warblers, the vireos, analysed in a general but not an intensive way by Hamilton (1962), present a similar picture to most European insectivorous passerine species. There are four subdivisions. In the subgenus *Vireo* (i) the 10 species in the *griseus* group forage in thickets, well over half of them are separated from each other by range and at least some of the rest by habitat; (ii) 3 of the 4 species in the *solitarius* group forage in trees and are separated from each other by range or habitat, and the fourth lives in scrub. In the subgenus *Vireosylvia*, (iii) the 4 species in the *olivaceus* group forage in trees in the lowlands, 3 being separated by range and the fourth by habitat; (iv) the 3 species in the *gilvus* group forage in trees in the highlands, two being separated from each other by range and the other by habitat. Hence almost all the species in the same subdivision are separated from each other by range or habitat, and this also applies to many of those in different subdivisions, while the rest are separated by whether they feed in the shrub layer or the crowns of the trees. Hamilton reported only two possible cases of 'co-occupancy'. First, the Yellow-throated Vireo *V.flavifrons* and the Red-eyed Vireo *V.olivaceus* coexist in eastern North America, but they are in different subgenera, which suggests that they may be so far apart as to have different feeding habits. Secondly, both the Jamaican White-eyed Vireo *V.(griseus) modestus* and the Blue Mountain Vireo *V.osburni* occur in the montane forest of Jamaica. However, the larger *V.osburni* has a much thicker shrike-like beak, so presumably differes in its main foods (Bond 1960).

Hawks in the genus Accipiter

Three species of hawks in the genus *Accipiter* coexist in eastern North America, the small Broadwinged Hawk *A.striatus,* the larger Cooper's Hawk *A.cooperii* and the much larger Goshaek *A.gentilis,* and in each of them the female is larger than the male, so that six forms of similar shape but different size hunt for prey in a similar way in the same area. Various morphological differences are associated with their differences in size, as shown in table 20 (from

Table 20. Differences in size, proportions and prey species in three sympatric species of *Accipiter* in North America

| | Broadwinged Hawk | | Cooper's Hawk | | Goshawk | |
| | *A.striatus* | | *A.cooperii* | | *A.gentilis* | |
	male	female	male	female	male	female
Mean weight (grams)	99	171	295	441	818	1137
Wing-length (mm)	141	168	204	233	297	315
Wing area (cm^2)	412	560	804	1064	1462	1976
Index of wing-muscle (mass in cm^3)	3	5	11	17	37	41
Mean weight of prey (grams)	18	28	38	51	397	522
Percentage of mammals in prey	2	3	16	19	39	66
Percentage of total kill made up of 7 commonest species in the prey		28		39		72

Notes: from Storer (1955, 1966). The wing area of *A.gentilis* was measured for only one individual of each sex, so may not be the true average. The wing muscles were not measured, and the index for muscle mass was the length × the width × the depth of the skeleton of the sternum as defined by Storer. The three species, and the two sexes of each, are of similar shape. A linear increase in wing-length corresponds to a cubic increase in weight, while wing area is related to body-weight as the square to the cubic function. The wing-muscle mass is proportionately greater in heavier birds as they have to fly faster to stay airborne. The degree of difference between the sexes is proportionately greatest in the smallest species and least in the largest.

Storer 1955). The surface area of the wing increases as the square, and the weight of the bird as the cube, of the wing-length, and since, for aerodynamic reasons, the larger the bird the faster it has to fly in order to stay up, the proportionate increase in the size of the wing muscles is greater than that in body-weight. Storer (1966) later showed that each of these six forms differs from the rest in its main prey. In general, the larger the hawk, the heavier the species on which it preys, the higher the proportion of small mammals in its prey, and the greater the tendency to concentrate on particular prey species. The last point is probably due to the fact that there are fewer species of larger than smaller prey available. Hence these three species, and the two sexes of each, form an excellent example of ecological differentiation through body-size and size of prey. There are only two coexisting accipiters in Europe, the Goshawk and the smaller Sparrowhawk *A.nisus* (winglength 199 mm in male and 236 mm in female, Niethammer 1938). It would be interesting to know whether this is because fewer small species are available as prey in Europe.

Arctic sandpipers

The high arctic provides unusual breeding conditions for birds, since insect foods are available for only a small part of each year, but may then be abundant. Four sandpipers (or peeps) in the genus *Calidris* which breed near Barrow, in northern Alaska, stay there for only 5 to 12 weeks each year (Holmes and Pitelka 1968). Their weights, length of beak, main habitats and main foods are set out in table 21. The Pectoral Sandpiper *C.melanotos* and Dunlin *C.alpina* have the same breeding habitat, move with their young to the same habitat after breeding, and take the same main foods throughout the summer. The only important difference between them is that the Dunlin occurs in similar numbers each year, whereas the Pectoral settles in much

134

Table 21. Overlap in breeding ecology of four arctic sandpipers in genus *Calidris*

	Pectoral	Redbacked (Dunlin)	Baird's	Semipalmated
	C.melanotos	*C.alpina*	*C.bairdii*	*C.pusilla*
Mean for males				
weight in grams	95	57	38	24
culmen in mm	30	33	22	18
Habitat				
June, first half July	⊢---- coastal plain ----⊣ and foothills		ridges and higher ground	near coast and along rivers
second half July	⊢-- lowland marshes --⊣		,,	,,
Main food				
first half June	⊢---- tipulid larvae ----⊣		⊢--chironomid larvae---⊣	
second half June	------- tipulid larvae -------			chironomid larvae
first half July	⊢-------------- tipulid adults --------------⊣			
second half July	⊢-- chironomid larvae --⊣		adult insects	chironomid larvae

Note: from Holmes and Pitelka (1968)

larger numbers in the summers when insects are plentiful than scarce. Baird's Sandpiper *C.bairdii* differs from them in habitat, and in its food in the first half of June and the second half of July, but takes the same main foods in the intervening four weeks. The Semipalmated Sandpiper *C.pusilla* is also separated in breeding habitat, it takes the same foods as Baird's in the first half of June, and the same species of food as the other two during the rest of the summer, though it prefers the smaller individuals.

Hence despite big differences in size and in length of beak, these four species overlap greatly in their diets, though two of them differ in habitat from every other. It may be asked why, if they take similar foods, they should have beaks of such different length, but Holmes and Pitelka noted that they are ecologically segregated outside the breeding season, to which their morphological differences are presumably adapted. If their numbers are determined by the availability of food in winter, there might not be enough individuals surviving the winter to fill the available breeding grounds, and under these circumstances ecological isolation might perhaps be relaxed. Nevertheless, they exercise some selection in their summer range, habitat, and diet. Moreover five other congeneric species have bred round Barrow, the White rumped Sandpiper *C.fuscicollis*, Knot *C.canutus*, Sanderling *C.alba*, Western Sandpiper *C.mauri* and Curlew Sandpiper *C. ferruginea*, and a sixth, the Rufous-necked Sandpiper *C.ruficollis* has occurred without breeding, and that these other species do not breed commonly suggests that they likewise exercise selection in their breeding areas and habitats. A similar ecological situation is perhaps found in arctic waders in the genus *Tringa* (see pp. 315-6).

Arctic auklets

Three marine auklets, the Crested *Aethia cristatella*, Least *A.pusilla* and Parakeet Auklet *Cyclorhnchus psittacula*, have been studied where they breed on St. Lawrence Island, Alaska (Bedard 1967). The Crested is much larger than the Least (weights 300 g and 90 g respectively), the Parakeet is of similar size to the Crested, but has a larger beak. All three prey on planktonic animals on the same feeding grounds. Between May and July, the Least and Crested eat a variety of crustacea, including hyperiids, mysids, gammarids and acridean larvae, but they differ markedly from each other in the size of their prey; and when particular prey species become abundant in August and September, the Least brings largely the copepod *Calanus finmarchicus* and the Crested largely the euphausiid *Thysanoessa* spp. to the young. Both species feed mainly in the early morning and the early afternoon, whereas the Parakeet Auklet feeds after the early afternoon and for the rest of the day. Over half of the Parakeet Auklet's diet consists of carnivorous macroplankton, especially fish and the large hyperiids (amphipods), which are larger than the main prey of the Crested; and as already mentioned, it has a larger beak.

Hence these three species take mainly different foods. Like the Atlantic auks, the three auklets are also separated by their nesting sites, the Parakeet on cliffs, the other two on talus slopes, where they nest among boulders of different size. Incidentally, only one Atlantic alcid, the Little Auk *Plautus alle,* feeds on plankton, as compared with five in the North Pacific.

Conclusion

North American studies amplify those for Europe in various respects. There are several instances of species separated primarily by range which overlap in a small area where they differ in habitat; but others, which have met only recently through man's modification of habitats, freely interbreed where they meet. Several species separated by range or habitat show interspecific territorial exclusion where they meet, but this does not appreciably influence their ecological isolation. The possible importance of such exclusion in two species of blackbirds, which breed together over a wide area, is uncertain. The many congeneric species of vireos are separated like typical European passerines, but five coexisting and similar-looking parulid warblers differ in feeding in unusually complex ways. Each sex of each of three species of hawks differs from the other five in size and size of prey. Four sandpipers of different size overlap in diet in the breeding season, but are segregated outside it. Three auklets are separated by both diet and nesting sites on the breeding grounds.

Chapter 9

African Studies

Usambara, Turacos,
Brood Parasites, Vultures

Usambara

After completing in 1944 my first analyses of ecological isolation in Darwin's finches, other island passerines and European passerines respectively, I sought similar information from elsewhere, but at that time the ecology of only one other area in the world was sufficiently known, Usambara in Tanzania, through the studies of R. E. Moreau, to whom I wrote.

Usambara consists of some 8000 sq km of extremely varied country, including lowland and highland rain forest (up to 2400 m), wooded grassland and savanna, semi-desert thorn country, a few swamps and various types of cultivated land. About 400 species of birds breed there, but Moreau (1948) analysed solely the largest families, all except one of them being passerine. Most of the birds concerned are insectivorous, but the bulbuls Pycnonotidae and thrushes Turdidae eat both insects and fruit, the sunbirds Nectariniidae insects and nectar, and the estrildine and ploceine weaver birds mainly seeds. The one non-passerine family included, the barbets Capitonidae, eat mainly fruit, also insects, and Moreau (pers. comm.) added information on the congeneric turacos Musophagidae, which eat fruit.

Moreau considered all the species in each family, but the summary in table 22, based on appendix 17, is restricted to congeneric species (with modern genera larger than those of Moreau). Only 2 per cent of these congeners are separated

Table 22. Ecological isolation in congeneric species in selected families in Usambara, Tanzania

Family	range	Number of instances of isolation by			
		habitat	feeding station	beak- or body-size	unknown (same habitat)
Bulbuls Pycnonotidae	2	11	3	9	
Shrikes Laniidae		12			
Helmet shrikes Prionopidae		3			
Flycatchers Muscicapinae		7			
Warblers Sylviinae		28	3	4	2
Thrushes Turdinae		11		1	
Sunbirds Nectariniidae		62		5	3
Estrildines Estrildidae		16		2	2
Whydahs Viduinae		1			
Weavers Ploceidae		51		6	20
Barbets Capitonidae	3	9			1
Turacos Musophagidae		1			
TOTAL	5	212	6	27	28
PERCENTAGE	2	76	2	10	10

Notes: based on appendix 17 (p. 322), derived from Moreau (1948). The analysis is of selected (not all) families, but includes all the larger families, mainly passerine.

by geographical range, but more could not be expected in so small an area, and separation by this means is not rare in Africa (cf. the tits *Parus* discussed earlier [pp. 45–48]). Just over three-quarters of the species are separated by habitat

Table 23. Habitat restriction in forest passerine birds in Usambara

Habitat	Number of species present
A. IN THE FOUR MAIN WOODLAND TYPES	
solely semi-arid with thorn trees	14
solely wooded grassland	11
solely riverine forest	5
solely evergreen forest (rain forest)	61
common to more than one of these	0
B. RAIN FOREST	
solely lowland	31
solely highland	24
both lowland and highland	5
C. RAIN FOREST	
solely interior	39
solely edge	20
both edge and interior	1

Notes: based on appendix 18 derived from Moreau (1948). The totals are of all the species in the families selected by Moreau which live in the woodland habitats concerned (i.e. not solely the congeneric species analysed in table 22 and appendix 16). One rain forest species is too little known for inclusion in parts B and C of the table.

(differences in altitudinal range included under habitat). At least 12 per cent are separated by feeding, and another 10 per cent might be, as they are found in the same habitat.

Apart from the means by which congeneric species might be separated, it is interesting to see the extent to which each forest species is restricted to one type of woodland. The summary in table 23, based on appendix 18, shows that none of the species in riverine forest, wooded grassland, semi-desert with thorn trees or rain forest respectively, is found in more than one of these four types of natural habitat (though some of them also occur in cultivated land). Further, of the 60 rain-forest species, only 5 are common to lowland and highland forest (2 of them being represented by different subspecies in each) and only 1 is common to the forest edge and the interior.

It is difficult to make comparisons with European forest, as the latter has been so much modified by man, but the figures in appendix 19 show that about two-fifths of the European forest birds are found in both broadleaved and coniferous woodland, a much higher proportion than are common to any two types of forest in Usambara. Further, though there are several types of broadleaved forest in Europe, most species found in one of them are also found in others, and the same holds for the birds in the various types of coniferous forest. The main exceptions are some of the finches with specialized diets discussed in chapter 5. Finally, the European forest birds are much less sharply divided than tropical species between those of the edge and the interior respectively, but partly for this reason, and due to the difficulties of forest clearance, I was unable to make a quantitative analysis of this point for Europe.

As already mentioned, 22 per cent of the Usambara congeners live in the same habitat; 10 per cent of them differ markedly from each other in size and presumably feeding, another 2 per cent differ in feeding stations, and while no means of separation were known to Moreau for the rest, biologists had not at that time appreciated the importance of differences in feeding stations, so at least most of them might well differ in feeding. The only putative overlap is in various ploceine weaverbirds which breed after floods have subsided and eat gramineous seeds, including rice, which might be in temporary superabundance. These birds disperse to unknown areas after breeding, so are perhaps separated by feeding outside the breeding season. On the assumption that the species not known to be segregated are separated by feeding, proportionately more congeneric passerines are segregated by feeding in Usambara than Europe (22 cf. 15 per cent). In both areas, segregation by feeding is particularly common in seed-eaters (44 per cent of the congeneric estrildines, viduines and ploceines in

Usambara, including the unknowns), but if all seed-eaters are excluded from both counts, the proportion remains higher (12 per cent) in Usambara than Europe.

Species diversity in the tropics

These findings help to show why there are more species of birds in the tropics than temperate regions, for Usambara, compared with Europe, has both more types of woodland, and also a much higher proportion of bird species restricted to only one type, or to either the edge or the interior of the forest. Further, though the habitats are more finely divided by the Usambara birds, the proportion of passerine species presumably separated by feeding is also higher there.

This rough analysis was prompted by the conclusions of MacArthur, Recher and Cody (1966) and MacArthur (1969). In tropical forest in Barro Colorado, Panama, there are $2\frac{1}{2}$ times as many resident bird species on each 5-acre (2 ha.) census plot as in temperate woodland in Vermont, USA. The number of species in each genus is fairly similar in the two areas, and the main increase is due to a greater number of genera in Panama, i.e. of birds with morphological differences, which are usually associated with differences in feeding rather than in habitat. The number of species in relation to the number of vertical layers of forest is nearly the same in the two areas, but only provided that the Panamanian forest is held to consist of four, and the Vermont forest of three, layers.*

*The four layers in tropical forest are at 0–2 feet, 2–10 feet, 10–50 feet and above 50 feet, those in temperate forest at 0–2 feet, 2–25 feet and above 25 feet, corresponding respectively to the ground, the shrub layer and the trees (of two heights in the tropics). The existence of these layers was discovered through, and justified by, mathematical analyses, not by observations on feeding stations. As they stand, the figures might suggest that the number of species in tropical as compared with temperate forest should be in the ratio of 4:3, not 5:2 as mentioned above, but the number of species increases logarithmically, not arithmetically, with the number of forest layers (MacArthur pers. comm.).

These findings indicate that birds subdivide the vertical feeding stations to a greater extent in tropical than temperate forest. The authors postulated further that more species can coexist in the same habitat in the tropics than in temperate regions because the high and seasonally much more uniform availability of different types of food enables each species to specialise to a much greater extent. Likewise Cain (1969) pointed out that whereas a European thrush can eat fruits only in autumn and winter, and at other times of year has to depend on animal foods, a tropical bird can find fruits all the year round. The larger proportion of bird species segregated by feeding in Usambara than Europe, mentioned earlier, fits the view that there are a greater proportion of food specialists in the tropics.

MacArthur also found that when his census areas were increased from 5 to 600 acres (2 to 243 ha), the number of species present is still $2\frac{1}{2}$ times as great in Barro Colorado as in Vermont, but for much larger areas the difference is greater, and Ecuador, for instance, has seven times as many birds as New England. The chief effect of enlarging the census area is to increase the number of habitats, so this presumably means that the tropics, as compared with temperate regions, have a greater diversity of habitats, and/or a greater number of bird species restricted to one kind of habitat (as just noted for Usambara).

Some gaps in the montane forest avifaunas of East Africa

Many of the mountains in East Africa with highland rain forest are detached from each other, with intervening savanna. On some of these 'island mountains' a species typical of highland forest elsewhere is absent and is replaced by a related species found elsewhere in a different kind of habitat (Moreau 1948, 1966). On most East African mountains, for instance, the flycatcher *Trochocercus albonotatus* is in highland forest and the paradise flycatcher

143

Terpsiphone (Tchitrea) viridis is in lowland or riparian forest. But on Mount Cholo in Malawi, *T.albonotatus* is absent and here, but only here *T.viridis* occurs in montaine forest. This particular example also suggests that competitive exclusion is not restricted to species in the same genus. Again, the warbler *Bradypterus mariae* is typical of highland forest, but is absent from Mount Hanang in N. Tanzania, where the related *B.cinnamomeus* occurs in highland forest, though elsewhere it lives in montane heath or scrub, not forest. On Mount Hanang, also, neither of the usual montane barbets *Pogoniulus bilineatus* or *P.leucomystax* is present, and the related *P.pusillus,* which elsewhere lives at lower altitudes, replaces them. Again, the usual thrush *Cossypha semirufa* and the usual insectivorous weaverbird *Ploceus insignis* of montane forest are absent from the forests of the Mbulu highlands just north of Mount Hanang, where they are replaced respectively by their congeners *C.heuglini* and *P.ocularis,* which elsewhere are lowland species of evergreen bush, not forest. In the same area, neither of the usual montane woodpeckers *Campethera taeniolaema* or *Mesopicos griseocephalus* is present, and here another woodpecker, *Dendropicus* (formerly *Yungipicus*) *obsoletus,* inhabits evergreen forest, though in the rest of its range it is restricted to open deciduous woodland.

In each of these instances, a species is present exceptionally in highland forest where, but only where, the species normally present in this habitat is absent. Presumably, therefore, the exceptional species is elsewhere excluded from this habitat by interspecific competition. Why, then, is the normal highland species absent from a few mountains? Although the highland forests concerned are now isolated by other types of vegetation, they were connected by similar forest in the cool period up to 18,000 years ago, and the Tanzanian forests concerned, on Hanang and Mbulu, are close to others where the normal species is present (Moreau 1966, including his fig. 42). Hence the normal species could presumably have reached these forests initially, and even if by chance they became extinct there later, they could hardly have found it difficult to reach them again. It is suggestive that instances of the type under discussion have been found in only three of the many highland forests in eastern Africa, Hanang and neighbouring Mbulu in Tanzania and Cholo in Malawi. There is also a partial parallel in Angola, where the grass warbler *Cisticola erythrops* and some other species which normally live outside forest are found in highland forest on Mount Moco (Moreau 1966, citing Hall 1960). But the Moco forest is patchy, the trees nowhere tall and dense; likewise the Hanang forest is 'ragged' (Moreau 1966, pp. 83, 92). Hence the ecological conditions are unusual, and I suggest that this is probably why unusual species are present, and that here, but nowhere else, they exclude the usual highland species. A similar controversy will be met later with respect to whether island species have broader habitats because other mainland species have failed to reach them, or because the ecological conditions on islands are such that the species present exclude other mainland species through competition. I favour the latter hypothesis.

Turacos Musophagidae

D. W. Snow commented in a lecture that it might be hard for wholly frugivorous birds of similar size to coexist. I therefore analysed the means of isolation between con-generic species in the main African family of wholly frugivorous birds, the turacos Musophagidae, based on Moreau (1958) and set out in appendix 20. The two species of *Crinifer* and the two of *Musophaga* occupy separate ranges, so do two of the three species of *Corythaixoides,* but the other two overlap in range and differ in habitat and probably in feeding. As shown in fig. 30a, 9 of the 10 species of *Tauraco* also have separate ranges, except for two which overlap in a small area where they differ in habitat. However, as shown in fig. 30b, the remaining species,

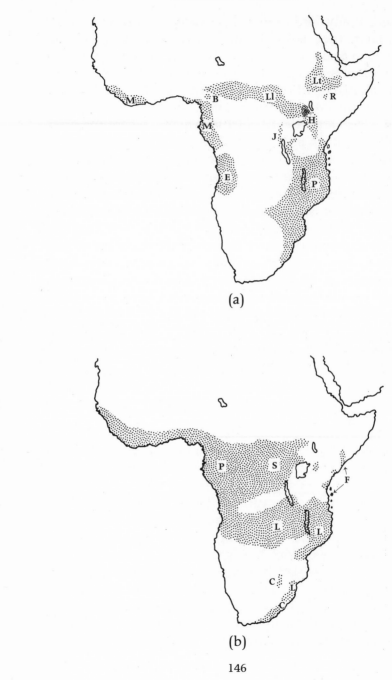

(a)

(b)

T.corythaix, overlaps in range with 5 of the other 9, but in all except one of these cases is known to be separated by habitat, and the other has not been studied. *T.corythaix* itself is by most authors (though not Moreau) divided into 5 geographical species in one superspecies which, if accepted, further increases the number of turaco species separated by range. The only coexisting turacos (except for two of *Corythaixoides* already mentioned) are in different genera, but their classification in separate genera recognises morphological, and hence presumably ecological, divergence; some of them differ markedly from each other in size, as do the two species of *Corythaixoides* in size of beak. Hence nearly all the congeneric species of turacos are separated by range, and all except one of the rest by habitat, justifying Snow's comment so far as this frugivorous family is concerned.

African brood parasites

The feeding habits of the tropical African birds which lay their eggs in the nests of other species are not sufficiently well known for discussion here, but these groups are

Fig 30. Ranges of turacos in genus *Tauraco* (from Moreau 1958).
(a) 9 geographically replacing species
    M   *T.macrorhynchus*
    B   *T.bannermani* ⎫
    E   *T.erythrolophus* ⎬ same superspecies
    P   *T.porphyreolophus*
    J   *T.johnstoni*
    H   *T.hartlaubi* ⎫
    Ll *T.leucolophus* ⎬ area of overlap, dense stippling
    Lt *T.leucotis* ⎫
    R   *T.ruspolii* ⎬ same superspecies
(b) *T.corythaix*, divided by most workers into 5 species:
    P   *T.c.persa*
    S   *T.c.schütti*
    L   *T.c.livingstonii*
    F   *T.c.fischeri*
    C   *T.c.corythaix*

considered because of the possibility that each species might parasitize different host species (termed 'alloxenia' by Friedmann 1967a, on whose studies this section is based). The most important bird family concerned is that of the cuckoos (Friedmann 1948, 1964, 1967b, 1968). The three African species in the genus *Cuculus* have wide ranges, which overlap extensively. *C.clamosus* parasitizes chiefly shrikes and *C.solitarius* chiefly thrushes; not enough is known to determine the main hosts of African *C.canorus*. The three African species of *Clamator* likewise have wide ranges and overlap extensively. *C.glandarius* parasitizes chiefly crows and starlings, *C.jacobinus* chiefly bulbuls and *C.levaillantii* chiefly babblers; but *C.jacobinus* parasitizes chiefly babblers in India, where *C.levaillantii* is absent. Two of the emerald cuckoos *Chrysococcyx* also have wide and overlapping ranges, in open country with trees; *C. caprius* parasitizes chiefly ploceid weaverbirds and *C.klaas* chiefly sunbirds and warblers. The main recorded hosts of a third species, *C.cupreus,* are likewise weaverbirds, sunbirds and warblers, but these records were not in its main habitat of forest, but in open country with trees near the forest edge, so the hosts concerned are probably not typical. Since *C.cupreus* differs in habitat from the other two, it probably differs in its main host species. The hosts of the fourth species, *C.flavogularis,* have not been found; it occurs in forest mainly south of *C.cupreus,* so might be separated by range rather than host species. The hosts of the two species of *Cercococcyx* in lowland forest are not known, and there is only one record for the third and montane forest species, in the nest of a forest thrush *Sheppardia.* Finally, too little is known to determine the main hosts of *Pachycoccyx audeberti,* which frequents open forest, but they include helmet shrikes and starlings. Hence each of the African cuckoos for which there are adequate records parasitizes mainly different host species, though Friedmann's lists show that there is wide overlap in the species parasitized occasionally.

The two European cuckoos, *Cuculus canorus* and *Clamator glandarius,* likewise have different main hosts, small passerines and corvids respectively. So do the four species of *Cuculus* in Japan (Royama 1963). Here, *C.canorus* parasitizes chiefly a reed warbler *Acrocephalus,* a shrike *Lanius* and a bunting *Emberiza, C.saturatus* parasitizes chiefly a leaf warbler *Phylloscopus, C.poliocephalus* parasitizes chiefly another type of warbler *Cettia* and a wren *Troglodytes,* and *C.fugax* parasitizes chiefly chats *Luscinia* and a flycatcher *Cyanoptila.* In Japan, therefore, each cuckoo parasitizes species in different genera, while in Africa each parasitizes species in different families, but in Japan, as in Africa, there is much overlap in the occasional hosts of each species. In contrast to these findings, Baker (1942) reported great overlap in the main hosts of the various Indian cuckoos, but it has become clear that his identifications of cuckoo eggs are unreliable (C. J. O. Harrison pers. comm.), so his records should be disregarded.

Another group of parasitic species in Africa are the honeyguides, of which various species overlap with each other widely in range, though some of them differ in habitat (in evergreen forest or dry bush country respectively). The hosts of only two species are well known, *I.indicator* parasitizing chiefly bee-eaters, hoopoes and starlings and *I.minor* parasitizing chiefly barbets (though only rarely the smallest barbets in the genus *Pogoniulus*). The few host records for other species are chiefly of woodpeckers for *I.variegatus,* of tinkerbirds *Pogoniulus* for *I. exilis,* of rock sparrows *Petronia* for *Prodotiscus regulus* and of white-eyes, small warblers and flycatchers for *P.insignis* (Friedmann 1955). Hence each of the honeyguides probably has different main hosts though, once again, they overlap in their occasional hosts. The hosts of various other species are not known.

The viduine weaverbirds are adapted for parasitizing solely estrildine weaverbirds (Friedmann 1960, Nicolai 1964). Each estrildine species has evolved striking and specific mouthparts, and is successfully parasitized solely by that viduine species in which the young have evolved similar mouthparts. Hence each viduine species has a different main host, but once again, there is overlap in their occasional (and unsuccessful) hosts. On the nomenclature used by Nicolai, the viduine *Steganura paradisaea* parasitizes the estrildine *Pytilia melba*, *S.orientalis* parasitizes a different race of *P.melba*, *S.obtusa* parasitizes *P.afra*, *S. togoensis* probably parasitizes *P.hypogrammica*, *Vidua macroura* parasitizes mainly *Estrilda astrild*, *Hypochera chalybeata* parasitizes *Lagonosticta senegala*, *Tetraenura regia* parasitizes *G.granatina* and *T.fischeri* parasitizes *G. ianthinogaster*. Hence each genus (or subgenus) of viduines parasitizes a different genus of estrildines. The hosts of the few remaining viduine species are not certainly known. Finally a ploceine weaverbird unrelated to the viduines, *Anomalospiza imberbis,* parasitizes warblers, especially grass warblers *Cisticola* (Friedmann 1960).

This survey shows that, in Africa, not only do the congeneric species of parasitic birds have different main hosts, but no two parasitic species in different genera or different families have the same main hosts. The only exception is that both the honeyguide *I.indicator* and the cuckoo *Clamator glandarius* often parasitize the starling *Spreo bicolor* in South Africa, but this is evidently a new habit for *C.glandarius,* which normally parasitizes corvids, and its eggs do not mimic those of the starling, though they do those of its corvid hosts.

While each parasitic species has different main hosts, there is wide overlap in the occasional hosts. That the latter are used only occasionally indicates that the individuals which parasitize them leave fewer offspring than those which parasitize the main hosts of their species, and since it is extremely unlikely that each parasitic species would have different main hosts by chance, such alloxenia (to use Friedmann's term) has presumably been evolved through natural selection. A revealing instance in this connection is that, in Africa, where *C.levaillantii* parasitizes chiefly babblers, the related cuckoo *C.jacobinus* parasitizes chiefly bulbuls, whereas in India, where *C.levaillantii* is absent, *C.jacobinus* parasitizes chiefly babblers. The parallels to this sort of situation among species with different habitats are readily interpreted in terms of competitive exclusion.

Various species of phytophagous insects may be found on the same food plant if each has a different set of insect parasites. But while it is generally held that the insect parasites play an important part in limiting the numbers of their insect hosts, no one has postulated that the bird parasites play a similar part in regulating the numbers of their host species. Hence it is not obvious how alloxenia might have been evolved in parasitic birds, except in the viduines, where it is essentially linked with the mimicry of specific mouthparts. One factor might be that the egg of a bird parasite would be unlikely to give rise to a fledgling if laid in a host nest which already contained the egg of another parasitic species but this does not provide a full explanation.

Vultures in the Serengeti

Seeing up to six species of vultures feeding round the same carcase on the Serengeti plains in Tanzania, one might think that there was a superabundance of food, and no effective competition. This is wrong however, for Kruuk (1967) has shown that the various species are at least partly separated from each other, with different specializations. The following account is based on his findings, with modifications made later by his student D. Houston (pers. comm.).

Much the commonest vultures in the Serengeti are the two griffons *Gyps rüppellii* and the White-backed *G. (Pseudogyps) africanus,* which feed especially on the muscles and guts of carcases. They differ from the other four in having long necks for reaching far inside corpses, a slender skull, a fairly thick beak, and a large, broad, gutter-shaped tongue with horny and backward-pointing teeth, which helps to grip soft meat. Two other large species, the White-headed Vulture *Trigonoceps occipitalis* and the Lappet-faced Vulture *Torgos tracheliotus,* eat mainly the skin, sinews and meat adhering to bones, generally tougher parts than those taken by the griffons, they commonly use their feet in gripping while tearing with their beaks, and their skulls are broad, their beaks very thick. The other two, the Egyptian Vulture *Neophron percnopterus* and the related Hooded Vulture *Necrosyrtes* (or *Neophron*) *monachus,* are much smaller with relatively slender beaks, and they feed chiefly on the small fragments left on the bones or on the ground by other scavengers. Hence they are the last to leave a carcase. They also feed much on village garbage and large insects. Hence each of these pairs is separated by feeding from the other two pairs.

As portrayed in fig. 31, one vulture similar to each of them is found in southern Europe, the Griffon *Gyps fulvus* being simply a geographical representative of *G.rüppellii,* the Black Vulture *Aegypius monachus* having a thick beak like *T.tracheliotus,* which it replaces geographically to the north, and the Egyptian Vulture *N.percnopterus* being the same species as in the Serengeti. But in Europe there is only one of each kind, so there remains the problem of how there come to be two of each in the Serengeti.

The two Tanzanian griffons are probably separated by habitat, *G.rüppellii* being found on high ground and nesting in cliffs, and the White-backed *G.(P.)africanus* in lowland savanna, nesting in trees, and these habitats adjoin in the Serengeti. The two small species are also separated by habitat, though in a different way, the Egyptian *N.percnop-terus* living in dry open plains and deserts and nesting in cliffs, and the Hooded *N.monachus* in wetter areas, including forest, wooded country and round villages, and nesting in trees. These two habitats also adjoin round the

Fig 31. Black, Griffon and Egyptian Vultures round a carcase. By Robert Gillmor.

Serengeti. Hence both members of each of these pairs meet in the Serengeti, though in most of Africa they are apart.

The two thick-beaked species are separated in a different way. The White-headed *T.occipitalis*, which is sparse, finds all its food for itself, and feeds solitarily, avoiding the other species. It is, however, highly efficient at finding carcases before the other species do so, and it often feeds on smaller carcases than they do. It is also possible that it captures living small mammals, which the others do not. In contrast, the large Lappet-faced Vulture is the most aggressive of all the vultures, it feeds primarily round large carcases and often robs the other species of meat.

Hence all six species are separated from each other, unlikely though this once seemed. A seventh species in Tanzania, and the fourth in southern Europe, is the Bearded Vulture or Lammergeier *Gypaetus barbatus*, which is largely separated by habitat, as it lives high up in precipitous mountains. It also has some peculiar habits, such as dropping large bones or tortoises from a height to break them, and swooping fast and low over living mammals on precipices to drive them over the edge.

Conclusion

In East Africa, most congeneric passerines are segregated by habitat, nearly all frugivorous turacos by range, and various vultures (in different genera) by habitat or the parts of the carcase which they eat. A montane forest species may be absent from a particular mountain, where its place is taken by a species found elsewhere in a different habitat. Each parasitic species, whether congeneric or not, has different main hosts, but overlaps with others in its occasional hosts. As compared with birds at high latitudes, tropical species are much more often confined to one type of woodland, there are more types of woodland, and probably, also, a higher proportion of species is separated by feeding; these points help to explain the greater number of species at low than high latitudes.

Chapter 10

Other Tropical Studies
Fruit-eaters, Honey-eaters,
Suburbs, Seabirds

Tropical fruit-eating birds

Often several, and sometimes many, species of birds can be seen feeding on the fruits of the same tree. At high latitudes, such an overlap in feeding is readily explained, because the birds in question eat fruit during only a brief period of superabundance when it is ripe, and in the rest of the year eat different foods. But in the tropics some fruits, of different species, may be ripe in every month of the year, as shown for the Melastomaceae in Trinidad by Snow (1962), and many species of birds can be seen eating fruit in the same tree. In less than two hours' observations in Colombia, for example, Willis (1966a) saw 28 species of birds, including 15 of tanagers, 9 in the genus *Tanagra*, feeding on Conostegia berries.* Willis asked: 'Do these species contravene the "competitive exclusion principle" ', and continued that, if the fruits in question are superabundant, the birds evidently have a succession of superabundant fruits through the year. Further, Snow (in prep.) has pointed out that whereas the insects sought by birds are well concealed, fruits have been evolved to attract birds, and this might give rise to a different type of ecological situation from any so far considered in this book.

Snow and Snow (in prep.) studied the foods and feeding

*He cited an instance in which 69 different species of birds fed on another superabundant food, the flowers of *Combretum farinosum* in Mexico.

habits of the 10 species (in 5 genera) of tanagers Thraupidae in Trinidad, details being set out in appendix 22. One of them, the rare *Tanagra trinitatis*, was seen eating solely insects, while another, *Tanagra violacea*, feeds exclusively on fruits, specializing on mistletoes and also taking aroid and cactus fruits which the others rarely if ever take, though it takes some fruits commonly eaten by other species. The remaining eight species eat much fruit, but also many insects, and some of them also eat flowers. They overlap greatly with each other in their fruit diets, but are separated ecologically through hunting for insects in different ways on different parts of the vegetation.

Thus of the three species of *Tangara*, one takes insects mainly from the leaves, one from thin twigs, and the other from thicker twigs. Of the two species of *Thraupis*, one takes insects from large and slippery leaves, such as those of palms, by clinging head downward, while the other feeds in more varied ways than any of the others, and in particular catches more insects in the air. Of the two species of *Tachyphonus*, one takes insects mainly from foliage in trees above 8 m from the ground, and the other mainly below 8 m and often on the ground. The closely related *Ramphocelus carbo* feeds differently again, hopping about looking for insects in the thick low canopy of bushes and in herbaceous vegetation. Finally the rare *Tanagra trinitatis* takes insects from the undersides of extremely thin twigs. Hence each of them differs from all the rest, and since the wholly frugivorous *Tanagra violacea* eats mainly different fruits from the rest, all ten species are segregated from each other. Each has adaptations to its particular way of feeding, for instance *Thraupis palmarum* has sharp curved claws for gripping slippery leaves, and the different species, especially those in different genera, differ from each other in overall size, length and thickness of beak, and proportionate length of tarsus and tail.

The Snows also analysed the feeding of five honey-creepers in the genera *Chlorophanes*, *Cyanerpes*, *Dacnis* and,

in a different family, *Coereba*, and found that in them, likewise, there is much overlap in the fruits and nectar taken,
but each differs from the rest in its hunting methods for
insects.

Since the honeycreepers and tanagers are isolated
primarily by their insect foods, it is of special interest to
determine what happens in birds which feed almost
entirely on fruits. Snow (pers. comm.) pointed out that
fruits are much less diverse than insects, both in size and in
where they occur, so that there are much smaller opportunities for ecological divergence among frugivorous than
insectivorous species. Hence, as cited in the previous
chapter, one would expect coexistence to be much harder
in wholly frugivorous than in insectivorous birds, and
Snow added that South American cotingas in the same
genus tend to have separate geographical ranges, and so do
congeneric birds of paradise.

Fig 32. Ranges of birds of paradise in genus *Paradisaea* (from
Gilliard 1969).
A   *P.apoda* (south-central and Aru I.)
B   *P.raggiana* (east, especially in south)
C   *P.minor* (north, including Japen I. and Misol in West Papuan
    Archipelago)
D   *P.decora* (D'Entrecasteaux I.)
E   *P.rubra* (Waigeu and perhaps Batanta in West Papuan
    Archipelago)
F   *P.gulielmi* (Huon Peninsula, above 670 m)
G   *P.rudolphi* (eastern mountains above 1350 m)

I have in appendix 21 set out the means of isolation of the congeneric birds of paradise and bower birds, based on Gilliard (1969). All the species are entirely frugivorous. In the birds of paradise, separation is by geographical range in one of the three species of *Manucodia*, the three species of *Ptiloris*, the two species of *Paradigalla*, the 5 species of *Astrapia*, the 4 species of *Parotia* and 6 of the 7 species of *Paradisaea*, whose ranges are mapped in fig. 32. Two species of *Manucodia*, two of *Epimachus* and two of *Paradisaea* are separated by habitat, one in lowland and the other in montane forest respectively, but these might almost be regarded as separated by geographical range in the conditions found in New Guinea. Only two species, in the unspecialized genus *Manucodia*, coexist in the same habitat and it is not known how they might be separated. Similarly in the bower birds, two of the three species of *Ailuroedius*, the three of *Amblyornis*, the three of *Sericulus* and the four of *Chlamydera* are separated by geographical range, while two species of *Ailuroedius* are separated by altitudinal range (and in part by geographical range), and two species of *Chlamydera* are separated by habitat in a very small area where their ranges overlap. The extent to which congeneric birds of paradise, bower birds and also turacos (see p. 146) are separated by range is highly exceptional in continental land birds, and is presumably due to limitations imposed by their fruit diet. Species in different genera often coexist, but they usually differ greatly in size.

Australian honeyeaters in the genus Melithreptus

Only a small part of Australia is tropical, but it is convenient to include in this chapter the only other Australian birds so far studied with respect to ecological isolation. The insectivorous meliphagids in the genus *Melithreptus* include four species on the mainland and two more in Tasmania, and the heads and beaks of the chief forms concerned are shown in fig. 33 (Keast 1968 and *pers. comm.*).

(i) Situation in southeast, where all three coexist.

(ii) Situation in southwest, where large *M.gularis* is absent and *M.lunatus* is much larger, *M.brevirostris* a little smaller than elsewhere.

(iii) Situation on Kangaroo Island, where *M.brevirostris* is sole species, and unusually large.

(iv) Situation in Tasmania, where *M.brevirostris* is absent and *M.lunatus* is replaced by the related but smaller *M.affinis*, and *M.gularis* by the related but larger *M.validirostris*.

Fig 33. Heads of *Melithreptus* species in Australia (from Keast 1968). (*M.albogularis,* which replaces the similar *M.lunatus* geographically in the north, is not shown.)

The broad situation on the mainland follows a pattern similar to that set out for many other genera in previous chapters. Two small species are separated mainly by range, *M.albogularis* in the north and *M.lunatus* in the south. A third small species, *M.brevirostris*, found only in the south, coexists with *M.lunatus* in forest, where it hunts chiefly on the upper branches and *M.lunatus* chiefly on the leaves. *M.gularis*, the fourth species, coexists with each of the others, but is much larger, so is presumably separated by feeding.

What makes these species of particular interest is a number of exceptions to this general picture. First, the northern *M.albogularis* and southern *M.lunatus* have a fairly extensive area of overlap in Queensland where they are separated by habitat, *M.albogularis* in tropical savanna woodland in the hills and *M.lunatus* in sclerophyll forest in the lowlands. Secondly, in southwestern Australia the foliage-gleaning *M.lunatus* and branch-feeding *M.brevirostris* are separated by habitat, *M.lunatus* in wetter forest and *M.brevirostris* in mallee scrub-forest, and in the mallee, *M.brevirostris* feeds mainly in the foliage. Thirdly, the large *M.gularis* is absent from Kangaroo Island in the west, where the only species present, *M.brevirostris*, has evolved a longer beak (by 15 per cent), and a somewhat longer tarsus and hallux, than elsewhere; presumably it to some extent combines the feeding niches divided between *M.gularis* and *M.brevirostris* elsewhere. Finally the large *M.gularis* is also absent from southwestern Australia, where the other small species, *M.lunatus,* has evolved a beak and tarsus 10 per cent longer than elsewhere, presumably for the same reason that *M.brevirostris* has done so on Kangaroo Island. *M.brevirostris* is also present in southwestern Australia, but is here separated by habitat as already mentioned (and is slightly smaller than elsewhere).

These are textbook examples of the effects of competitive exclusion: two species with largely separate ranges differ in habitat where they overlap, one species expands its feeding stations in the absence of another with which it usually coexists, and in two areas a small species has evolved a longer beak in the absence of a larger species with which it normally coexists.

In Tasmania, the small *M.affinis* is a geographical representative of mainland *M.lunatus*, and the large *M.validirostris* of mainland *M.gularis*. But the difference in beak-length between these related pairs of species is much greater in Tasmania than on the mainland. This is associated with divergence in their feeding habits, as the small *M.affinis* has become more warbler-like, and the large *M.validirostris* not only hunts for food on branches, but probes in crevices in bark, linked with the absence from Tasmania of the usual trunk-feeding species of the forests and woods of Victoria, the Australian treecreepers *Climacteris* and nuthatches *Neositta*. As discussed in later chapters, broader feeding stations and bigger differences in size between congeners are characteristic of island birds.

Suburban Singapore

The colonization of farmland, gardens and towns by European birds occurred so long ago that it is not possible, now, to trace where most of the species came from, as discussed for finches in chapter 5. The same applies to long-settled parts of the tropics such as India, which has a rich variety of garden species. In contrast, Ward (1968) found only 31 native species yet established in the relatively new suburbs of Singapore. Originally Singapore, like the rest of Malaya, was covered with evergreen forest, except for an extremely narrow coastal strip of mangroves, sandy ridges with halophytic trees and shrubs, and cliffs. Yet, as shown in table 24, as many as 25 of the 31 suburban species have come from coastal habitats, and three others breed both on the coast and inland. None of them have come from highland rain forest or montane scrub, and only two, both flowerpeckers *Dicaeum*, live in lowland rain forest, but these flowerpeckers also live in open coastal vegetation, and in the rain forest are above it in the open habitat of the forest canopy (P. Ward pers. comm.).

Similarly, most species in the suburbs of Ibadan in Nigeria have come from savanna (Elgood and Sibley 1964), and the 22 which these authors termed 'forest birds' live in

Table 24. Origin of town and garden birds of Singapore

Natural habitat in Malaya	Number of species from this habitat in Singapore
COASTAL	
(i) mangroves	11 [6 of them also in (ii) and 1 in (iii)]
(ii) sand ridges (halophytic trees and shrubs)	21 [6 of them also in (i) and 2 others in (iv)]
(iii) cliffs	3 [1 of them also in (i)]
INLAND	
(iv) lowland rain forest (canopy)	2 [both also in (ii)]
(v) montane rain forest	0
(vi) montane Rhododendron scrub	0
(vii) inland cliffs	1 [also in (iii)]
OTHERS (not forest)	3
NONE (introduced)	9

Notes: from Ward (1968). The 6 town species which live in both mangroves and sandy ridges are in the genera *Treron*, *Lalage*, *Malacocincla*, *Rhipidura*, *Leptocoma* and *Zosterops*; the one in mangroves and cliffs is in *Halcyon*; the 4 only in mangroves are in *Otus*, *Hemiprocne*, *Hirundo* and *Aplonis*; the 2 in sand ridges and rain forest are in *Dicaeum*; the 13 only in sand ridges are in *Turnix*, *Streptopelia*, *Centropus*, *Caprimulgus*, *Merops*, *Dendrocopos*, *Aegithina*, *Pycnonotus*, *Orthotomus*, *Anthus*, *Anthreptes* and *Munia* (2 spp.); the one on sea and inland cliffs is in *Apus*; the one only on sea cliffs is in *Collocalia*; the others not in forest are *Amaurornis* (in swamps) and *Halcyon* and *Cypsiurus* (in open country); and the introduced species are in *Columba*, *Oriolus*, *Corvus* (2 spp.), *Orthotomus*, *Passer* and *Padda*.

riverine forest or the forest edge, not in the interior of forest; no species from the last habitat have colonized Ibadan (Elgood pers. comm.). Suburban habitats consist of open spaces with trees and bushes, so look like savanna, which covers large parts of Africa, so it is understandable that Ibadan should have been colonized by species of savanna or the forest edge. But suburban gardens do not look like mangrove forest or the sandy ridges of Malaya,

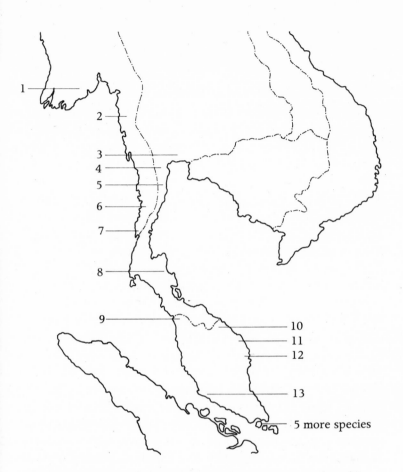

Fig 34. Southern limit of birds of open country spreading south from Burma (from Ward 1968).

1. parrot *Psittacula krameri*
2. sparrow *Passer domesticus*
3. bulbul *Pycnonotus aurigaster/cafer*
4. owl *Tyto alba*
5. kite *Milvus migrans*
6. starling *Sturnus nigricollis*
7. crow *Corvus splendens*
8. hawk *Accipiter badius*
9. tree-pie *Crypsirhina temia*
10. bulbul *Pycnonotus blanfordi*
11. bulbul *Pycnonotus jocosus*
12. sparrow *Passer flaveolus*
13. barbet *Megalaema haemacephala*

The 5 that have already reached Singapore are the oriole *Oriolus chinensis,* crow *Corvus macrorhynchos,* tailor-bird *Orthotomus sutorius,* and the mynahs *Acridotheres tristis* and *A. fuscus.* An isolated population of the crow *Corvus splendens* is also in Singapore, but might have originated from escaped cage-birds.

which anyway form a tiny fraction of the original vegetation of the country.

Nine bird species now living in Singapore were not originally present in Malaya. Some of these are escaped cage-birds or were deliberately introduced by man, but others, such as the mynahs *Acridotheres tristis* and *A.fuscus*, spread of their own accord from much further north, where they are native in open country, their progress being facilitated by the recent clearing of forest by man. As shown in fig. 34, a number of other species are currently spreading south from Burma into Malaya in areas cleared of forest and have covered a varying distance.

The avifauna of Singapore is not yet stable. Of particular interest are those species which have recently replaced others with similar feeding habits; presumably the later of the two took longer to reach, or to adapt itself to, suburbia, but once it did so was the more efficient. Thus the kingfisher *Halycon smyrnensis* has greatly declined and the related *H.chloris* has colonized and increased. The native munia *Munia maja* was once common, but has now been largely replaced by the related *M.punctulata*, which was perhaps introduced by man. The native tailorbird *Orthotmus sericeus*, still common on the coast, has been largely replaced in the suburbs by the common species of Indian gardens *O.sutorius* (Kipling's 'Darzee'), which has spread down from the north. The dove *Geopelia striata*, once numerous, has been largely replaced by another, *Streptopelia chinensis*. It is not certain whether the marked decline of the Magpie Robin *Copsychus saularis* is linked with the big increase of the two mynahs which, as already mentioned, have spread down from the north; they are in different families, but feed in a similar way. Again, two swifts *Apus affinis* and *Cypsiurus parvus*, have greatly decreased, while two others, *Collocalia francica* and *Hemiprocne longipennis*, have much increased, presumably due to changes in the insect life, but possibly helped by interspecific competition.

Some other changes do not appear to be linked with those in other birds. The Java Sparrow *Padda oryzivora* decreased with the decrease in spilt grain about buildings. The fantail flycatcher *Rhipidura javanica* and the two flower-peckers *Dicaeum* have decreased for unknown reasons. The oriole *Oriolus chinensis* established itself in the nineteen-twenties, and the House Crow *Corvus splendens* (perhaps introduced by man) some twenty years later. Other gaps remain to be filled. For instance, there is not yet a garden flycatcher, a predator on small passerines, or a scavenging raptor, though all three are found in long-established tropical towns. Moreover an accipitrine hawk and a scavenging kite are currently spreading south through the Malay Peninsula, as shown in fig. 34. That Ibadan has 75 garden species, about twice as many as in Singapore, also suggests that the latter has more species yet to come, either because there are vacant niches, like that for a hawk, or because two species might be able to divide a niche now occupied by only one.

In Singapore today, one is seeing an accelerated version of what presumably happens when a new volcanic island has been formed, but with this difference, that a volcanic island has first to be colonized slowly by plants and insects, whereas in Singapore man has provided a relatively rich garden habitat within a few years. There is inevitably a timelag between the creation of a land area and its discovery by birds, and if the habitat is of a strange type, birds may not at first recognize it as suitable. Further, those best pre-adapted to the new ecological conditions may not be the first to arrive or to recognize the area as suitable. Even after they have arrived, time is needed for competitive exclusion to sort the species out, and when its results are clear, the less efficient species will have to evolve behaviour to keep away from where they will be excluded by competition. The present unstable situation in Singapore well illustrates one of the main themes of this book, that ecological isolation may take a long time to evolve.

165

Tropical seabirds

The members of the British Ornithologists' Union Centenary Expedition to Ascension Island, in the Atlantic, found a big overlap in diet, not only between congeneric terns, tropic birds and boobies respectively, but between terns in different genera and even between species in these three different families, as shown in appendix 23 (Stonehouse 1962). Yet terns, tropicbirds and boobies differ so much in size, shape and hunting methods that one would have supposed that they must be adapted to different types of prey, and the same applies to the big difference in size between the two species of tropicbirds, and to the smaller differences between the various terns. A similar big overlap in prey species was recorded in the seabirds of the Barents Sea and the Farne Islands (see pp. 124, 321–2), so it might be wondered whether seabirds constitute a general exception to the principle of competitive exclusion. Although, however, the Brown Booby *Sula leucogaster* and White Booby *S. dactylatra* breeding on Ascension Island take mainly the same species of prey, the Brown often feeds close inshore and the White mainly out of sight of land, so they hunt in different feeding areas (Dorward 1962).

The problem has been carried considerably further by the Ashmoles (1967) and Ashmole (1968b) in a later study on Christmas Island in the Pacific. Here are found the same four species of terns which breed on Ascension, together with three others. Of these, the Greybacked Tern *Sterna lunata* was not studied, and food samples were not obtained from the Crested Tern *Thalasseus bergii*, but the latter feeds only close to the beaches within the lagoon and along the coast, where none of the others feed to an appreciable extent, so it is separated from them. The other five feed

Fig 35. Beaks of five tropical terns (top to bottom: Sooty, Brown Noddy, Fairy, Black Noddy and Blue-grey Noddy). By Robert Gillmor.

Table 25. Size of Christmas Island terns and size of their prey

Species	Wing-length (mm)	Length of culmen (mm)	Area of beak in cross-section (mm²)	Mean vol. (in ml) of largest 2% of prey		Most important fish in prey
				fish	squid	
Sooty Tern *Sterna fuscata*	284	42	18	11	17	*Exocoetidae* *Scombridae*
Brown Noddy *Anous stolidus*	284	42	21	14	13	*Exocoetidae*
Fairy Tern *Gygis alba*	236	39	15	7	8	*Blenniidae* *Exocoetidae*
Black Noddy *Anous tenuirostris*	226	43	15	4	10	*Exocoetidae* *Blenniidae*
Blue-Grey Noddy *Procelsterna cerulea*	180	25	7	0·5	1	*Gempylidae*

Notes: from Ashmole (1968b) Tables 1 and 2. Note that the Black Noddy has a long but narrow beak. The ranking order for size of prey is the same if the most frequent size of fish and squids in the prey is used.

largely on small fish driven to the surface periodically by schools of tuna, though they take mainly different prey from the tuna itself. All five take many of the same species of fish and squids, which at first sight supports the overlap in diet recorded on Ascension, but fuller study showed that they differ in important ways. The heads of the five species are shown in fig. 35 and various details in Table 25.

The two largest species, of similar size to each other, are the Sooty Tern *Sterna fuscata* and Brown Noddy *Anous stolidus*, and they have very similar diets. In both of them the mean size of prey is similar, about half the diet consists of fish and half of squids, and the commonest fish are flying fish Exocoetidae, with tunas Scombridae next, the latter being more important to the Sooty than the Brown Noddy. This suggests an almost complete overlap, but the Sooty feeds up to hundreds of miles from land and the Brown Noddy within fifty miles, hence their feeding areas are largely different. The Sooty Tern has relatively narrow wings adapted for flying long distances rapidly, it can stay on the wing indefinitely, so need not return to land at night, and its chick is adapted to starve for several days and to ingest a large meal when it comes. The Brown Noddy has proportionately wider wings and tail, which presumably enable it to manoeuvre more efficiently than the Sooty, and this might well give it the advantage if they feed together, but it cannot fly so fast and returns to land each night, so cannot reach most of the areas hunted by the Sooty.

These differences recall those between the Swift *A.apus* and Swallow *Hirundo rustica*, in England. Both species feed on airborne insects. The Swift has a long narrow wing, so flies fast and far, and it can store much food in its throat, while its chicks can starve for several hours and can ingest large meals; but it manoeuvres less well than the Swallow, so feeds mainly in the open and high up. The Swallow has a proportionately broader wing and can feed lower down, where insects are more numerous, as its greater power of manoeuvre enables it to avoid trees and other obstacles, hence it does not need to fly far from its nest and feeds the young on smaller meals and more often.

The next largest tern on Christmas Island is the Fairy Tern *Gygis alba*. Its diet is similar to that of the two larger species, but includes some smaller prey, and the most frequent fish are species of Blenniidae. Although it hunts above the tuna by day, it is also able, unlike the other terns, to take fish at the surface in the half light at dawn and dusk. Probably correlated with this, it has an unusually large eye and its plumage is pure white, which makes it much less visible to a fish on the surface than is the Noddies' dark plumage; but the latter both protects the skin from the sun and does not wear out so quickly as white plumage. Presumably as an alternative means of protection from the sun, the Fairy Tern has evolved dermal melanin, so the skin is black, not white as in noddies.

The Black Noddy *Anous tenuirostris* feeds in the same way as the Brown Noddy, and flying fish are likewise its most frequent prey. But it is smaller, and the smaller prey which form most of its diet comprise only a small proportion by volume (though not by number of individuals) of the prey of the Brown Noddy. Unlike the latter also, it often feeds within a few miles of the coast by day. Its long thin beak gives it the advantage in taking highly mobile small fish.

Finally the Blue-grey Noddy *Procelsterna cerulea*, which is closely related to *Anous*, is much smaller. It often feeds only a few miles offshore, frequently with the Black Noddies, but takes mainly smaller prey, notably small fish in the family Gempylidae, planktonic crustacea and marine water striders *Halobates*, which are not taken by the other terns, presumably because it would be uneconomic for them to hunt them. These are the smallest prey available to seabirds in the area.

The size of these five terns in relation to the size of their prey has been set out in table 25. The mean volume of their largest prey provides a good indication of their size preferences, but their ranking order would be the same if the most frequent size of fish and squids in their prey were used instead (as in Ashmole's fig. 1, not reproduced here). The size of prey is correlated with the winglength of each species, but not with the length of beak, largely because the small Black Noddy has the longest beak. As Ashmole pointed out, the length of the beak is not necessarily a good measure of the size of prey which a bird can take, and in terns its thickness, and hence strength, is important; a long slender beak is adapted for catching mobile prey, and a strong thick beak for catching larger prey.

Three seabirds in other families were briefly studied. The Phoenix Petrel *Pterodroma alba* has a deep beak with sharp cutting edges, and feeds to a much greater extent than any of the other species on squids, including large dead ones. The Christmas Island Shearwater *Puffinus nativitatis* presumably hunts like other shearwaters by swimming under the surface, so takes prey unavailable to terns at the time. The Redtailed Tropicbird *Phaethon rubricaudus* hunts solitarily for dispersed fish, many of which are larger than those taken by the terns and petrels, and though flying fish are especially taken, the next most favoured species are in three families rarely taken by the other seabirds studied.

Hence while the seabirds of Christmas Island take many of the same species of fish, they are largely separated from each other by differences in the distance from the island at which they feed, the time of day at which they hunt, the depth of water in which they catch their prey, and the size of their prey. Although, however, each species is well adapted to its particular way of feeding, each is not equally successful, if success is measured by the number of individuals breeding on the island. Thus there are about one million adult Sooty Terns but only some 1900 adult Brown Noddies, so the capacity of Sooties to travel much further from the land evidently makes a much greater volume of food available to them than to the Brown

Noddies. Again, there are about 15000 adult Black Noddies, but only 3300 Blue-grey Noddies, so presumably the larger prey of the Black Noddy provide much more food than the smaller prey of the Blue-grey Noddy; and there are only about 1800 adult Fairy Terns. Presumably the scarcer species are able to persist because the adaptations of the abundant species carry disadvantages as well. The longer and narrower wings of the Sooty Tern enable it to fly further but to manoeuvre less well than the Brown Noddy; the longer and narrower beak of the Black Noddy makes it more efficient at catching small active fish, but less efficient at catching larger fish than the Brown Noddy; the white plumage of the Fairy Tern wears out more quickly, but makes it less visible to fish on the surface, than the brown plumage of the noddies; and so on.

Similarly in Galapagos waters, the Bluefaced Booby *Sula nebouxii* feeds close inshore, the White *S.dactylatra* a moderate distances offshore, and the Redfooted *S.sula* far out to sea (Nelson 1968); and of the three storm petrels, *Oceanites gracilis* feeds inshore and the two species of *Oceanodroma* far out, *O.castro* by day and *O.tethys* by night (Harris 1969). Probably, therefore, separation by feeding area and time of day is common in seabirds.

Conclusion

Tropical fruit-eating birds and tropical seabirds have provided apparent examples of a big overlap in diet between coexisting congeneric species, the fruit-eaters because they have a succession of superabundant fruits, and the seabirds because they bring the same species of prey to their young. In fact, the tropical species which eat both much fruit and many insects are segregated by their hunting methods for insects, and those which eat wholly fruit are separated primarily by range, while the tropical seabirds are largely separated by the distance from land or the times of day at which they feed, and by the mean size of their prey. Australian *Melithreptus* provides some classic instances of ecological isolation and character displacement. Suburban Singapore illustrates the rapid replacements that may take place when a new habitat is provided.

Chapter 11

Adaptive Radiations
in Archipelagoes

Galapagos Finches and Hawaiian Sicklebills

The Galapagos

The Galapagos archipelago, as Darwin wrote in the 'Voyage of the Beagle', 'is a little world within itself', and it was through consideration of the size differences between some of Darwin's finches, the endemic subfamily of Geospizinae, that I first came to appreciate the importance of ecological isolation in birds. This group has undergone a miniature adaptive radiation in the archipelago, and it is of special interest for this book to see how the different species are segregated ecologically from each other at their relatively early stage in speciation and evolution. The present account is based on my earlier book (Lack 1947), together with the many new facts discovered by Bowman (1961), and some minor revisions of these writings (Lack 1969), criticisms of details in this last paper not being referred to here unless of special importance.

The Galapagos, shown in fig. 36, are equatorial volcanic islands of relatively recent origin, some 600 miles west of Ecuador. The low ground is arid, with tree cactuses and rather small deciduous trees, the high ground, found only on the larger islands, has cloud forest, with a transition zone between. Apart from Darwin's finches, only five passerine species breed in the islands, each of which feeds in a different way from the rest. These are a martin *Progne*, mockingbird *Nesomimus*, parulid warbler *Dendroica* and two tyrant flycatchers, a larger *Myiarchus* and smaller

Pyrocephalus. None of these feeds like any of Darwin's finches except the parulid warbler, which is decidedly larger (by 10 mm in winglength) than the rather similar warbler-finch *Certhidea*.

Fig 36. Map of Galapagos.

Isolation of the five genera

Darwin's finches are dull-coloured short-tailed birds about the size of sparrows (mean wing-length 50–90 mm), which eat seeds or insects and, with two exceptions, have rather thick beaks. It is generally agreed that all 14 species evolved from a common ancestor, and they are so like each other except in beak that this evolution was evidently recent. Hence the 5 genera, as well as the congeneric species, are considered here. Three of these genera, the heads of which are shown in fig. 37, consist of a single species each. One of them, *Pinaroloxias inornata* lives on Cocos Island, 600 miles northeast of the Galapagos, and is separated geographically from the rest, which are restricted to the Galapagos. It has a rather thin and somewhat decurved beak and eats a variety of foods, including insects (Slud 1967).

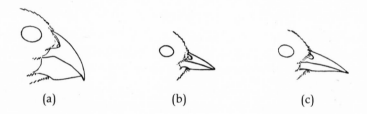

(a)   (b)   (c)

Fig 37. Heads of three monotypic genera of Darwin's finches (from Swarth 1931).
(a) the vegetarian *Platyspiza crassirostris*, (b) the warbler finch *Certhidea olivacea* and (c) the Cocos finch *Pinaroloxias inornata*. Natural size.

The vegetarian tree finch *Platyspiza crassirostris* eats mainly buds and leaves, which the others rarely do, and also fruit, and has a beak adapted for crushing along its whole length, in shape rather like that of the European Bullfinch *P.pyrrhula* (see p. 79). The warbler-finch *Certhidea olivacea* has a rather slender, straight, tweezer-like beak, like that of a warbler, for picking small insects off leaves and fine twigs, which the other Geospizinae rarely if ever do. These two genera are clearly separated by diet from each other and from the other two genera, the seed-eating ground finches *Geospiza* and insectivorous tree finches *Camarhynchus,* each of which includes several species. The only possible overlap in diet might be between the warbler-finch *Certhidea* and the parulid warbler of the Galapagos, already mentioned.

The ground finches Geospiza on the main islands

The ground finches *Geospiza* typically feed on the ground on seeds and, with one exception, have the finch-like beaks characteristic of seed-eating birds. Five of the 6 species breed on the central Galapagos islands, and their habitats, main foods and size are summarized in table 26 and their heads are shown in fig. 38. Four of them breed in the arid and transitional zones, but the fifth, *G.difficilis*, in the humid zone, so it is separated by habitat, but probably also by feeding, as it spends much time digging in the litter of the forest floor, which the others do not do, and has a straighter and slightly longer beak than the seed-eating species.

Of the four species in the arid zone, the cactus finch *G.scandens* is clearly separated from the rest in diet, since though it at times eats seeds on the ground, it depends on the flowers and fruits of the prickly-pear *Opuntia*, which the others rarely if ever take. Moreover it breeds solely where *Opuntia* is found and is absent, for instance, from large parts of Chatham where this tree is absent, though found where it is present (personal observation). It has a relatively long thin beak, well adapted for probing in *Opuntia* flowers.

The other three species of *Geospiza* depend primarily on seeds and differ markedly in size from each other. The small *G.fuliginosa,* with its proportionally small and narrow beak, eats mainly small and soft seeds. The medium-sized *G.fortis* takes rather larger and harder seeds than *G.fuliginosa*, though with some overlap. The large *G.magnirostris*, with its proportionally large and deep beak, though it takes many of the larger seeds taken by *G.fortis*, includes a higher proportion of the larger and especially of the harder

177

Table 26. Ecology of *Geospiza* species on central Galapagos Islands

	G.difficilis	G.scandens	G.fuliginosa	G.fortis	G.magnirostris
Habitat zone for breeding	humid	arid and intermediate	arid and intermediate	arid and intermediate	arid and intermediate
Size of seeds	?	fairly small	small	medium and large	large and medium
Hardness of seeds	?	medium	soft	fairly hard, medium and soft	hard and medium
Other main foods	?	flowers and fruits of Opuntia	some insects	—	—
Wing length in mm	72	70	64	72	84
Culmen from nostril in mm	10·3	12·9	8·4	11·5	15·9
Depth of beak in mm	9·4	8·8	8·0	12·5	20·5

Notes: habitats from Lack (1947) (and for *G.difficilis* also Lack, 1969), foods from Bowman (1961, 1963) and measurements from Lack (1945b), these last being of males on James, those on other islands being similar, and females being similar to males but a little smaller. A small proportion of insects has been recorded in the diets of the four species of the arid zone, this being highest in *G.fuliginosa* and smallest in *G.magnirostris*.

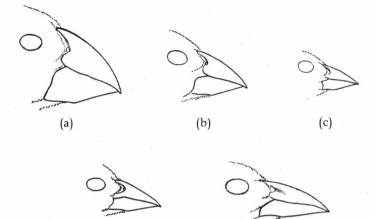

Fig 38. Heads of ground finches *Geospiza* on central Galapagos islands (from Swarth 1931).

(a) large *G.magnirostris,* (b) medium *G.fortis,* (c) small *G.fuliginosa,* (d) *G.difficilis* of humid zone and (e) cactus-eating *G.scandens.* Natural size.

ones. For instance, Bowman (1961) found the hard seeds of *Bursera* and *Croton* in about 80 per cent of the *G.magnirostris* examined, but in only 15 per cent of the *G.fortis* examined. Bowman demonstrated corresponding adaptations in the size of the jaw muscles of these three species. Clearly, they take mainly different foods, but with some overlap. In view of what is now known of various seed-eating birds in Europe and tropical Africa (Ward 1965), the seeds eaten by Darwin's finches are probably scarcest shortly before the time at which the new crop is formed, i.e. near the end of the dry season. If so, one might expect the potential competition between them to be most severe, and that each species would differ most clearly in its diet from the others, at this time. Hence their diets as analysed throughout the year by Bowman might show greater overlap than they would if the analysis had been restricted to a critical season (if such exists).

The medium ground finch *Geospiza fortis*, has an exceptionally variable beak for a passerine bird. Indeed the smallest individuals overlap slightly in length (but not depth) of beak with the small *G.fuliginosa* and the largest overlap in both length and depth of beak with the large *G.magnirostris* (Lack 1945). Further, Bowman found that smaller individuals of *G.fortis* tend to take smaller and softer seeds than larger individuals, and overlap in diet with *G.fulignosa*, whereas the larger individuals overlap in diet with *G.magnirostris*. This suggests to me (though Bowman disagreed) that the limits of beak-size in *G.fortis* have been determined through interspecific competition, the lower limit through competition with *G.fuliginosa* and the upper through competition with *G.magnirostris*. On this view, one might expect *G.fortis* to include larger individuals on islands where the large *G.magnirostris* is scarce, absent, or exceptionally large, than on islands where it is common and of normal size, and this is so, as shown for the depth of beak in fig. 39.*

*For the culmen, the upper limit in *G.fuliginosa* is 10·2 mm, the lower limit in *G.magnirostris* is 13·8 mm, while that of *G.fortis* varies between 9·8 and 14·2 mm. The corresponding figures for the depth of beak are *G.fuliginosa* 9·5 mm, *G.magnirostris* 15·8 mm, *G.fortis* 10·0–16·6 mm. But the upper limits for length and depth in *G.fortis* are only 12·6 mm and 14·1 mm respectively on the northern islands of Abingdon, Bindloe and James, where *G.magnirostris* is common, as compared with 14·2 and 16·6 mm respectively on Seymour, Indefatigable, Duncan, Albermarle, Chatham and Charles, where *G.magnirostris* is sparse, absent or (on Charles, now extinct) extremely large. The lower limit of beaks-size in *G.fortis* is similar on all these islands, so it is reasonable to correlate the variations in the upper limit with the situation in *G.magnirostris*. Bowman (1961 pp. 264–6) argued, instead, that *G.fortis* is more variable on those islands with a more diverse flora, but this does not explain why the increased variability is solely at the large end of the size-range, or why the other species of *Geospiza* on these islands do not show a greater variability than elsewhere (see his table p. 265). My explanation is tentative however, and food studies are needed to test it.

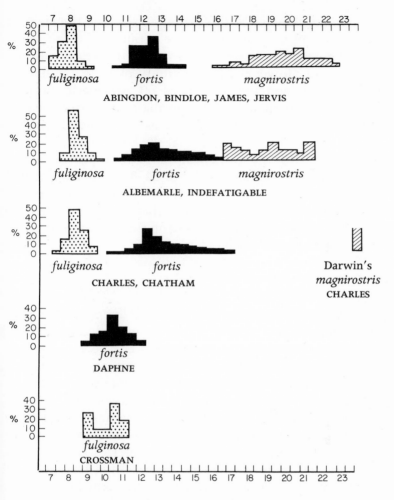

Fig 39. Variations in beak-depth of small, medium and large ground finches on different islands (from Lack 1947).

Above : note larger size of *G.fortis* on Albemarle, Indefatigable, Charles and Chatham, where *G.magnirostris* is sparse, absent or unusually large, than on northern islands (top line) where *G.magnirostris* is common.

Below : note intermediate measurements of *G.fortis* on Daphne and *G.fuliginosa* on Crossmans in absence of the other species.

181

It would likewise be interesting to see how the medium *G.fortis* might vary in the absence of the small *G.fuliginosa*, but on none of the main islands is *G.fortis* present without *G.fuliginosa*. However, on two groups of islets, Daphne off Indefatigable and the Crossmans off Albemarle, populations occur which are intermediate in measurements between these two species in size, as also shown in fig. 39. Relative proportions suggest that the Daphne birds belong to *G.fortis* and the Crossman birds to *G.fuliginosa* (for measurements see Lack 1947, pp. 84–86). The existence of such intermediate forms where only one of the two species is resident reinforces the view that the size difference between *G.fortis* and *G.fuliginosa* on the large islands is due to competitive displacement.

Geospiza on the small outlying islands

The small outlying Galapagos islands, Culpepper, Wenman, Hood and Tower, consist solely of an arid zone and are inhabited by only two, or on Tower three, species of ground finches, as set out, with their measurements, in table 27, while the heads of these forms are shown in figs. 40 and 41. The medium-sized *G.fortis* and the cactus-finch *G.scandens* do not occur on the outlying islands, the small *G.fuliginosa* breeds solely on Hood, and the large *G.magnirostris* solely on Tower and Wenman. A fairly large species, *G.conirostris*, which is not found elsewhere, breeds on at least three of them. Finally *G.difficilis*, which on the central islands is confined to the humid zone, breeds in the arid zone on Culpepper, Wenman and Tower, thus providing an example of a difference in habitat between subspecies of the same species.

On Tower, where the small ground finch *G.fuliginosa* is absent, a small form of *G.difficilis* appears to fill a similar niche, as it is of similar overall size and size of beak, though its beak is slightly longer, straighter and proportionately thinner, suggesting that it in part takes different foods.

Perhaps it digs in the ground, a habit recorded in most other forms of *G.difficilis*. The large *G.magnirostris* is also present on Tower, and so is a narrow-beaked form of the fairly large *G.conirostris*, the beak of which is a thicker version of that of the cactus-eating *G.scandens*.

On Hood, the small *G.fuliginosa* is present, and is typical in size, even though the medium *G.fortis* is absent (cf. p. 180). The only other ground finch is a deep-billed form of *G.conirostris*, and it is tempting to suppose that it takes foods which on Tower are divided between the narrow-beaked form of *G.conirostris* and the heavy-beaked *G.magnirostris*, since it is intermediate between them. Whether this is so or not, it also has a unique feeding habit, scratching up gravel with its legs for food underneath and pressing its beak against a boulder as a brace (DeBenedictis 1966).

On Wenman, as on Tower, occurs the large *G.magni-rostris*. The only other common species is an unusually long-beaked form of *G.difficilis*. This has been seen digging in the ground for food, like other forms of its species, but it also feeds on *Opuntia*, like the cactus finch *G.scandens* on the central islands, and like this species it has an unusually long beak. It takes a wider range of foods, however, since it also pecks in bark, presumably for insects, and more remarkably, it eats carrion and the blood of moulting boobies *Sula* spp. (Gifford 1919, Bowman and Biller 1965).

On Culpepper occurs the same cactus-eating form of *G.difficilis* as on Wenman. The only other ground finch is an extremely variable form, probably of *G.conirostris*, which ranges in size from the Tower form of *G.conirostris* almost up to *G.magnirostris*, found on nearby Wenman, so perhaps combines their food niches.

Detailed observations on food have not been made on these outlying islands, but the beaks of the *Geospiza* species present on any one island are so different that each clearly takes mainly different foods from the one or two others present with it. The species on the same island differ

Table 27. Ecological isolation in *Geospiza* species on outlying Galapagos islands (mean measurements are of wing and culmen × beak-depth, in mm.)

	Tower	Hood	Wenman	Culpepper
Large ground finch	*G.magnirostris* (86, 16·5 × 21·2)	} *G.conirostris* (80, 15·4 × 16·0)	*G.magnirostris* (86, 15·7 × 20·4)	*G.conirostris* (82, 15·0 × 16·5)
Cactus feeder	*G.conirostris* (77, 14·4 × 13·0)	(?)	} *G.difficilis* (72, 10·7 × 8·3)	} *G.difficilis* (73, 11·3 × 9·0)
Small ground finch	*G.difficilis* (63, 9·4 × 7·9)	*G.fuliginosa* (64, 8·8 × 8·3)		

Notes: measurements, all of males, are from Lack (1945b pp. 142–147). *G.conirostris* on Hood has not been observed eating *Opuntia*, its beak is intermediate in size between that of the two larger species on Tower, and *Opuntia* occurs on Hood. Recently, Curio and Kramer (1965) collected a specimen of *G.conirostris* on Wenman (with beak depth 11·0 mm and wing 71 mm, hence like Tower form) and M. P. Harris (pers. comm.) saw a pair feeding a fledgling there, but it seems unlikely that this species will have been overlooked by the earlier collectors. Probably, therefore, it is a recent colonists and liable to extinction, though the possibility of its being a rare permanent resident cannot be excluded.

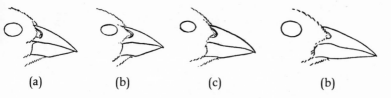

Fig 40. Resemblance of *G.difficilis* on Tower to *G.fuliginosa*, and on Culpepper to *G.scandens* (drawings from Swarth 1931).
(a) *G.fuliginosa* on Hood, (b) *G.difficilis* on Tower, (c) *G.difficilis* on Culpepper and (d) *G.scandens* on James. Natural size.

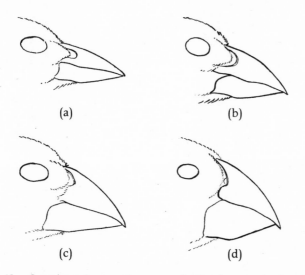

Fig 41. *G.conirostris* on Tower and Hood compared with *G.scandens* and *G.magnirostris* (from Swarth 1931).
(a) thick-billed form of *G.scandens* on Bindloe, (b) *G.conirostris* on Tower, (c) *G.conirostris* on Hood and (d) *G.magnirostris* on Tower. Natural size.

from each other to a greater extent than do those on the larger central islands. The evidence also suggests that the Tower form of *G.difficilis* is convergent with *G.fuliginosa* and the Culpepper and Wenman form with *G.scandens* (though also taking foods taken by a small ground finch elsewhere), and that the Hood and Culpepper forms of

Table 28. Ecology of *Camarhynchus* species on central Galapagos islands

| | Subgenus *Camarhynchus* | | Subgenus *Cactospiza* | |
	C.parvulus	*C.psittacula*	*C.pallidus*	*C.heliobates*
Habitat zone for breeding	humid and transition	humid and transition	humid and transition	coastal mangroves
Feeding station	twigs	branches and large twigs	branches, trunks	branches, trunks and leaves
Feeding methods	searching and picking	twisting twigs, pulling bark	excavating, probing with twig	excavating, probing with twig
Size of insects	small	fairly large	fairly large	fairly large
Wing-length (mm)	64	74	75	72
Culmen from nostril in (mm)	7·0	9·8	12·6	10·5
Depth of beak (mm)	7·4	11·2	9·3	8·2

Notes: upper sections based on Bowman (1961, 1963), but on Curio and Kramer (1964) for *C.heliobates*. The measurements of *C.parvulus*, *C.psittacula* and *C.pallidus* are from James, those from other islands being a little different and those of females a little smaller, while those of *C.heliobates* are from Albemarle and Narborough (all from Lack 1945b). *C.parvulus* is rather larger (wing 65 mm, culmen 8·0 mm, depth 7·9 mm.) on Chatham, where *C.psittacula* is absent, than on the other islands.

186

G.conirostris may, in part, take foods divided between *G.magnirostris* and the narrow-beaked form of *G.conirostris* on Tower.

Camarhynchus on the large central islands

The insectivorous tree finches in the genus *Camarhynchus* breed especially in the humid zone of the larger islands, and not at all on the small outlying islands discussed in the previous section. The ecology of the four species on the central islands is summarized in table 28 and their heads are shown in fig. 42. They have relatively thick beaks, short in the subgenus *Camarhynchus* and longer in the subgenus *Cactospiza*.

The small insectivorous tree finch *Camarhynchus parvulus* searches the leaves and twigs for insects on their surface, and also strips off small pieces of bark for insects underneath like the Blue Tit *Parus caeruleus*, which has a rather similar beak and a similar, indeed greater, agility in exploring small twigs. Bowman (1961) found that the large

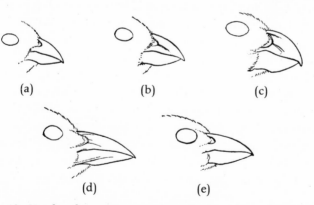

Fig 42. Heads of insectivorous tree finches *Camarhynchus* in central Galapagos (from Swarth 1931).

(a) small *C.parvulus*, (b) medium *C.pauper* (only on Charles), (c) large *C.psittacula* (form on Indefatigable and James), (d) woodpecker finch *C.pallidus* (James form) and (e) mangrove finch *C.heliobates*. Natural size.

insectivorous tree finch *C.psittacula*, which has a proportionately larger and deeper beak, with a strongly decurved culmen, spends less time on the small twigs and proportionately more time stripping off bark, including bark from larger twigs and branches than those searched by *C.parvulus*, and it also tears small twigs apart in its hunt for concealed insects, for which its strong beak is well adapted. As a result, it takes mainly larger larvae and other insects than *C.parvulus* and in part takes them from different places, but there is some overlap in their diets. In tits, likewise, the larger species feed on larger branches and take larger insects than the smaller (see p. 24).

The woodpecker finch *C.(Cactospiza) pallidus* is of similar overall size to *C.psittacula* and lives in the same habitats, but has a longer and much straighter beak. It spends part of its time hunting for insects under bark on branches, like *C.psittacula*, but also digs trenches in decaying wood, rather than tearing off bark, in which it is helped by its chisel-shaped beak. It also probes deep into crannies, by holding a small stick lengthwise in its beak, and it climbs and feeds on vertical trunks which *C.psittacula* does not do. Hence while Bowman found that part of its prey (including phalaenid and ostomid larvae) is the same as that of *C.psittacula*, much else is different, including large dipterous larvae which live in decaying wood and can be reached only by trenching and probing. The remaining species, the mangrove finch *C.heliobates*, is in size, and in size and shape of beak, very similar to *C.pallidus*, like which it takes insects from under bark and probes with a stick in its beak (Curio and Kramer 1964), but it is restricted to coastal mangroves, so is separated by habitat.

Camarhynchus on outlying islands

The situation on the southernmost island of Charles differs from that on any other island since, in addition to the small *C.parvulus*, which is common, and the large *C.psittacula*, which is very scarce, a similar medium-sized species *C.pauper* is numerous, which feeds in a similar way and presumably takes prey of intermediate size. In measurements (see note to table 29, and fig. 42), it is nearer the large *C.psittacula* than the small *C.parvulus*, and since *C.psittacula* is very scarce, its diet is probably closer to that of the larger than the smaller species. Whereas the large, medium and small ground finches in the genus *Geospiza* are found together on all the main islands, the parallel situation in *Camarhynchus* is restricted to this one island, which I formerly supposed might be because it originated there in relatively recent times, but now wonder whether there might be some ecological peculiarity on this island.*

The southeasternmost island of Chatham is at the opposite extreme to Charles, as here there is only one species in the subgenus *Camarhynchus*, namely the small *C.parvulus* which, as might be expected, is unusually large.

Although the large insectivorous tree finch *C.psittacula* has a short decurved beak and the similar-sized woodpecker finch *C.pallidus* has a long straight one, Bowman found, as already mentioned, that they feed partly in the same kind of way on similar prey. Hence it is suggestive that on the northern islands of Abingdon and Bindloe, where the woodpecker finch *C.pallidus* is absent, the endemic form of *C.psittacula* has a more elongated beak than elsewhere, which suggests that it may there take some of the foods taken on other islands by *C.pallidus*. In addition, on Bindloe, it taps the pads of the prickly pear *Opuntia* for water and eats the pulp, which it does not do elsewhere (Bowman 1961 p. 267).

The converse situation holds on the southeasternmost island of Chatham, where *C.psittacula* does not breed, for the Chatham form of the woodpecker finch *C.pallidus* has a

---

C.psittacula is the same size on Charles as it is on Indefatigable and James where *C.pauper* is absent, though larger than on Albemarle, where *C.pauper* is also absent. Hence there is no good evidence for character displacement.

189

Table 29. Variations in measurements of *Camarhynchus psittacula* and *C.pallidus* in presence and absence of the other

Island	Other species	Average measurements of males in mm			
		wing	culmen (from nostril)	beak-depth	ratio of culmen to beak-depth
		MEASUREMENTS OF *C.psittacula*			
	C.pallidus				
James	present	74	9·8	11·2	0·9 ⎫
Indefatigable	present	74	9·6	10·7	0·9 ⎬
S. Albemarle	present	69	8·5	9·3	0·9 ⎭
Abingdon	absent	71	10·1	10·4	1·0 ⎫
Bindloe	absent	71	10·5	10·5	1·0 ⎭
		MEASUREMENTS OF *C.pallidus*			
	C.psittacula				
Chatham	absent	72	10·8	8·9	1·2
James	present	75	12·6	9·3	1·4 ⎫
Indefatigable	present	73	12·1	9·1	1·3 ⎬
S. Albemarle	present	72	11·2	8·1	1·4 ⎭

Notes: measurements from Lack (1945b). The righthand column indicates the shape of the beak, a higher figure meaning a proportionately longer beak. *C.pallidus* is also absent from Charles, where *C.psittacula* is of the same size as on Indefatigable, but on Charles it is found alongside *C.pauper* (wing 70, beak 9·0 × 8·8 mm).

shorter beak than elsewhere, recalling in shape that of *C.psittacula* on Abingdon and Bindloe; this suggests that on Chatham *C.pallidus* takes some of the foods taken on other islands by *C.psittacula*. The measurements of the forms concerned have been set out in table 29, and their heads are shown in fig. 43. Hence in *Camarhynchus*, as in *Geospiza*, the beak tends to be modified in some of the forms on outlying islands, correlated with the absence of another species, and this suggests that its shape and size on the central islands have been determined through competition for food with these other species.

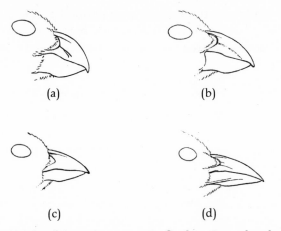

Fig 43. Heads of insectivorous tree finches *Camarhynchus* on outlying islands (from Swarth 1931).

(a) *C.psittacula* on James, where *C.pallidus* present, (b) long-beaked *C.psittacula* on Bindloe, where *C.pallidus* absent, (c) short-beaked *C.pallidus* on Chatham, where *C.psittacula* absent and (d) *C.pallidus* on James, where *C.psittacula* present. Natural size.

Conclusion on Darwin's finches

In the adaptive radiation of Darwin's finches in the Galapagos, the four genera differ markedly in feeding methods and shape of beak from each other. Most of the ground finches in the genus *Geospiza* also differ from each other in the foods which they take, with correlated adaptations in the size and shape of their beaks, though one species is also separated in part by habitat. In the tree finches in the genus *Camarhynchus*, likewise, nearly all the species are separated from each other by differences in feeding and the size and shape of the beak, though here also, one species differs in habitat.

The way in which each species is isolated from its congeners has been set out in table 30. This shows that separation is by feeding in 71 per cent of instances, a far higher proportion than in continental passerine birds. In Europe, however, separation by feeding is rather common in seed-eating birds, hence the proportion of species separated in this way in the genus *Geospiza* is not so excessive. This does not apply, however, to the tree finches in the genus

Table 30. Means of ecological isolation of congeneric Darwin's finches

(G by range, — no contact, H by habitat, F by feeding)

A. GEOSPIZA ON CENTRAL ISLANDS

1	*G.difficilis*	1				
2	*G.fortis*	HF	2			
3	*G.fuliginosa*	HF	F	3		
4	*G.magnirostris*	(H)F	F	F	4	
5	*G.scandens*	(H)F	F	F	F	

B. GEOSPIZA ON SMALL OUTLYING ISLANDS

1	*G.conirostris*	1			
2	*G.difficilis*	F	2		
3	*G.fuliginosa*	F	G	3	
4	*G.magnirostris*	GF	F	—	

C. CAMARHYNCHUS

1	*C.heliobates*	1				
2	*C.pallidus*	H	2			
3	*C.parvulus*	HF	F	3		
4	*C.pauper*	—	—	F	4	
5	*C.psittacula*	HF	F	F	F?	

Camarhynchus, which are insectivorous, yet in this genus also, most species are separated by feeding.

One reason for the paucity of species separated by habitat in the Galapagos is the paucity of habitats, for there are only three distinctive types of woodland, that of the arid zone, that of the humid zone, and, locally, that of mangroves. The similarities in their plumage and breeding behaviour suggest that the species of Darwin's finches diverge from each other in relatively recent times. In mainland birds, differences in feeding habits and beak comparable with those in Darwin's finches are found between different genera, most of which have later diverged into different species with separate habitats. It might therefore be suggested that more of Darwin's finches would have been separated by habitat if their evolution had not been so recent. This seems unlikely, however, in view of the paucity of habitats. Moreover the Hawaiian sicklebills, which constitute a more mature evolution, are likewise separated mainly by feeding, not habitat, as discussed in the next section.

One point illustrated better in Darwin's finches than most other birds is the tendency, in the absence of one species, for another to take a wider range of foods, but the evidence for this is as yet based solely on the size and shape of the beaks of the forms concerned, and detailed food studies are much needed. All such examples come from very small or outlying islands, presumably because here the foods available for the finches are reduced, and in such circumstances, one might expect a single generalized species to oust two specialists (MacArthur and Levins 1967).

The Hawaiian sicklebills Drepanididae

The Hawaiian archipelago lies some 2000 miles from California on one side and from the Marquesas on the other, so is far more isolated than the Galapagos. Linked with the heavy rainfall, its vegetation is much richer, though still impoverished compared with a continental forest, and various groups of plants and insects have undergone striking adaptive radiations (Zimmerman 1948). Few land birds have colonized, including only seven passerine forms, a thrush *Phaeornis*, a warbler *Acrocephalus,* an Old World flycatcher *Chasiempsis,* two Australian meliphagids *Chaetoptila* and *Moho,* a crow *Corvus,* and the ancestor of the sicklebills, and there is only one species of each of these on any one island, except for two species of thrush on Kauai, and the many sicklebills.

Twenty-two species of sicklebills were alive in the nineteenth century, but various of them are now extinct. Most of them are rather larger than Darwin's finches and much more brightly coloured. Many have thin decurved beaks, and a tubular tongue with a brush tip, adapted for feeding on nectar. Most of these also eat some insects, while other species are primarily adapted for taking insects in various ways and yet others, with thick beaks, eat mainly seeds or fruits. The group has diverged to a much greater extent than Darwin's finches. Formerly many more genera and species were recognized than now, and I have here followed the taxonomic revision by Amadon (1950), and for ecology Perkins (1903). The measurements and diets of the congeneric species have been summarized in appendix 24, with the monotypic genera listed in a footnote.

The species of *Loxops* have fairly short, thin and slightly decurved beaks recalling those of white-eyes *Zosterops*. Three species coexist on most of the islands, but though they are similar in overall size, they differ considerably in size and shape of beak, and also in feeding. *L.coccinea* has crossed tips to its relatively short beak, which is used for opening buds to get insects, its main diet. The short-tailed *L.maculata* creeps over the branches and uses its rather longer beak to obtain insects from crevices in bark. *L.virens* differs from the other two in having a complete tubular tongue and eats much nectar, though also insects. Hence these three species are separated by feeding.

On each of the islands of Hawaii and Kauai, there is a

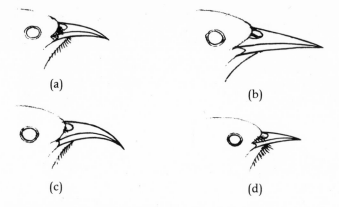

(a)

(b)

(c)

(d)

Fig 44. Character displacement in *Loxops virens* in Hawaiian archipelago (drawings from Rothschild 1893).

(a) small form of *L.virens* on Hawaii, which coexists with (b) the large *L.sagittirostris*, (c) large form of *L.virens* on Kauai, which coexists with (d) the small *L.parva*.

further species closely related to *L.virens* and coexisting with it. The heads of the forms concerned are shown in fig. 44. The further species on Hawaii, *L.sagittirostris*, is much larger with a heavier and straighter beak than *L. virens*, from which it also differs in diet, as it eats caterpillars and adult insects taken from leaves. The form of *L.virens* on Hawaii is similar to, but slightly smaller than, those on the islands where it is the only species in its superspecies. On Kauai, in contrast, the other coexisting species, *L.parva*, is much smaller than *L.virens*, with a straighter beak, and it feeds on nectar in shallow flowers and on insects from leaves, rarely removing bark. In contrast, the Kauai form of *L.virens* is much larger than any other form of this species, and feeds mainly on insects taken from the trunks and branches of trees, using its heavy beak to dig in moss and bark (Amadon 1947). The difference between the two island forms of *L.virens*, according to whether it coexists with a larger or smaller second species, provides a good examples of character displacement (p. 10). (The measurements are included in appendix 24.)

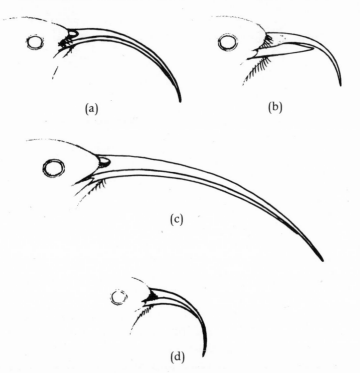

Fig 45. Geographical replacement of species in same superspecies
of *Hemignathus* in Hawaiian archipelago (from Rothschild 1893).
   (a) *H.obscurus,* which coexists on Hawaii with (b) *H.wilsoni*
(in *H.lucidus* superspecies), (c) *H.procerus* (in *H.obscurus* superspecies), which coexists on Kauai with (d) *H.lucidus.*

The species of *Hemignathus* are somewhat larger than
those of *Loxops,* with much longer and more decurved
beaks, and they climb on trunks and branches like creepers
or woodpeckers. There are two coexisting superspecies,
each with one main species and a second different-looking
species restricted to one island; their heads are shown in
fig. 45. *H.obscurus* (and the related *H.procerus* on Kauai)
take nectar, especially from lobelias, and also probe for
insects in cracks in decaying wood and the stems of tree
ferns; the lower mandible is about four-fifths as long as the
upper. In the second superspecies *H.lucidus* (and the

related *H.wilsoni* on Hawaii), the lower mandible is only half as long as the upper (and in *H.wilsoni* is straight, not curved), and it is used separately from the upper mandible for hammering and levering off bark, especially for weevils Curculionidae. Hence the two superspecies differ from each other in beak, feeding methods and diet (Amadon 1947). The marked differences in the beaks of the two species in each superspecies, shown in fig. 45, have not been explained.

The species of *Psittirostra* are somewhat larger than those of *Hemignathus,* but their beaks are much shorter and thicker, being adapted for taking fruits or seeds. One species is widespread. Four others are, or were, found solely in Hawaii where, as set out in appendix 20, at least three are specialized for different species of fruits or seeds, and all of them have beaks of markedly different size. One further species is found solely on the small remote islands of Laysan and Nihoa, so is separated geographically from the rest, and from all other sicklebills except one (*Himatione sanguinea,* now extinct on Laysan). Fitting with this, it takes a wide variety of foods, including adult and larval insects, seeds, starch roots, seabirds' eggs and dead birds, hence it is a typical 'all-purpose' bird on a remote island.

Finally the two long-billed species of *Drepanis* are simply well-marked geographical forms; they feed, or fed, on nectar from lobelias and on insects.

The means of isolation of the congeneric species of sicklebills have been summarized in appendix 25. The proportion segregated by feeding is 73 per cent, almost the same as in Darwin's finches, and since only one of the genera concerned, *Psittirostra,* is specialized for eating seeds and fruits, it may be concluded that differences in feeding are characteristic of passerine birds on remote islands. The few Hawaiian sicklebills which are not separated by feeding either replace each other geographically or have no contact, and none of them differ from each other primarily in habitat.

Two other cases

The only other cases in which a passerine genus on a remote archipelago has there evolved into two species are the thrushes *Phaeornis* on Kauai in the Hawaiian archipelago, and the finches *Nesospiza* in the Tristan da Cunha group. On most of the Hawaiian islands there is only one species of thrush *P.obscurus*, but on Kauai it coexists with a smaller species, *P.palmeri. P.obscurus* eats mainly fruit but some insects, and feeds especially in large trees, though also in bushes, while *P.palmeri* eats mainly insects, but some fruit, and feeds especially in small trees and the underbrush (Perkins 1903). There is a little evidence for character displacement as the forms of *P.obscurus* on Lanai and Molokai are intermediate in size between the two species on Kauai, and were seen feeding in both trees and the underbrush.*

The two species of *Nesospiza* in the Tristan group differ markedly in size, and while the smaller feeds on a variety of seeds and insects, the larger specializes on the seeds of the tree *Phylica* and of *Nertera,* and was not recorded taking animal food (Hagen 1952).**

*Mean winglengths of males, with numbers measured in brackets, are for *P.palmeri* c. 90 mm (only 3 females or juveniles), *P.obscurus* on Kauai 105 mm (7), Lanai 94 mm (6), Molokai 96 mm (8) and Hawaii 103 mm (8). *P.palmeri* probably has a slightly longer beak than any form of *P.obscurus*. Measurements, kindly sent by Miss M. LeCroy per D. Amadon, are of the specimens in the American Museum of Natural History.

**Mean measurements of males, from Hagen 1952, are for *N.acunhae* wing 79 mm and culmen 14·2 mm on Nightingale, and wing 85 mm and culmen 15·4 mm on Inaccessible, and for *N.wilkinsi* wing 95 mm and culmen 19·3 mm on Nightingale and wing 89 mm and culmen 18·2 mm on Inaccessible. Hence they differ more markedly in size on Nightingale than Inaccessible. *N.acunhae* occurred formerly on Tristan itself, but is now extinct, and presumably *N.wilkinsi* might also have done so, but it was never recorded there.

Conclusion

In both the Galapagos finches and the Hawaiian sicklebills, most congeneric species live in the same habitat and are separated by feeding, with associated differences in size and shape of the beak. This holds not only for the seed-eating genera *Geospiza* and *Psittirostra,* but also for the insectivorous *Camarhynchus* and *Hemignathus,* and the partly insectivorous and partly nectar-feeding *Loxops.* It also holds in the only other known examples of passerine speciation within a remote archipelago, the thrush *Phaeornis* in Hawaii and the finch *Nesospiza* on Tristan. The frequency of ecological isolation by feeding on remote archipelagoes is partly due to the paucity of species separated by habitat, linked with the reduced diversity of habitats for birds on such islands. In addition, since continental passerine *genera* differ in size, beak and feeding habits to a similar extent to these island congeneric *species,* it represents a more primitive level of evolutionary and ecological divergence, linked with the reduced ecological diversity on islands. In Darwin's finches, the Hawaiian sicklebills and the Hawaiian thrushes, there are examples of subspecific differences in beak-size correlated with the presence or absence of a closely related species (character displacement).

Chapter 12

A Widespread Family on Remote Islands

The White-eyes Zosteropidae

As shown in the previous chapter, the frequency of each type of ecological isolation is very different in insular birds like Darwin's finches or the Hawaiian sicklebills from what it is in continental passerine birds, and in particular, a high proportion of the island birds are separated by size and feeding. One further passerine family, the white-eyes Zosteropidae, is typical of tropical islands and will be analysed here. These birds did not, like Darwin's finches or the sicklebills, evolve within one archipelago, but instead, have colonized many archipelagoes independently from the mainland. Many other continental passerine birds have, of course, colonized islands, but the extent to which white-eyes have done so is unique, and they are also remarkable in that sympatric species are common on islands but rare on the mainland.

White-eyes are widespread on the mainland of tropical Africa, southern and eastern Asia and Australia, and also on the islands off these continents, including those of the Indian Ocean, Indonesia, Melanesia, and parts of Micronesia. A typical species in the main genus *Zosterops* is warbler-like, with a wing-length of 55–60 mm, i.e. slightly shorter than that of a European leaf warbler *Phylloscopus*, in comparison with which it has a slightly longer and slightly decurved beak, also a brush tongue adapted for taking nectar. It has greenish upper parts, yellow or sometimes grey underparts, and a white ring round the eye. The varied diet includes insects, fruit, small seeds and nectar. Many species live in primaeval forest, others in secondary growth or cultivated land, and they occur from coastal mangroves to bushes above the treeline on high mountains. Subspecies of the same species, and related species, tend to be larger on islands than the mainland, and larger at higher than lower altitudes. A few island species are much larger than the rest, and are coloured rather differently, but they are obviously members of the family Zosteropidae, which is well defined, so are included in the present discussion, even though they are usually put in separate genera.

The taxonomy of the group has proved complex, especially owing to the difficulty of deciding which of the forms that replace each other geographically on islands are closely related and in which of them the resemblances are due to a common basic pattern combined with convergent evolution. There is the further question of which closely related island forms should be treated merely as subspecies and which as full species in the same superspecies. Fortunately the main difficulties in the Indo-Australian region have been resolved by Mees (1957, 1961, 1969), and the differences between his treatment and that of Mayr (1967) in Peters' Check-List are extremely small. There is also good agreement on the African forms (Moreau 1967, Hall and Moreau *in prep.*). Following Mees, there are in the Indo-Australian region 50 species of *Zosterops* and another 18 species in 9 other genera, while in Africa there are 12 species of *Zosterops* and 4 in another genus. These are set out, with their ranges, habitats, mean wing-lengths and mean beak-lengths in appendix 26, and the text is restricted to points of general interest. This chapter could not have been written without the monograph by Mees, a model of its kind, and I am also extremely grateful to him for reading an earlier draft of the manuscript.

Zosterops is found throughout Africa south of the Sahara except for the desert regions and, at the other extreme, most of the Congo forest. Most of the lowlands are occupied by a single species, but this is replaced geographically by a different species in the northeast and another in the south. In addition, a fourth and larger species, *Z.poliogastra*, occurs in many isolated populations in the East African mountains, where it is separated by altitude from the lowland species. Similarly on Mount Cameroon in West Africa, a large white-eye in the genus *Speirops* replaces the lowland species at higher altitudes. Except in these two areas, no two species of white-eyes occur in the same part of Africa.

The situation is different on the African islands, some of which have only one species, but others two, one of which is much larger than a typical *Zosterops*. In the Gulf of Guinea, a *Zosterops* of normal size and a larger species in the genus *Speirops* occur on Fernando Po, where they are separated by altitude, and on Sao Tomé and Principe, where they are not. On the fourth and much smaller island of Annobon, there is only one species which, except in being larger, looks very like the species of *Zosterops* on Principe and Sao Tomé, so may have evolved from it. It is intermediate in size between the *Zosterops* and *Speirops* on the other islands, and has a beak as long as that of *Speirops*. This is probably an example, similar to those in various of Darwin's finches, in which one species on a small remote island replaces two more specialized forms found together on larger islands.

In the Indian Ocean, there are two species of *Zosterops* on each of Grand Cormoro, Reunion and Mauritius, one of them being larger than the other or, on Mauritius, with a longer beak but not larger in overall size. On Grand Comoro, as on Fernando Po, the larger species is at a higher altitude than the smaller, and there might be partial separation by habitat on the other two islands. Each of the smaller islands in the Indian Ocean has only one species, which in some cases is the same size as the lowland species on the African mainland, but on Marianne, Mahé, Anjouan and (in beak) Mayotte, is larger, like the Annobon white-eye. The extremely large island of Madagascar has only one species, though it provides much more diverse habitats than the other islands, but in this respect it resembles continental Africa, in which there is normally only one species in any one area, as already mentioned.

Mainland Asia and Australia

On the mainland of southern and eastern Asia, as on the mainland of Africa, there are three lowland species that replace each other geographically. Apparently, however, two of them breed where their respective ranges meet in Szechwan and Yunnan, and the third also occurs there in winter. There is no additional species in the high mountains of Asia. But two species occur in the Malay Peninsula, one in mangroves, cultivated land and the forest edge in the lowlands, and the other in heavy forest above 700 m. Unusually, the highland species is not larger than the lowland one. The same lowland species *Z.palpebrosa* ranges from sea level up to 2000 m in India, while on Java and Sumatra it is restricted to middle altitudes (Stresemann 1939). This is an unsual variation in habitat in subspecies of the same species (although there are parallels in other genera), and it may be attributable to the presence or absence of other species of *Zosterops*. The other Malayan species, *Z.everetti,* also varies in altitudinal range, though to a smaller extent, as it is submontane in Borneo and occurs in both lowlands and hills in the Philippines.

The two species on the mainland of Australia are separated by range except for a small area of overlap in the middle of the west coast, where they are separated by habitat. A third species occurs only on certain northeastern offshore islands, so is also separated by range.

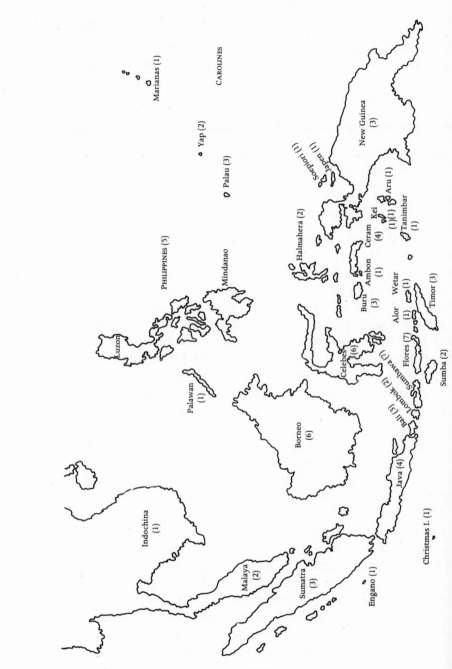

Marianas (1)

CAROLINES

Yap (2)

Palau (3)

Sarawak (1)

Japen (1)

New Guinea (3)

Halmahera (2)

PHILIPPINES (5)

Mindanao

Aru (1)

Ceram Kei
(4) (1)(1)
Tanimbar
(1)

Ambon
(1)

Buru Wetar
(3) (1)

Luzon

Alor
(1)

Timor (3)

Celebes (6)

Flores (7)

Palawan
(1)

Sumbawa (1)

Borneo
(6)

Lombok (2)

Sumba (2)

Bali (3)

Indochina
(1)

Java (4)

Malaya
(2)

Sumatra
(3)

Engano (1)

Christmas I. (1)

204

In contrast to the situation on the mainland, many of the islands of the Malay Archipelago and the Western Pacific have several coexisting species, as shown in figs. 46 and 47.

Sunda Islands

Sumatra has 3 species of Zosteropidae, with one more on an outlying island, Java has 4, with 2 more on outlying islands, while the smaller islands of Sumbawa and Flores have 7 each, this being the largest number of species of white-eyes on one island anywhere in the world. As there are 57 resident passerine species on Sumbawa and 72 on Flores (Rensch 1931), the Zosteropidae comprise one-eighth and one-tenth respectively of their passerine avifaunas.

Some species in this area are separated by range. Many others are separated by altitudinal range, which is often, and perhaps always, linked with a change in the type of forest; and on Java, Sumbawa and Flores there are not simply a lowland and a highland species, but a third at intermediate levels. Some lowland species are separated by habitat, being respectively in forest and in cultivated or other open ground with trees. Finally, some species coexist in the same habitat but differ markedly in size, the larger being in separate genera. Hence all the white-eyes of the Sunda islands are segregated from each other, most of them by range, altitude, or size (and presumably feeding).

Borneo, New Guinea, Moluccas and Celebes

Borneo has 6 species of Zosteropidae, two of which replace each other geographically, with a seventh on some outlying islands. Some of these species are separated by altitudinal range, including one at mid-levels, and others coexist but differ in size. The latter include two endemic monotypic

Fig 46. Malay archipelago and adjoining islands, and the number of species of white-eyes Zosteropidae on each.

genera, one larger than *Zosterops,* while the other is the only genus of white-eyes much smaller than *Zosterops.*

The mainland of New Guinea, though much larger with more diverse habitats than Borneo, has only 3 species of *Zosterops.* But there are 6 more on outlying islands, where the species endemic respectively to Great and Little Kei, Soepiori, Tagula and Goodenough show how frequently species in this genus replace each other geographically. More remarkably, the four white-eyes on Ceram, only 150 km from New Guinea, include an endemic monotypic genus and another endemic species, while the three on Buru, only 70 km from Ceram, include a different endemic monotypic genus and another endemic species. In Celebes, each of the three main peninsulas has a different lowland species, though these are not separated by sea; the other species on this island are separated by habitat, altitude, or size (and presumably feeding), the last being in a separate genus.

Philippines and Solomons

Five species occur in the Philippines, where they are separated by range, altitude, habitat or size. Fourteen occur in the Solomons, the greatest number in one archipelago, but most of them replace each other geographically and there are at most two on one island. Some species are separated by extremely short sea-gaps, for instance on Ganonga and Vella Lavella, 5 km apart, and on Gizo and Kulambangra, 6 km apart; and endemic subspecies of the same species on Tetipara, Rendova and New Georgia are only 1·7 and 2 km apart, respectively (Mayr 1942 p. 227). This suggests that white-eyes are highly sedentary, a conclusion belied by their exceptional success in colonizing distant islands in the Pacific Ocean.

Four of the Solomon Islands have two species each. Kulambangra and Bougainville have a smaller lowland and rather larger montane species, while low-lying and remote

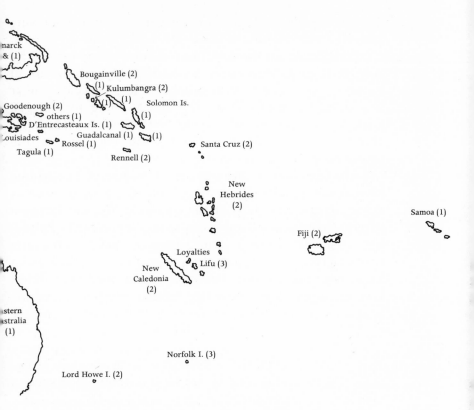

Fig 47. Islands in western Pacific, with number of species of white-eyes Zosteropidae on each.

Rennell and also Santa Cruz have one typical and one much larger species in a separate genus, which presumably differs in feeding. That on Santa Cruz is the largest of all the white-eyes. On both islands, there are only 10 other passerine species, so that Zosteropidae comprise one-sixth of the passerine avifauna.

Other Pacific islands

The Fijis, New Hebrides, Banks, New Caledonia and most of the Loyalties have two species each, one of which probably differs in habitat from the other. But Lifu in the Loyalties has three species, one of which is very large; its colouring suggests that it evolved from the widespread *Z.lateralis*. Likewise there are (or were) two species differentiated by size on Lord Howe and three on Norfolk Island, one of which is *Z.lateralis* and the others are large species similar to it and presumably derived from it. There are only 10 indigenous passerine species on Norfolk Island (Mathews 1928), hence white-eyes comprise 30 per cent of the passerine avifauna.

The remarkable situation in the white-eyes of the Caroline Islands is summarized in table 31, with a map of the islands in fig. 48 and drawings of the species in fig. 49. The two largest islands, or island-groups, Ponape and the Palaus, have a small, a medium and a large species, whereas the smaller Truk and Yap have only two species, a small and a large, and remote Kusaie has only one, the medium species. It is reasonable to suppose that smaller and remote islands have less diverse ecological resources than larger and nearer islands. If the resources are such that the island can support only two white-eyes instead of three, one could expect on general grounds that the middle-sized species would be the one excluded, while if they are such that only one species can survive, one might expect it to be the middle-sized one, which means that the latter has an interrupted range. (However, on the Marianas to the north, only the small species occurs.). The large species on the Palaus looks very different from those in the genus *Rukia* on Yap, Truk and Ponape, and though Mayr treated it as congeneric, I agree with Mees that it should be in a separate genus *Megazosterops* and is probably of independent origin. Hence it provides an example of geographical replacement at the generic level.

The frequency of each type of isolation

If each of the three continents is analysed separately and the results are combined, there are 15 possible contacts between mainland species of *Zosterops,* and separation is by

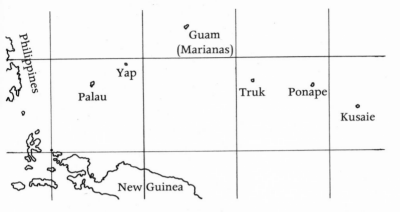

Fig 48. The five main islands of the Carolines.

Table 31. White-eyes on the Caroline Islands

| | Main islands in sequence from west to east | | | | |
	Palau	Yap	Truk	Ponape	Kusaie
Area (km²)	490	100	100	330	108
Distance (km) from next island to west	880	450	1500	640	500
Z.conspicillata (mean winglength [mm])	56	57	56	55	—
Z.cinerea (mean winglength [mm])	66	—	—	61	64
Rukia spp. (mean winglength [mm])	—	73	81	71	—
Megazosterops sp. (mean winglength [mm])	82	—	—	—	—

Notes: Mayr (1967) merged *Megazosterops* in *Rukia,* but I fully support Mees (1969) in keeping it separate and in considering it of independent origin. The further east the island, the more remote it is from the mainland and from the other archipelagoes with resident white-eyes. Measurements of beak-length in appendix 25.

range in 80 per cent, and by habitat (including altitude) in the rest. Hence as compared with the continental genera analysed in previous chapters, the proportion separated by

Zosterops conspicillata

Zosterops cinerea

Rukia longirostra

Megazosterops palauensis

range is extremely high. The same holds on islands, but here the precise figure is greatly affected by the methods of analysis.

There are three main difficulties. The first two are best illustrated by examples. If in one archipelago there are several species, each on a different island, then if each is scored against every other, the number of instances of separation by range is 3 for 3 species, 6 for 4 species, 10 for 5 species, and so on, so it rises disproportionately; probably, therefore, it is best to count each such species only once each. Secondly, if there are two islands, each with a different lowland and a different highland species, the two lowland species are obviously separated from each other by range, and so are the two highland species, but should the lowland species on the one island be regarded as differing in range or in habitat from the highland species on the other? Thirdly, since there are often different species on adjoining archipelagoes, the number of instances of separation by range is lower if each archipelago is analysed separately, and the results from each are then added, than if several archipelagoes are combined in one analysis, and it is hard to determine how best to delimit the areas analysed. The first two of these difficulties also arise on continents, but so few species are there separated by range that how they are scored makes little difference to the total; on islands, however, so many species are separated by range that the method of analysis makes a big difference.

Assessing each species against every other, giving a difference in range priority over a difference in habitat, and analysing separately each area set out in a separate paragraph in Appendix 26, out of 280 possible contacts between island Zosteropidae, separation is by range in 62 per cent, but if each geographically separated species is scored only once the figure is 35 per cent (out of a total of 163 potential

Fig 49. The White-eyes of the Caroline Islands (cf. table 31). By Robert Gillmor.

contacts); the others are by habitat in 36 per cent, feeding in 28 per cent and unknown in one per cent of instances. These figures are for all species in the family. For the genus *Zosterops* alone, out of 163 potential contacts on the first method of analysis, separation is by range in 69 per cent, which for 93 contacts on the second method of analysis, it is by range in 46 per cent, habitat in 44 per cent, feeding in 8 per cent and unknown in 2 per cent. Had larger geographical areas been analysed, e.g. the whole of Indonesia, or the whole of Melanesia, or for that matter the whole Indo-Australian region, the proportion of species separated by range would have been much greater.

Separation by range

Whatever the method of scoring, far more species are separated by range in *Zosterops* than in any other widespread genus of birds, and to answer an obvious objection, this is certainly not because the systematists who have worked on *Zosterops* use a different concept of the species from those who have studied other genera. The two most recent revisers of the group, Mayr and Mees, are highly experienced and are in striking agreement with each other.

Each continental species of *Zosterops* has a wide range, and the high proportion of them separated by range is not due to there being many such species, but to the rarity of other means of separation. That so few continental white-eyes live in the same area and differ in habitat or feeding is presumably because they are restricted to a particular ecological niche by the many other passerine birds present. Otherwise, it would be puzzling that more species coexist on islands, where the ecological diversity is reduced, but so is the number of other passerine species.

Many of the island white-eyes separated by range are in the same superspecies, and the large number of such species separated by range is partly due to the readiness with which white-eyes evolve into new species on islands. Another factor is the paucity of species separated by habitat or feeding on very small or remote islands, which is presumably due to the limited ecological resources. Islands of moderate size often have several species of white-eyes, and here not only the lowland species, but also those of montane forest, and those in other genera, are often represented by different species in the same superspecies on adjoining islands or archipelagoes.

It is remarkable that such restricted ranges are found in a passerine genus which has been more successful than any other in colonizing remote islands. Indeed, colonization is still continuing. Thus *Z.chloris* has established (or rather re-established) itself on one of the islands between Sumatra and Java whose avifauna was presumably destroyed by the eruption of Krakatau (Mees 1969 p. 351). More strikingly, *Z.lateralis* evidently crossed some 1600 km of sea from Australia to establish itself in the South Island of New Zealand about the year 1830 (Mees 1969 p. 68) and it reached Norfolk Island, some 1300 km from Australia and 720 km from New Zealand, in the year 1904. As *Z.lateralis* lives mainly in open woodland or cultivated land with trees, rather than in primaeval forest, it might not have found a suitable habitat in New Zealand in earlier times.

Of these two recent colonists, *Z.chloris* inhabits low coral islands, *Z.lateralis* open country, and as pointed out by Mees, it is primarily the forest white-eyes whose ranges are so restricted. As already mentioned, only 5 km of sea separate some species in the same superspecies in the Solomons. Birds could fly across in a few minutes, so presumably they have evolved sedentary habits because each population is better adapted to its own island than any other, and is excluded by competition with the resident forms on the other islands. Presumably the same applies where the species which replace each other geographically are in different superspecies, like those on the three peninsulas of the Celebes.

In a few instances, species in different genera replace

each other geographically. Thus the large *Rukia* is found on various of the Caroline Islands, but is replaced by the equally large *Megazosterops* on Palau. Again *Lophozosterops* is found on various of the larger islands in the Banda Sea but is replaced by *Madanga* on Buru.

Separation by habitat

Separation by habitat without a difference in altitude is rare in *Zosterops*, and the known instances are of one species in primaeval lowland forest and another at the forest edge or in cultivated land, sometimes also in coastal mangroves. In Singapore, as shown earlier (p. 161), the cultivated land has been colonized by coastal birds, not those of closed forest, and the same could well hold in island *Zosterops*. Probably, therefore, the modern replacement of forest by cultivated land has resulted in the coastal species spreading inland and to higher altitudes, so that they are now less clearly separated by habitat from the forest species than in earlier times.

Separation by altitudinal range is frequent in white-eyes, as in many other tropical forest birds. Often, and perhaps always, it is linked with a difference in the type of forest. The species at higher altitudes is usually somewhat larger than that in the lowlands, of which many examples are set out in appendix 26, but there are exceptions (for instance *Z.montana* on some islands).

Differences in size

As just noted, highland species of *Zosterops* are usually a little larger than lowland species, but the difference tends to be small, and the situation is exceptional on Mount Cameroon and Fernando Po, where the highland species is so different as to be in a separate genus. Almost everywhere else that a large white-eye in a different genus is found with a typical species of *Zosterops,* they live in the same habitat, and though critical observations on diet have not been made, it is here presumed that they differ in feeding. Separation by size is found in 28 per cent of the Zosteropidae, which is only a little lower than the proportion separated by habitat, and is much higher than that separated by feeding in most continental passerine genera. It therefore confirms the finding for Darwin's finches and the Hawaiian sicklebills that separation by size, and presumably feeding, is especially common in island birds. In the genus *Zosterops* itself, however, the proportion separated by size is much lower than in the family as a whole. This is because most, though not all, of the unusually large white-eyes are sufficiently different in other ways to be put in separate genera.

Large species in the genus *Zosterops* coexist with one or two smaller species on (a) Lifu, Lord Howe and Norfolk, (b) Grand Comoro, (c) Mauritius and Reunion and (d) Palau, Ponape and Kusaie, and these four groups of species look so different in colour or other ways and, except for (b) and (c), are so distant from each other, that they presumably evolved independently. The same holds for the large species in separate genera on, respectively, the Guinea islands *(Speirops)*, Lesser Sundas *(Heleia)*, Borneo *(Chlorocharis)*, Ceram *(Tephrozosterops)*, Buru *(Madanga)*, Rennell and Santa Cruz *(Woodfordia)* Palau *(Megazosterops)* and the other Carolines *(Rukia)*, each of which is restricted to one, or at most a few adjoining islands, while one further genus *Lophozosterops,* is rather widespread in Indonesia. Hence large size has been independently evolved about twelve times in white-eyes, always where a typical species of *Zosterops* is also present and normally on an island, hence presumably through two successive colonizations from the mainland or another island, followed by competitive displacement. The only example on the mainland is the *Speirops* on Mount Cameroon, but this genus is also found on the Guinea islands, so could have evolved there.

Most of these large species are found on remote and

fairly, but not very, small islands with few other passerine birds. The smallest of these islands is Lord Howe, only 13 sq km in area, which has 6 other indigenous passerine species (Mathews 1928). Norfolk Island, $2\frac{1}{2}$ times as large with 7 other indigenous passerine species, is the smallest island with three white-eyes, but one of these arrived only in 1904, after man had cultivated part of the former forest. Yap and Truk, each of about 100 sq km, Ponape and Palau, three to five times as large and with three species, and Principe, 127 sq km in area, are also remote, with, at most, 10 other passerine species. Others of the islands concerned are between 900 and 2500 sq km in area and have from 9 to 16 other passerine species, namely Sao Tomé, Fernando Po, Mauritius, Reunion, Grand Comoro, Rennell and Santa Cruz. Less remote, much larger (between 9000 and 18,000 sq km) and with many more passerine species, are Flores, Sumbawa, Ceram and Buru. Finally Borneo is much larger still (over 700,000 sq km) with a much richer avifauna than any of the other islands involved, and it also has the only genus of white-eyes, *Oculocincta*, which is smaller than *Zosterops*.

Very small and remote islands normally have only one species of white-eye, and though some of these species are similar in size to typical mainland white-eyes, many of them are larger, as for instance on Annobon (Gulf of Guinea), some of the Seychelles and Comoros, Christmas Island (Indian Ocean), Great and Little Kei (Banda Sea), Rossel (Louisiades), Samoa, and Kusaie (Carolines). Further, where the same species of *Zosterops* is found on both large and small islands, the subspecies on small islands tend to be the larger, as in *Z.lateralis* on islets off Australia (Mees 1969). This is a widespread trend in island songbirds, and the usual explanation is that it enables them to take a wider range of foods from the limited number available on small islands; but while this seems reasonable, it has not been tested in the field.

Once an island form of *Zosterops* has become larger than

its mainland ancestor, there is the chance that a later colonist of normal size will be sufficiently different ecologically to persist alongside it, after which they will tend to diverge further. That this has been the sequence of evolution where two island white-eyes now coexist is supported by the fact that the larger of the two is always the more different from *Zosterops* in other ways, so presumably was the first to arrive. (The small *Oculocincta* on Borneo is a special case.)

That only one species of white-eye is normally found on very small or remote islands is presumably because, with limited ecological resources, one generalised species tends to exclude two specialists (cf. MacArthur and Levins 1967), and it is suggestive that the species concerned is often intermediate in size between the small and large species on adjoining islands, for instance on Annobon compared with Sao Tomé, and on Kusaie compared with Truk. The only very small and remote islands with two or three white-eyes are Lord Howe and Norfolk Islands. Some of the islands with a single endemic white-eye evidently provide extremely limited ecological conditions, as shown by the paucity of other passerine species, for instance there is only one other passerine species on Annobon, and also on Christmas Island (Indian Ocean). Part of the success of *Zosterops* on islands is evidently due to its ability, perhaps helped by its varied diet, to survive in conditions where most other passerine species cannot do so. Hence it may normally be one of the first species to colonise a new island, and will be able to evolve adaptations to the local conditions as the island vegetation matures, and prior to the arrival of other passerine genera from the mainland.

That only one species of white-eye is likewise found on certain large islands, such as Madagascar, is presumably because the ecological diversity is offset by the presence of many passerine birds in other genera, as argued for the similar situation on the continents. Why Borneo, with 8 species, should be an exception is not known.

There is the further question of why a few islands should have three, and not just two, species of white-eyes separated by size. In the Carolines, only the two largest islands have three species, presumably because these provide greater ecological diversity than the others. On Norfolk Island, the third species colonized only recently, perhaps due to man modifying the forest by cultivation, as already mentioned. Sumbawa and Flores (with *Heleia* and *Lophozosterops*) and Ceram (with *Tephrozosterops* and *Lophozosterops*) are exceptional in having large white-eyes in two different genera (in addition to *Zosterops* itself), and while they provide more varied ecological conditions than smaller islands, one might have thought that this advantage would be offset by the presence of a greater number of passerine species in other genera.

As discussed in previous chapters, there is a general tendency in congeneric forest birds for smaller species to forage on thinner and higher branches, and for those with smaller beaks to take smaller prey, than their larger congeners. To what extent this holds among the island white-eyes is not certain, but the proportionate differences in winglength and beak between species coexisting on one island are so different in different cases that the nature of their ecological separation must likewise differ greatly.

The biggest proportionate difference in winglength (and hence overall size) between two coexisting white-eyes is 45 per cent on Truk, it is 28 per cent on Yap, about 22 per cent on Buru, Rennell and Santa Cruz, and around 20 per cent on various other islands, but smaller on others, only 5 per cent on Reunion, while on Mauritius the species with the much longer beak actually has a slightly shorter wing than the other.

There are similar big variations with respect to beak-length. Grant (1968) postulated that species-pairs on islands (in any passerine genus) differ in beak-length by at least 15 per cent, but Mees (1969) pointed out that the difference is less than 5 per cent between the two larger species of

Zosterops on Norfolk Island. Similarly it is just under 5 per cent between the two larger white-eyes on Ceram, and about 11 per cent between the pairs on Principe, Mauritius and Buru respectively. It is actually less than 1 per cent between the pair on Fernando Po but these, unusually, are separated by habitat. At the other extreme, the difference is 75 per cent between the largest and the medium-sized species on Ponape (where the smallest species in overall size has a slightly longer beak than the medium-sized species), it is just over 60 per cent on Truk, 57 per cent on Santa Cruz, and between 30 and 15 per cent on many other islands.

Overall size and beak-length are not the only ways in which larger white-eyes differ from smaller coexisting species. While those on Norfolk and Lord Howe are enlarged, but otherwise typical, *Zosterops,* the large genus on Yap, Truk and Ponape has a relatively long and slender beak, that on Palau has a much heavier beak, also broad wings and soft plumage, and that on Rennell and Santa Cruz has a thick beak and short tail; in addition, most of them differ markedly in colour or pattern of plumage from each other. At least most of these differences are presumably adaptations to rather different ways of life.

Hence except that the larger member of each pair is the more different from a typical *Zosterops* in other ways, these large white-eyes show no uniform trend. The essential point is that each of them has presumably evolved sufficient differences to be ecologically isolated from the smaller species living with it, but the ways in which it has done this are different on different islands, doubtless due to differences in the habitats, the available foods, and the other passerine species present. Sometimes, perhaps, the difference is mainly in feeding stations, with a related difference in body-size, sometimes in size of prey, with a related difference in beak-length, sometimes in feeding methods, occasionally in habitat and often, doubtless, in differing combinations of these factors. Hence the statement that two coexisting island species 'differ in size' is a bald summary of a complex ecological situation.

Although this type of evolution is common in white-eyes, it is rare in other mainland passerine genera that have colonized islands. Examples are known, for instance, a typical and an unusually large species of sunbird *Nectarinia* are found on Sao Tomé, and another pair on Principe (Amadon 1953a). In contrast, this type of evolution is common in the endemic genera of the Galapagos and Hawaii, which suggests that its prevalence in white-eyes may be because they have often been the first passerine species to colonize oceanic islands in the Indian and Pacific oceans, as Darwin's finches and the sicklebills presumably were in the Galapagos and Hawaiian Islands respectively. One might then ask why the white-eyes have not differentiated much further, like these other two groups, but they have presumably been checked in this by the colonization of other mainland genera, adapted to different niches, before they could evolve to fill them.

Conclusion

There is normally only one species of *Zosterops* on each small and remote island (presumably owing to limited ecological resources), and also in any one continental area (presumably because other passerine genera restrict white-eyes to a particular niche). But there are often between two and seven species on fairly small and fairly remote islands, presumably because they provide more diverse conditions than small remote islands, while as compared with a continental area, their reduced ecological diversity is more than offset by the reduced number of other passerine genera present. White-eyes are evidently among the first species to colonize new islands in the Indo-Australian region, and colonization is still continuing. Nevertheless, many island species are separated from each other by narrow seas, presumably through competitive exclusion, because each is best adapted to its particular island, in which case there will be strong selection for sedentariness.

Where more than one white-eye is found on the same island, they are often separated by habitat (including altitude) or by size (and presumably feeding), but most white-eyes separated by size are in different genera. Unusually large white-eyes have evolved solely where another species of normal size is present, hence presumably through competitive displacement, and usually on islands with few other passerine species. This has happened about twelve times independently, but except in being larger, the resulting species look very different from each other, so have evidently diverged ecologically in different ways from the small species with which they coexist.

Chapter 13

A Tropical Archipelago
The West Indies

One further example may be given of ecological isolation in insular land birds, this time not in a particular group, like Darwin's finches or the white-eyes, but in all the congeneric species in an archipelago. The most suitable islands for this purpose are the West Indies, because the avifauna is not too depauperate and includes many congeneric species, and the birds are relatively well known through the work of Bond (1960), with detailed studies of Puerto Rico by Wetmore (1927), and Hispaniola by Wetmore and Swales (1931) and Wetmore and Lincoln (1933). I had brief field trips in Puerto Rico (guided by C. and K. Kepler) and Jamaica (helped by I. Goodbody and others) in June 1969. I am further indebted to the Keplers for extremely helpful notes and criticisms. The wading and water birds are not considered here.

The West Indies, shown in fig. 50, lie between 12° and 28°N. Puerto Rico is about 700 km from South America and Jamaica over 600 km from Central America, but adjacent islands are fairly near each other, and the Bahamas are only 100 km from Florida, Cuba is about 200 km from both Florida and Mexico, and Grenada less than 150 km from South America, hence the archipelago has often been colonised by mainland species. The lowlying Bahamas have much pine forest, the Greater and Lesser Antilles are mountainous with rain forest, with scrub forest on lowland limestone. Former rich lowland forest has been largely replaced by cultivation, with the loss of various land birds, mainly non-passerines. Nearly all of the few introduced birds are restricted to cultivated land.

The means of ecological isolation in congeneric species have been summarized in appendix 27 for passerines and appendix 28 for other land birds. The classification is that of Bond (1960), but differences in Peters' Check-List and in recent papers on particular genera have been noted. Later workers have often reduced island species to sub-species, which affects the total of species separated by range. Further, as noted for *Zosterops* in the previous chapter, in genera with many species separated by range,

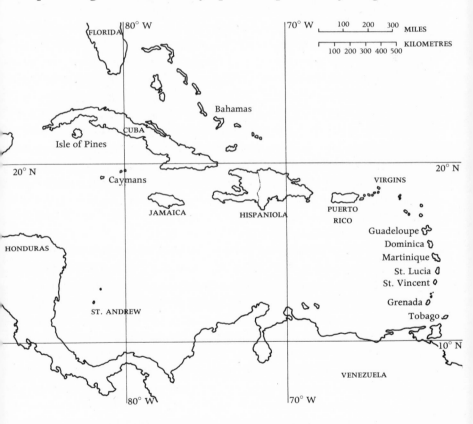

Fig 50. Map of West Indies.

the total is absurdly high if each is scored against every other, so it is probably best to score each such species only once.

Passerine species separated by range

On Bond's classification and with each species counted against every other, 72 per cent of the congeneric passerines are separated by range, (out of 127 possible cases), but on the revised classification of certain groups noted in appendix 27, and with each geographically separated species counted only once, the figure is 59 per cent by range, with 21 per cent segregated by habitat, 17 per cent by food and 2 per cent unknown (out of 81 possible cases).

Hence the proportion of species separated by range is extremely high, as in insular *Zosterops*. It arises chiefly because many island forms in the same superspecies are sufficiently distinctive to be classified as full species, whereas on continents most geographically replacing forms intergrade with each other and are considered as sub-species. But this is not the whole explanation, for in the West Indies some species which replace each other on different islands are in different superspecies. The distribution of the crested flycatchers *Myiarchus* is shown in fig. 51, according to the classification of Lanyon (1967), in which *M.sagrae*, *M.stolidus*, *M.antillarum* and *M.oberi* are in the same superspecies, but *M.nugator* is in a different one, derived from a different mainland species. Each island has only one species except Jamaica, where there are three. While nearly all the West Indian passerine congeners which replace each other geographically are on different islands, there are two species of *Teretistis* warblers in eastern and western Cuba respectively, and two species of palm tanagers *Phaenicophilus* in eastern and western Hispaniola respectively.

No two islands provide identical conditions, so some of the species which replace each other on different islands

Fig 51. Ranges of crested flycatchers *Myiarchus* in West Indies (from Lanyon 1967).

There is only one species on each island except Jamaica, where there are three. *M.sagrae, M.stolidus, M.antillarum* and *M.oberi* are in the same superspecies, but *M.nugator* is derived from a different mainland species.

probably differ in ecology as well. This is obvious in a few instances. Thus the forest warbler on Puerto Rico, *Dendroica adelaidae*, is there restricted to dry lowland forest, whereas that on Jamaica, *D.pharetra*, is restricted to wet montane forest (but *D.adelaidae* is in both lowland and montane forest on St Lucia). Again, the Puerto Rican representative of the white-eyed vireos, *V.latimeri*, is restricted to dry lowland forest, but that on Jamaica, *V.modestus*, is in both dry lowland and wet montane forest.

In a number of instances, species in different genera replace each other on different islands (not being congeners, they are omitted from appendix 27). For instance, the Golden Swallow *Kalochelidon euchrysaea* on Jamaica and

225

Hispaniola is replaced by the Bahama Swallow *Callichelidon cyaneoviridis* in the Bahamas; there is a forest species of *Dendroica* warbler on almost every island except St Vincent, and only here the Whistling Warbler *Catharospeza bishopi* is found; the widespread West Indian Bullfinch genus *Loxigilla* is absent from Cuba and Grand Cayman, on which the Cuban Bullfinch *Melopyrrha nigra* occurs; the Blue-hooded Euphonia *Tanagra musica* on Hispaniola, Puerto Rico and the Lesser Antilles is replaced by the Jamaican Euphonia *Pyrrhuphonia jamaica* on the latter island; the Redlegged Honeycreeper *Cyanerpes cyaneus* occurs solely on Cuba, while the Bananaquit *Coereba flaveola* is found on all the West Indies except Cuba. Some of these pairs seem closely related, others not; in either case, their being put in different genera means that they differ morphologically, and hence to some extent in feeding habits or other ecology. That, nevertheless, they replace each other on different islands suggests that the ecological situation is so simple that it allows only one species of each of these types to exist on any one island. The same was pointed out long ago by Mayr (1933) for the flycatchers in the Tonga archipelago, where there is a species of either *Pachycephala* or *Clytorhynchus* on each island, but never of both. Further study is needed to show whether all the pairs just listed for the West Indies are true ecological replacements. For instance, the Jamaican swallow lives in montane forest, the Bahaman on low ground, and the Cuban honeycreeper is in a different family from the Bananaquit.

Passerine species separated by habitat

Few of the congeneric passerine birds differ in habitat. Some of these are in dry lowland forest and wet highland forest respectively, but since rich lowland forest has largely gone, it is not easy to say whether the humidity or the altitude provides the critical difference. Some lowland species have penetrated into the highland forest in man-made clearings, including a hummingbird discussed later, and the mockingbird *Nesomimus polyglottos*.

On most islands where it occurs, the flycatcher *Elaenia martinica* is the sole species in its genus, and it lives in both forest and open country. But on Grenada another species, *E.flavogaster*, is in open country, and here *E.martinica* is restricted to forest, presumably through competitive displacement. Both species also occur on St Vincent, where their separation is less clearcut than on Grenada, perhaps because *E.flavogaster* is a recent arrival (Crowell 1968). Some other differences in habitat on different islands are not linked with the presence or absence of a congeneric species, notably those in the *Dendroica* warblers and white-eyed vireos mentioned earlier (p. 225).

Passerine species separated by feeding

Congeneric passerine species are separated by feeding in only 17 per cent of instances, a much smaller proportion than in Darwin's finches, the Hawaiian sicklebills and the island white-eyes; but most of these latter live on remote islands with few other land birds, so that their opportunities for evolving differences in feeding may be unusually good. In the West Indies, many mainland genera with different feeding habits are established, and this is presumably why there has been no adaptive radiation of birds within the archipelago. Each of the few endemic genera consists of, at most, two or three geographically replacing species, and the only endemic passerine family, the palm chats *Dulidae*, consists of a single species on Hispaniola.

On Jamaica, three species of *Myiarchus* flycatchers and two thrushes in the genus *Turdus* are separated partly by habitat and partly by size, and presumably feeding. Two widespread vireos are likewise separated by size, but unusually for forest birds separated in this way, the large Black-whiskered Vireo *V.altiloquus* forages in the canopy and the various island representatives of small white-eyed vireo in the bushes, but the latter have thin twigs, so the correlation between body-size and size of branch still holds. A third and unusually thick-billed vireo, *V.osburni*, occurs with the other two on Jamaica, but nowhere else, and presumably differs in feeding. Remarkably, two species of West Indian Bullfinch, *Loxigilla portoricensis* on Puerto Rico and *L.noctis* on the Lesser Antilles, which are separated mainly by range, coexisted on the small island of St Kitts, where the larger species, *L.portoricensis*, was larger than on Puerto Rico, presumably through character displacement (it is now extinct).

Hummingbirds

The congeneric hummingbirds of the West Indies are included in appendix 28. Nearly all of them are separated from each other by range on separate islands, but two species of *Anthracothorax* on Puerto Rico are separated by habitat, one high and the other low, except that the lowland species has moved into highland forest in manmade clearings. This latter is the only species of *Anthracothorax* on Hispaniola, where it occurs in both lowlands and highlands, indicating that its avoidance of highland forest on Puerto Rico is due to displacement by its congener there.

As shown in table 32 and portrayed in fig. 52, hummingbirds in 10 different genera live in the West Indies, though usually there are only 3 (in a few cases 2 or 4) on any one island. Hence there is much geographical replacement by species in different genera. Some of these replacing species are similar in size, so even though they are in different genera, they perhaps have similar diets, but others are

Fig 52. The hummingbird genera of the West Indies (cf. Table 32). By Robert Gillmor, based on plates by D.Eckleberry in Bond (1960).

Trochilus polytimus

Chlorostilbon ricordii

Mellisuga helenae

Anthracothorax mango

Calliphlox evelynae

Cyanophaia bicolor

Orthorhyncus cristatus

Sericotes holosericeus

Eulampis jugularis

Glaucis hirsuta

Table 32. Geographical replacement of hummingbirds in West Indies. Only generic names are given, a number in brackets after it denoting the species if there is more than one in the genus. On second line, habitat is specified if it helps in segregation, and the figures are mean winglength and culmen in mm for males by Ridgway (1911).

Islands	Small	Medium	Large
Bahamas	*Calliphlox* [open, 39, 16]	*Chlorostilbon* (ii) [woods, 53, 17]	
Cuba	*Mellisuga* (i) [28, 10]	*Chlorostilbon* (ii) [53, 17]	
Jamaica	*Mellisuga* (ii) [37, 10]		*Anthracothorax* (i) [low, 74, 27] *Trochilus* [low and high, 66, 21]
Hispaniola	*Mellisuga* (ii) [open, 35, 10]	*Chlorostilbon* (iii) [heavy forest high, 56, 17]	*Anthracothorax* (ii) [low and high, 68, 24]
Puerto Rico (except extreme east)	*Chlorostilbon* (i) [49, 14]		*Anthracothorax* (ii) [low, 63, 23] *Anthracothorax* (iii) [high, 64, 24]
Virgins and extreme E. Puerto Rico	*Orthorhyncus* [49, 10]		*Sericotes* [61, 22]
Lesser Antilles (except Grenada)	*Orthorhyncus* [49, 10]	Additional on Dominica, Martinique *Cyanophaia* [high, 60, 17]	*Sericotes* [low, 61, 22] *Eulampis* [high, 74, 23]
Grenada	*Orthorhyncus* [49, 10]		*Sericotes* [low, 61, 22] *Glaucis* [high, 66, 35]

230

Notes: The species are as follows: *Calliphlox evelynae, Mellisuga* (i) *helenae,* (ii) *minima, Chlorostilbon* (i) *maugaeus,* (ii) *ricordii,* (iii) *swainsonii, Orthorhyncus cristatus, Cyanophaia bicolor, Anthracothorax* (i) *nango,* (ii) *dominicus,* (iii) *viridis, Trochilus polytmus, Sericotes holosericeus, Eulampis jugularis* and *Glaucis hirsuta.* One futher species, *Anthracothorax prevostii,* is on Old Providence and St. Andrew. Of these, the genera *Mellisuga, Orthorhyncus, Cyanophaia, Trochilus, Sericotes* and *Eulampis* are endemic, and so are the species of *Calliphlox, Chlorostilbon* and *Anthracothorax* (except *prevostii*). *Anthracothorax dominicus* has been recorded from the Virgin Islands, but not recently (Robertson 1962, Kepler pers. comm.), so may well not be resident, and *Eulampis jugularis* has been recorded from Grenada, apparently as a casual visitor.

rather different in overall size or length of beak, so presumably fill partly different ecological niches, even though they have mutually exclusive ranges.

As shown in fig. 53, each island has only one small hummingbird, which may belong to one of four different

Fig 53. Geographical replacement of small hummingbirds in West Indies.

There is only one small species on each island, derived from one of four separate genera (see Table 32 for details).

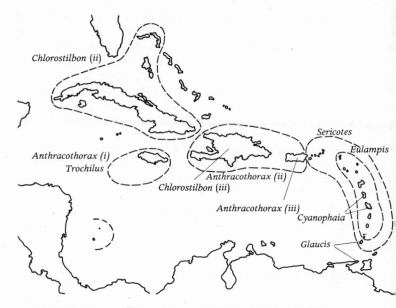

Fig 54. Ranges of larger hummingbirds in West Indies.
There are two larger species on most islands, but only one on the Bahamas, Cuba, and the Virgins, and three on Dominica and Martinique. The extreme east of Puerto Rico has the same species as the Virgins, and not those in the rest of Puerto Rico (see table 32 for details).

genera. The species of *Mellisuga* on Cuba and Jamaica are smaller than the rest, indeed they are the smallest birds in the world. *Orthorhyncus* in the Lesser Antilles has a beak of similar length to *Mellisuga*, but is a larger bird, and the other two genera, in the Bahamas and Puerto Rico respectively, are larger in both beak and overall size; hence these birds probably fill rather different niches. While *Chlorostilbon* provides the smallest species on Puerto Rico, the other species in this genus, on the Bahamas, Cuba and Hispaniola, are of medium size and coexist with a small species in another genus.

This completes the species in the Bahamas and Cuba which, with the Virgins, are the only main islands with only two species. The situation elsewhere is more complex, as shown in fig. 54. Hispaniola has a small, a medium and a large species, while Jamaica, Puerto Rico and various of the Lesser Antilles have a small and two large species, the two latter being in most cases separated in lowland and montane forest respectively. But these large species belong to five separate genera, and are rather different in size, as can be seen from table 32. There is a further complication on Dominica and Martinique, where there is also a medium-sized species in the highlands, making four on each of these small islands.

I at first thought that these geographical replacements might be due to the historical factor of which mainland species happened to have arrived, but the records summarized by Bond (1960) show that several species have been recorded outside the islands where they reside, hence even if the situation originated through the chances of first colonization, it is presumably reinforced nowadays by competitive exclusion. This is shown most clearly on Puerto Rico, where the northeast of the island is inhabited, not by the species on the rest of Puerto Rico, but by the two characteristic of the Virgin Islands. Here, there is no sea gap, and replacement must be due to interspecific competition; and if the latter applies in this instance, it is likely to apply to the many similar instances involving separate islands.

The other land birds

On Bond's classification, and with each species assessed against every other, the number of instances in the other (non-passerine) land birds of separation by geographical range is 78 per cent, but if two genera of swifts and two of woodpeckers are merged, as by recent revisers, and if each geographically isolated species is scored only once, the figure is 65 per cent, while 11 per cent are isolated by habitat and 5 per cent by feeding, with 18 per cent unknown.

Hence the proportion of congeneric species separated by range is again high, due primarily to the existence of insular species in the same superspecies in parrots, lizard cuckoos, todies, woodpeckers, swifts and hummingbirds. There is a woodpecker in the genus *Centurus* or *Melanerpes*, but not both, on each island, and while these genera look very similar and are merged by many workers, they are sufficiently distinctive to suggest that the West Indian birds were derived from one of two mainland species. The woodland owls are also represented by different genera on different islands in the Greater Antilles, but they differ so much in size that they cannot be ecological equivalents, even though they replace each other.

Few congeneric non-passerine land birds are separated by habitat, but there are lowland and highland, or dry and wet, forest species in pigeons, parrots, todies and the hummingbirds already mentioned. The Todidae are an endemic family, but except for two separated by habitat on Hispaniola, there is only one species on each island. That two endemic species of *Amazona* parrots occur on Jamaica, and another two on Dominica, perhaps means that a second species was formerly present on other islands but has been exterminated. Few congeneric non-passerine land birds are separated by feeding, but coexisting species which differ in size are found in raptors and swifts. The means of separation are unknown in various pigeons, the ecology of which may have been disturbed by both cultivation and heavy shooting.

The small number of species on each island

As in other archipelagoes, each West Indian island has far fewer species of resident land birds than has an area of equivalent size on the nearest mainland. I had hoped that the ways in which the congeneric species are isolated ecologically from each other might throw light on this paucity, but most of them are separated by range on different islands, whereas the question here is what determines the number on one island. Congeneric species on the same island are separated by either habitat or feeding in roughly similar proportions, which does not reveal anything of importance in the present context, though the situation is in contrast with that on the much more remote Galapagos or Hawaii, where differences in feeding are paramount.

What most struck me in my brief visit in 1969 was that in two days on Puerto Rico and three on Jamaica, I saw 87 per cent of all the passerine species, although I made no special effort to see them all. This is a far higher proportion than it would be possible to see in the same time on the mainland, and suggests that most of the island species are common and widespread. Further, in contrast to mainland tropics, many forest species occur both in semi-arid lowland forest and in humid montane forest (which is not apparent from the information in appendices 27 and 28 because most of the birds concerned have no congeners on the same island).

I have summarized the situation for the passerine and various other land birds on Jamaica, Puerto Rico and St John (Virgin Islands) in appendix 29, based on Bond (1960), the Keplers (1969 and pers. comm.) and Robertson (1962) respectively. The comparison is restricted to forest birds, because most open habitats are due to, or have been much modified by, cultivation. Out of 54 species of forest birds analysed, just a little over a half are found in both semi-arid lowland and humid montane forest or, on St John, in both dry and moist lowland forest. In the only comparable figures so far published for tropical forest on the mainland, those for Usambara (see pp. 140, 325-6), no species (out of 89) live in both arid lowland woodland and rain forest, indeed each of the three arid types of woodland has a separate avifauna; and only 8 per cent (out of 60 species) live in both lowland and montane rain forest. Further, only three of the West Indian species appear to be restricted to forest edge, while another two are restricted to the edge in highland but not lowland forest, which (scoring each of the latter as $\frac{1}{2}$), makes 7 per cent in all; and there is no evidence that any forest species avoids the edge. In Usambara, on the other hand, one-third of the 60 rain forest species are restricted to the edge, nearly two-thirds to the interior, and only one is common to both. Clearly, the West Indian species are much less restricted in habitat than those of Usambara. In addition, Usambara provides a wider range of forest types than the West Indies.

That the Puerto Rican forest birds have broader habitats than those in Panamanian forest was earlier postulated by MacArthur, Recher and Cody (1966), because they found that in Panama, but not Puerto Rico, larger census areas, which bring in a greater diversity of habitats, include more bird species than smaller census areas. These authors attached more weight to their demonstration that the number of species in each vertical feeding layer is comparable in the two areas only provided that Puerto Rican forest is held to consist of two layers (the ground and the trees) and Panamanian forest of four (the ground, the shrubs and two heights of trees). This assumption was justified for, and by, their mathematical analyses, but was not checked for the feeding stations of particular species. It will be recalled that one Puerto Rican vireo feeds in the shrub layer and the other in the canopy, so at least one species is segregated in the shrub layer, but broadly their findings must mean that each forest species has a wider range of feeding stations in Puerto Rico than Panama.

Schoener (1965) found that the mean difference in beak-length of congeneric species is greater in West Indian than mainland land birds, which suggests that the island species need to differ from each other to a greater extent in size to achieve ecological isolation (actually some, but not all, of the passerine species concerned are isolated by habitat in the West Indies). The main reason for the size difference, shown by Grant (1969), is that the mainland species resident in the West Indies are not a random sample of those on the mainland, but are those which differ greatly from each other in size of beak, and hence presumably in feeding.

This holds in the buzzards (hawks) *Buteo*, ground doves *Geotrygon*, crows *Corvus*, vireos *Vireo* and warblers *Dendroica*, though I would add that the differences in overall size between these species might be as important as their differences in beak-length, so far as competition for food is concerned.

The size difference between the small Broadwinged Hawk *Buteo platypterus* and large Redtailed Hawk *B.jamaicensis*, both found on Cuba and Puerto Rico, is greater than that between the mainland species of *Buteo*, but simply because the intermediate Redshouldered Hawk *B.lineatus* occurs with them on the mainland but not in the West Indies. Similarly, as mentioned earlier (p. 133), there are three species of *Accipiter* hawks in North America, the small Sharpshinned *A.striatus*, medium *A.cooperii* and large Goshawk *A.gentilis*. But in Cuba there are only two, the small Sharpshinned and an endemic species *A.gundlachi*, which is closely related to *A.cooperii* but much larger. Hence in *Accipiter*, as in *Buteo*, three species have been reduced to two, but whereas in *Buteo* the medium species has dropped out (cf. *Zosterops* in the Caroline Islands, p. 208), in *Accipiter* the large species is absent and the medium species has evolved into a larger bird.

A further indication that habitats or feeding stations may be broader in the West Indies than on the mainland is the fact that two species which replace each other on different islands may occur in the same country on the mainland. This holds in the honeycreepers *Cyanerpes cyaneus*, in the West Indies only on Cuba, and *Coereba flaveola*, on all the main islands except Cuba, in the nightjars *Caprimulgus cayennensis*, in the West Indies solely on Martinique, and *C.rufus* solely on St Lucia, and in the swifts *Chaetura brachyura*, in the West Indies solely on St Vincent, and *C.cinereiventris* solely on Grenada. All six of these species coexist in northern South America, including the offshore island of Trinidad. In addition, woodpeckers in the genera (or subgenera) *Melanerpes* and *Centurus* replace each other geographically in the West Indies, species of *Melanerpes* on Guadeloupe and Puerto Rico and of *Centurus* on the other Greater Antilles and the Bahamas, but species in both genera occur in some countries in central America.

A striking example of a broad ecological niche in the West Indies is provided by the Trembler *Cinclocerthia ruficauda*, an endemic monotypic mimid genus in the forests of the Lesser Antilles. This species feeds in a variety of ways which, on the mainland, are subdivided between species in different families of forest birds that are absent from the archipelago (Zusi 1969).

Hence West Indian land birds, as compared with their mainland counterparts, have broader habitats and broader feeding stations and the coexisting species differ more markedly from each other in size, and hence presumably in feeding. In their equilibrium theory of island biogeography, MacArthur and Wilson (1963, 1967) postulated that the number of species resident on an island is a balance between the rates of colonization and extinction respectively, and that the rate of colonization is lower on remote than near islands, while the rate of extinction is higher on small than large islands, thus accounting for the correlation between the number of species present with (a) isolation and (b) island-area respectively. It has also been argued, and I formerly did so myself, that it is because various mainland species have failed to reach islands that the species present have broader habitats and feeding stations, as in the West Indies. But in fact birds wander much more widely and more often than once thought. Hence in my view the most reasonable explanation of these various points is that, with a reduced diversity of ecological resources on islands, fewer bird species with broader habitats or ecological niches tend to oust greater numbers of more specialized species (Lack 1969). If this explanation is correct, it is necessary to suppose that the diversity of the ecological resources for birds on islands is correlated both with island-area, which is widely agreed, and also with the degree of isolation. The latter probably holds good, since the number of plant genera present is correlated with the degree of isolation (Balgooy 1969).

Size differences between sexes

In the Hispaniola woodpecker *Melanerpes (Centurus) striatus*, the beak of the male is 27 per cent longer than that of the female, with no overlap in measurements, and the horny tip of the tongue is half as long again (Selander 1966). Since the male is only slightly larger in overall size (by 5 per cent in winglength, a standard amount in birds), the beak difference is evidently a special adaptation. It is as great as that between many species segregated by feeding, and in fact the two sexes feed in largely different ways, the male probing much more often than the female, and the female searching and gleaning much more often than the male. *M.striatus* is related to the Golden-fronted Woodpecker *M.aurifrons* of the mainland, in which there are only slight differences between the sexes in these various respects, as shown in table 33.

A big difference in beak-length between the two sexes was found in another 5 out of 31 American woodpeckers examined by Selander and Giller (1963), three of these being in the West Indies, on Cuba, Puerto Rico and the Lesser Antilles respectively, (but in the Jamaican woodpecker the two sexes are similar); the other two are on the peninsula of Baja California, and none were found on the main continent. A search by Selander (1966) for big differences in beak-length between the sexes in other kinds of birds produced one more example from the West Indies, the Trembler *Cinclocerthia ruficauda* in the Lesser Antilles, and five more on oceanic islands, namely three Hawaiian sicklebills, a sunbird on Sao Tomé, and the extinct New Zealand Huia *Heteralocha acutirostris*, in the last of which the sexes were known to feed differently (cf. Lack 1944). The only others probably linked with feeding are in pursuing raptors such

Table 33. Differences between the sexes in beak and feeding in a West Indian and a mainland woodpecker

	Hispaniola woodpecker *Melanerpes striatus*		Golden-fronted woodpecker *Melanerpes aurifrons*	
Mean length (in mm)				
culmen	33	26	33	30
(and limits)	(30–37)	(22–29)	(30–37)	(28–34)
Horny tip of tongue	15·6	10·2	15·0	13·6
Wing	121	116	130	128
(and limits)	(113–128)	(109–127)	(124–136)	(124–133)
Percentage of foraging time spent				
probing	35	9	17	10
searching and gleaning	25	58	50	50
pecking, excavating	33	25	28	27
other ways	8	9	5	13

Notes: from Selander (1966, tables 2, 3, 6 and 7). In addition to the recorded feeding differences, the male may have spent proportionately longer than the female picking on live as compared with dead branches, and on larger as compared with smaller branches, but these differences were not statistically significant.

as falcons and *Accipiter* hawks (see p. 133), and in a few wading and waterbirds. Hence in land birds such a difference is found mainly on remote islands where, as just suggested, fewer species with broader niches tend to oust more specialists, and perhaps under these conditions, it is sometimes advantageous for the two sexes to divide, rather than share, the available food resources.

Land birds on other archipelagoes

I could find no other tropical archipelago about which enough is known to make a similar analysis of the land birds, except for some with extremely few congeneric species. I have however set out the information for the congeneric passerine birds on four other archipelagoes in Appendix 30. This does not provide information of special interest, though the high proportion of species separated by either range or size on the Guinea islands may be noted.

Conclusions

Most congeneric land birds in the West Indies have separate ranges; and various species in different genera also replace each other geographically. Such species are not always very alike, and a few of them coexist elsewhere but evidently they are alike enough to prevent their living in the same area, under the conditions of reduced ecological diversity found on islands. Most species which replace each other geographically are on separate islands, but two hummingbirds replace two or three others within Puerto Rico, two tanagers replace each other within Hispaniola, and two warblers within Cuba. In these, and probably also the cases involving separate islands, geographical replacement is evidently due to competitive exclusion, not to inability to cross a geographical barrier.

The paucity of species in the West Indies compared with the mainland is associated with those present having broader habitats and feeding stations, and being separated by bigger differences in size or size of beak, than their mainland congeners. It is suggested that this is because, with limited ecological resources on islands, fewer species with broader ecological niches tend to displace more species with narrower niches. In a few species, there are big differences in size of beak and feeding between the two sexes.

Chapter 14

Conclusions

The aim of this final chapter is to summarize the general biological conclusions of the book, together with some not yet mentioned, and in the concluding section to suggest briefly the other fields of biology influenced by the findings on ecological isolation. Page references are to earlier chapters in the book.

A. RANGE

Although new species of birds originate from geographically isolated subspecies, their present ranges have been determined primarily by events taking place after speciation. Even insular species would spread into each other's ranges unless restricted by competitive exclusion. Species inhabit separate but adjoining ranges when (i) they have such similar ecology that only one of them can persist in any one area and (ii) each is better adapted than the other to part of their combined ranges. Separation by range is particularly common in wholly frugivorous congeners, because they eat fruits of similar size, and also in the land birds of oceanic archipelagoes, because the reduction in ecological diversity on islands means that even rather dissimilar species may exclude each other. Even species in separate genera, such as the flycatchers of the Tonga islands and the hummingbirds of the West Indies, replace each other geographically, usually on separate islands, but at times on different parts of the same island (pp. 225, 228).

The frequency of geographical separation in land birds on archipelagoes might be helped by the readiness with which new species are formed there, but this is certainly not the main factor involved.

Separation by geographical range is uncommon in congeneric species in other types of birds, but examples are found in most groups. Some of the genera concerned are specialized. Others, such as the continental white-eyes *Zosterops,* or the African and North American tits *Parus,* have varied diets, and they are probably restricted in their ecological niches by the presence of birds in other families rather than by their own feeding habits as such, for species in both these genera frequently coexist elsewhere, white-eyes on islands and tits in the palaearctic.

Adaptations to range differences

There is circumstantial evidence that some species which replace each other geographically have adaptations to their respective ranges. As mentioned earlier (pp. 89–91), the Brambling *Fringilla montifringilla* replaces the Chaffinch *F.coelebs* in Fenno-Scandia; and it is somewhat larger, with a proportionately smaller beak. These trends conform with Bergmann's rule of increasing body-size, and Allen's rule of proportionately smaller external parts, in warmblooded animals in the colder parts of their range, and though these rules were formulated for subspecies of the same species, they might well apply to closely related species. Precisely what these trends are adaptations for is not clear, but they are so widespread that they are generally accepted to be adaptive. Also, C. M. Perrins (pers. comm.) commented that, in cold weather in England, the Brambling seems much more at ease than the Chaffinch, its feathers being less fluffed out, for instance. Further, the zone of overlap between the two species in Fenno-Scandia has been shifting north with the recent warmer summers, providing further indirect evidence that their boundary is determined by their respective adaptations to cold, and in the zone of overlap, the Brambling breeds higher up and the Chaffinch in the valleys.

Similarly Brünnich's Guillemot *Uria lomvia* replaces the Common Guillemot *U.aalge* to the north in the Atlantic (p. 122) and is likewise larger, with a proportionately smaller beak. Thus the southernmost and northernmost races respectively of *U.aalge* have mean winglengths of 198 and 212 mm, *U.lomvia* of 205–226 mm, and the exposed culmen is 24 per cent of the winglength in an intermediate race of *U.aalge*, 21 per cent in the northernmost and 16 per cent in *U.lomvia* (Salomonsen 1932). Hence *U.lomvia* carries further the trends found in different subspecies of *U.aalge*.

The northern Whimbrel *Numenius phaeopus* and southern Curlew *N.arquata* are another pair which replace each other geographically in Fenno-Scandia (p. 122), and as in the Brambling and Chaffinch, their zone of overlap is currently shifting north, and in it the Whimbrel occurs at the higher altitudes. The Whimbrel also has the smaller beak. Unexpectedly, it is the smaller in overall size, which Salomonsen (1955, citing Hemmingsen 1951) attributed to its living in warmer winter quarters than the Curlew. He gave further examples, notably in the Ringed Plover *Charadrius hiaticula*, in which northern breeding races are smaller than more southern ones, but spend the winter in warmer regions, to which their sign is presumably adapted.

While such general, and presumably adaptive, trends occur in continental birds, no evidence yet exists to show whether congeneric species which replace each other on islands are differentially adapted. But though this may seem hard to credit, some of the birds in question have been recorded on other islands but do not persist there, so they are presumably excluded through competition.

Apart from a shift north in the zone of overlap between species which replace each other in Fenno-Scandia, one species has rarely been observed in the process of replacing another in natural habitats, probably because the latter are stable. But in the man-made habitat of suburban Singapore, one species has frequently replaced another, for instance in munias and kingfishers (p. 164). What has been observed, strictly, is the appearance of one species and the disappearance of a similar species, but it would be straining coincidence too far to suggest that, in every such case, the two events were unconnected.

Other factors limiting range

Except in frugivorous birds and the land birds of islands, the range of a species is rarely limited by that of a similar species. A few species are limited to areas where their preferred food occurs, but these are food specialists such as crossbills *Loxia* (pp. 86–87), and most birds have such varied diets that this factor can be ruled out. Moreover even specialized feeders may have different diets in different areas. Thus the Nutcracker *Nucifraga caryocatactes* depends on the seeds of Arolla pine *Pinus cembra* in east and central Europe, where its range coincides with that of this tree, but the Swedish race is adapted to hazel nuts *Corylus avellana*. Many species seem limited in range by the range of their preferred habitat, desert, forest and so on, and often by a particular type of this habitat. Thus one African tit is restricted to those parts of Africa with brachystegia woodland and another to acacia savanna (pp. 45–7), but though it is true to say that they are limited by the range of their habitats, their habitat restriction is, in these and various other cases, due to competitive exclusion.

The range of certain other species seems limited by climate. In Europe, in addition to the species-pairs already mentioned, various southern species have extended further north, and various northern species have retreated, with the current warmer summers in the north; others, including the Redbacked Shrike *Lanius collurio*, have retreated from western Europe with the current wetter and cooler summers there (summarized in Lack 1954). Again, the northern limit of the Dartford Warbler *Sylvia undata* in Europe is probably set by winter cold, since many of them die in severe weather. But in at least some of such cases, climate acts indirectly. For instance, the Redbacked Shrike depends on large insects, which are scarce where the summers are wet and cool. Warning should be added that, with ingenuity, one can find some combination of temperatures and other climatic factors to fit almost any line on a map, but it does not necessarily follow, where the line in question demarcates a species' range, that this range is determined by the climatic factors in question. Birds are warm-blooded, and it may be doubted whether climatic factors ever limit the range directly.

The limits of the range of many other species do not coincide with those of their particular foods or habitats, or with the boundaries of a particular species, or with obvious climatic factors. For instance, the European warblers typical of reeds *Phragmites*, or the passerines typical of pine forests, do not occur everywhere in Europe where these types of vegetation are found. Again, the range of a species often extends to different climatic limits in different areas. Hence an extremely important factor limiting range has evidently been left out of account, and this, I suggest, is the complex of other bird species present, acting in conjunction with the available habitats and foods.

Such exclusion may be of two main types. First, a habitat may be rich but a particular species may not be able to survive there because the many other species present prevent it from finding a suitable niche. It was suggested that this is why, for instance, so few species of tits coexist in tropical Africa, and why so few migratory palaearctic forest species occur in evergreen forest in Africa in winter. In such cases, there is no reason to think that each species is excluded by a particular ecological counterpart; probably, the complex of species present divide up the feeding stations in such a way that a potential newcomer would have to displace several of them from part of their niches to obtain enough food. Again, it is hard to believe that various warblers common in northern and central Europe would be absent from the Mediterranean region if the warblers characteristic of the latter had been absent, and in this case, likewise, there is no reason to think that each is excluded by a particular counterpart.

Secondly a habitat may be poorer in one area than another. In Austria, for instance, there are six species of acrocephaline warblers (p. 100), but westward the number in marshy habitats becomes progressively smaller, and is eventually reduced to one, the Sedge Warbler *A.schoeno-baenus*, which in places occupies, in addition to its own habitat, those occupied elsewhere by the Marsh or Reed Warblers *A.palustris* or *A.scirpaceus* (p. 291). The absence of some of these species from western Europe is not due to difficulties of dispersal, for they occur on migration, and some of them have bred occasionally. I suggest, instead, that if the diversity of ecological resources in marshland is reduced, fewer species with broader habitats or feeding stations tend to exclude a greater number of specialists. On this view the western limit of the ranges of the warblers concerned is set by competitive exclusion by species with which they can coexist elsewhere, acting in conjunction with a less rich environment. However, the possibility that marshland is more impoverished further west in Europe has not been tested. A similar view will explain the smaller numbers of species on islands than the mainland, and on smaller or more distant than on larger or nearer islands (pp. 234–9, see also Lack 1969).

These views are speculative, but the restricted ranges of many species cannot easily be attributed to the effects of habitat, food, climate or competitive exclusion by one similar species, and though it is dangerous to argue from ignorance, it is hard to believe that, if many of the existing species were absent, others would not have wider ranges. While this explanation has been introduced here to explain limitations in range, it also applies to limitations in habitat and feeding (a point which will not be repeated in the later sections of this chapter on habitat and feeding).

Migrants with two homes

Most species of birds reside in one area throughout the year, but fully migratory species occupy two, their breeding range, presumed to be where they on average raise most young, and their winter range, where they presumably have the greatest chance of individual survival. The potential areas involved need not be similar, hence a small wintering area might in itself restrict the summer range. As shown in fig. 55, Kirtland's Warbler *Dendroica kirtlandii*, winters solely in the Bahama Islands, and probably this is responsible for its restricted summer range, which is much smaller than the area occupied by its preferred breeding habitat, secondary growth of the Jack Pine *Pinus banksiana* (Van Tyne 1951). The small breeding ranges of Bachman's Warbler *Vermivora bachmani* and the Bristle-thighed Curlew *Numenius tahitiensis* may likewise be due to the few individuals which can survive in their restricted winter quarters (Amadon 1953).

While this situation is clearest in species with small wintering areas, the size of the population that can survive the winter probably influences the size of the breeding area in many, if not all, other migrants. For instance, the Waxwing *Bombycilla garrulus* fluctuates greatly in numbers, probably due to fluctuations in the berry crops on which it depends in winter, and as shown in fig. 56, it breeds further west and south than usual in Fenno-Scandia only in years when it is abundant (Cornwallis 1961). Presumably the latter areas are less suitable for breeding except when the permanent range is densely occupied.

As shown in chapter 6, many more congeneric passerine migrants are separated by range in winter than summer. This presumably means that it is harder for them to coexist

249

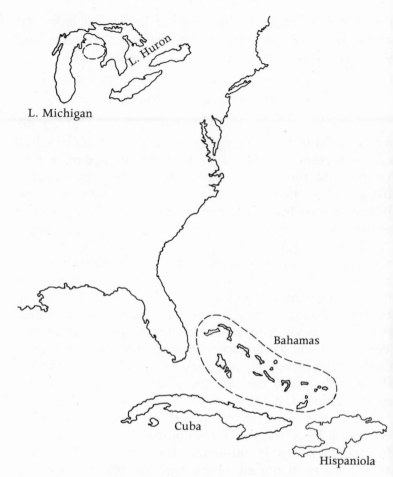

Fig 55. Summer and winter range of Kirtland's Warbler *Dendroica kirtlandii* (from Van Tyne 1951).

Summer range is restricted to a small area between Lakes Huron and Michigan, winter range to the Bahamas.

in the same area in winter than summer, perhaps because the ecological resources available to them in Africa are more restricted, either due to the nature of the habitat, which is often arid, or to the presence of indigenous African species. The difference is similar to that between the tits of Europe and Africa at all times of the year (p. 49).

250

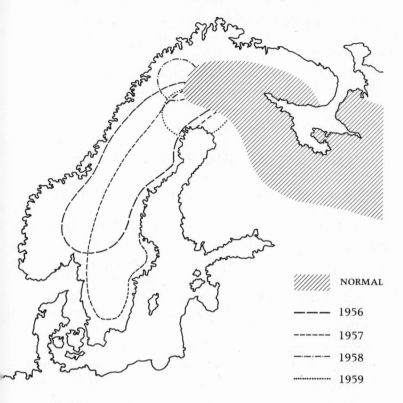

Fig 56. Westward limit of breeding range of Waxwing *Bomby-cilla garrulus* in years of differing abundance (from Cornwallis 1961).

Proximate factors limiting range

The main ultimate factor limiting the breeding range is the failure of those individuals which try to breed outside it. For instance, 14 species have bred occasionally in Britain between 1940 and 1966 but without persisting (Parslow 1966–67), and a similar list could be compiled for any other well-watched area. Likewise migratory species occasionally spend the winter outside the normal range of their species. Given that each species is adapted to its range, those individuals which have evolved behaviour to keep within it will, on average, leave more offspring than those which settle outside it. Moreover those with such behaviour have a greater chance of evolving adaptations to local conditions, which will further increase the advantage to them of keeping within the normal range.

To settle in the range of their species, resident birds need only to stay near where they are born. But migrants have to find their winter range and to return to their summer range. The young of a few species, including cranes, swans, geese, some terns and gulls, migrate with their parents or other adults, and so learn the wintering grounds partly by tradition. But the young of most species migrate unguided; nevertheless most of them reach the specific wintering grounds and return to the specific breeding grounds. The means of navigation come outside the scope of this book, except to note that such elaborate behaviour would not have been evolved unless each species survives better in its normal winter and summer ranges than elsewhere.

For successful return in spring, the young of most migrant species probably have to learn the whereabouts of the breeding area before they leave it in their first autumn. To test when they learn it, Löhrl (1959) took young Collared Flycatchers *M.(F.)albicollis* from the nest, reared them by hand, and then took them 90 km south of where they hatched, and released them in an area where this species does not normally breed. The proportion of males, about one-fifth, which returned to the area where they hatched, was extremely similar in the controls and in those released in the experimental area either 10–12 days or 4–5 weeks after fledging, but none released six weeks after fledging returned there. Hence the young evidently learn the area of their birth within just over a month of leaving the nest. Similarly, young Australian muttonbirds *Puffinus tenuirostris* learn the whereabouts of their natal area some time before they leave the nest burrow for the last time (Serventy 1967).

B. HABITAT

A difference in habitat is much the commonest means of ecological isolation between congeneric species in continental passerine birds and is widespread in most other groups. That is why it was formerly argued that new bird species might arise through isolation by habitat, a view now abandoned. In European birds there is every gradation between species separated by their breeding ranges, like the two starlings *Sturnus vulgaris* and *S.unicolor* (p. 116), *via* species with mainly separate breeding ranges and a small area of overlap where they differ in habitat, like the two nightingales *Luscinia* (p. 107), or with a larger area of overlap, again with a difference in habitat, like the two treecreepers *Certhia* (p. 113), to species which are separated mainly by habitat but have partly separate ranges. The existence of such a series suggests that a difference in habitat between two species may take a long time to spread from its area of origin. Alternatively, overlap with a difference in habitat might be possible only in an area of greater ecological diversity than elsewhere.

Adaptations to habitat

Segregation by habitat, or by range, is not normally brought about by fighting. Some species defend mutually exclusive territories where their ranges or habitats adjoin, but the function of this behaviour is to secure for each male a territory without other males of either species, and such border zones are relatively so narrow that the species in question are clearly kept apart by selecting their respective habitats or ranges. The only known exception, not satisfactorily explained, is that the larger and commoner of two American blackbirds excludes the smaller and more widespread species throughout their common range, in a way found otherwise only in competition for sites by hole-nesting species (pp. 128-9).

The evidence that congeneric species are adapted to their differences in habitat depends mainly on the existence of general trends. For instance, the species of tits *Parus* and nuthatches *Sitta* which live in coniferous forest are normally whiter below and greyer (less blue) above, with proportionately longer and thinner beaks, than their congeners in broadleaved forest; moreover parallel beak differences are found in different subspecies of the same species of tit in the two habitats (pp. 34–35, 63). Again, the shorter hind claw of the Tree Pipit *Anthus trivialis* than the Meadow Pipit *A.pratensis* accords with a general difference in this respect between arboreal and terrestrial species, as shown in fig. 57. The difference is smaller between the Wood and Skylarks *Alauda arborea* and *A.arvensis*, but the Wood Lark, though it lives among trees, perches on them much less often than the Tree Pipit. Above the level of congeneric species, adaptations to habitat, for instance the webbed toes of swimming birds or the long legs of wading birds, are too obvious to need discussion here.

Fig 57. Foot of Meadow Pipit *A.pratensis* (left) and Tree Pipit *A.trivialis* (right). By Robert Gillmor.

The competition resulting in separate habitats presumably took place long ago in natural environments, and the main evidence for its existence there is that species occasionally breed outside their normal habitat but do not persist. Replacement has, however, been observed in habitats modified by man. For instance, suburban Singapore was colonized by an indigenous coastal tailorbird

Orthotmus, but later a congeneric species found in Indian gardens spread south and has replaced it in the suburbs, though not on the coast, so the two species are now separated by habitat in Singapore (p. 164). Again, the Rock Sparrow *Petronia petronia* formerly occurred in both fields and towns of the central and western Canary Islands, but the Spanish Sparrow *Passer hispaniolensis* established itself in the town on Gran Canaria in the mid-19th century and later on the other islands, after which the Rock Sparrow disappeared from the towns, but not the countryside, so the two species are now separated by habitat in the archipelago (p. 10). The same has happened recently in Madeira, where formerly there was only the Rock Sparrow, and the Spanish Sparrow has similarly displaced another sparrow, *Passer iagoensis,* from towns but not the countryside of the Cape Verde Islands, where the two now live in adjacent habitats (Bannerman 1963, 1965, 1968, see also comment in Lack 1969).

Variability in habitat

The habitat of a species is not necessarily fixed. This is shown by the fact that some species have different habitats in different parts of their range, including the Ring Ousel *Turdus torquatus* in pine forest in central Europe and above the treeline in northwestern Europe (p. 306), the Rock-Alpine Pipit *Anthus spinoletta* in the alpine zone in central Europe and on the seashore in northwestern Europe (p. 287), and the thrush *Turdus poliocephalus* in montane forest in the Solomons and in the lowlands on Rennell (Mayr 1931). Again, while three-fifths of the birds of the isolated mountain blocks of Pantapui in Venezuela are derived from other mountains, the rest are most nearly related to species in the adjacent tropical lowlands, which have spread into the upper tropical, and even the lower subtropical, zone on Pantapui, some sufficiently long ago to have become endemic genera, others probably so recently that they are not even endemic subspecies (Mayr 1965, Mayr and Phelps 1967).

In natural habitats, such changes are probably slow, but some species have quickly adapted themselves to habitats modified by man, where buildings are used by cliff-nesters, fields by steppe birds, hedgerows by scrub birds or those of the forest edge, the suburbs of Ibadan by savanna species and those of Singapore by birds of coastal scrub. Some species have moved into habitats which, to human eyes, look unlike their natural ones, such as the Lesser Redpoll *Carduelis flammea* (pp. 74–75) and Wood Pigeon *Columba palumbus* (p. 312) in rural England. Various species, perhaps most, first colonized the man-modified habitats of Europe in winter, probably in search of food, and have modified their breeding behaviour only later (e.g. the Bullfinch *P.pyrrhula* p. 75, and the Blackbird *Turdus merula* as summarized in Lack 1966). Changes are still in progress, not only in new habitats like Singapore, but even in long-established European towns, which the Black Redstart *Phoenicurus ochruros* and Serin *S.serinus* colonized only in the 19th century and the Collared Dove *Streptopelia deca-octa* in the 20th century.

Just as some species, when abundant, extend their breeding range, so others may at such times extend their habitats. For instance, in the years when their arrival coincides with warm southerly winds, various southern migrants breed in Sweden in unusually large numbers, including the Thrush-Nightingale *L.luscinia*, Redbacked Shrike *Lanius collurio*, Wood Warbler *Phylloscopus sibilatrix* and Garganey *Anas querquedula*, and they then spread into parts of their habitat not occupied in other years and which look atypical for them to an ornithologist (Svärdson 1949). This is presumably because the individuals concerned, mainly in their first breeding year, avoid the most suitable parts of the habitat when these are crowded. Other instances in Fenno-Scandia, linked with annual fluctuations in numbers in the Great Spotted Woodpecker *Dendro-*

copos major, and with a long-term increase in the Lapwing *V.vanellus,* were cited by Hildén (1965).

Similarly, in Holland, three species of tits and the Chaffinch settled to breed at a higher density, and fluctuated less in numbers, in mixed woods than pine woods, which Kluijver and Tinbergen (1953) and Glas (1960) ascribed to their filling up the richer woods first, the remaining individuals being excluded and settling in the poorer. While, however, this explanation seems correct in broad outline, the populations in the richer woods fluctuated more than might have been expected on this view and, in particular, did not rise to a more or less fixed number before the poorer woods were colonized. In the same connection, when territory-owning Great Tits were experimentally removed from a wood near Oxford, their place was taken by individuals coming in from adjacent hedgerows, where they breed less successfully; before the experiment, the hedgerow birds were evidently excluded from the wood by the individuals in occupation (J. Krebs pers. comm.).

There are supporting observations on other birds. In the American Redstart *Setophaga ruticilla,* sub-adult males may occupy less favourable habitats than mature males (Ficken 1967). Again, when Orians (1961) experimentally removed breeding male Red-winged Blackbirds *Agelaius phoeniceus,* some of them were replaced by sub-adults, which had presumably been excluded from the favourable breeding areas until then. After a big reduction in their numbers after an extremely cold winter in England, Wrens *T.troglodytes* bred chiefly in woods and beside streams, and only later, as their numbers rose, did they spread back to gardens and orchards, and lastly to hedgerows (Williamson 1969). Finally, when the breeding population of the White Stork *C.ciconia* declined in south Germany, perhaps due to winter losses, the decrease seemed to be steeper in poorer upland than richer lowland habitats (Zink 1963, extended in Lack 1966), but as in the tits studied by Kluijver and

Tinbergen, there was not a precise quantitative relationship between the numbers in the more and less favourable habitats.

Nesting sites

Many species breed within their feeding grounds and have simple nesting requirements which are met without competition. Competition is observed mainly in hole-nesting and cliff-nesting species, such as hirundines, swifts, tits, and various seabirds, which breed in relatively safe sites, often away from their feeding grounds. In some of these groups, each species occupies a different type of site from every other in the same area, presumably through competitive exclusion, while in tits there is partial separation (pp. 26–27). Many of the species concerned fight for holes, the larger usually ousting the smaller, but the latter being able to use smaller holes than the larger. Hirundines are separated by their nest architecture, which is adapted to a different type of site in each European species, as shown in fig. 19 (p. 95). A parallel to the occupation of more or less suitable habitats, just discussed, was observed for nesting sites during a population decline of the Peregrine Falcon *Falco peregrinus* due to toxic chemicals, when safer sites were occupied for more years than less safe ones, presumably because some individuals moved to the better sites after their previous owners had died (Rice 1969).

Habitat selection

Hand-raised ducklings enter the water spontaneously, and whereas young Shovelers *Anas clypeata* and Tufted Ducks *Aythya fuligula* then seek reeds, young Eiders *Somateria mollissima* avoid them and seek rocks, in conformity with the habitats of their respective species (Fabricius 1951). Hence innate factors influence habitat selection. Again, Klopfer (1963) took young Chipping Sparrows *Spizella passerina* from the nest before they had seen the type of foliage which this species frequents, which in southeastern USA is pine (Burleigh 1958). Ten such hand-raised young, later given the choice of sitting among pine or oak branches in an aviary, spent nearly three-quarters of their time in pine, and 9 out of the 10 spent more than half their time in it, so they presumably have a hereditary preference for it. However, when other hand-raised young were initially provided with oak branches and foliage in the aviary, and only later were given the choice of pine or oak, they spent just under half their time in pine and only half the individuals spent more than half their time there. Hence their hereditary preference was modified, though not eliminated, by the type of vegetation where they lived soon after fledging.

Again, Reed Buntings *Emberiza schoeniclus* nest at a higher density in marshes than on dry ground near water where predation is higher. As shown in Table 34, the proportion of young which changed their breeding habitat was higher in those hatched on dry than marshy ground, presumably owing to a hereditary preference for marshy

Table 34. Nesting habitat in relation to natal habitat in Reed Bunting *Emberiza schoeniclus*

Natal habitat	No. of ringed young which returned	No. which bred in marshes	No. which bred in drier ground	Percentage which changed from natal habitat
Marshes				
(a) males	12	8	4	33 } 26
(b) females	7	6	1	14 }
Dryer ground				
(a) males	24	12	12	50 } 46
(b) females	22	9	13	41 }

Notes: from Bell (1968). On 5·8 ha of marsh there were 4·3 pairs per ha and on 18·6 ha of drier ground 2·5 pairs per ha, so marshland was preferred, as on other areas.

sites, but the natal habitat had some influence on the habitat chosen, especially in the females, which chose the nesting site (Bell 1968).

Any innate factor in habitat selection is likely to relate to simple features. This might help to explain why the differences in habitat between congeneric species are usually simple, such as the presence or absence of scattered trees for larks and pipits, or the presence or absence of standing water for acrocephaline warblers; and many other species are separated in broadleaved and coniferous trees respectively.

Further probable evidence for the influence of the natal site on habitat selection is found in the Mistle Thrush *Turdus viscivorus* and Curlew *Numenius arquata* in Germany. Once these species had started breeding in, for them, the new habitat of cultivated land, they expanded rapidly in numbers, presumably because their young returned to breed in cultivated land, though part of this effect might have been due to a tendency to return to the natal area, as distinct from the natal habitat. The same might have applied in an experiment in which the young of a coastal race of the Herring Gull *Larus argentatus* were transferred inland and subsequently bred there (Hildén 1965, citing work by Peitzmeier and Drost respectively; much else in Hildén's comprehensive review is relevant to this section).

In flock-feeding birds, habitat selection by the juveniles may also be influenced at a later stage by their tendency to join existing flocks, which will normally be in the habitat of the species. Similarly in colonial birds, habitat selection is influenced by the tendency for birds seeking breeding sites to join established colonies, probably because the presence of a colony is a good indication that a site is safe (Lack 1966, 1968). However, colonial birds breeding for the first time sometimes start new colonies, usually in sites typical for their species, so the presence of other individuals is not their only guide. Finally, as noted earlier in this section, habitat selection is influenced by the density of birds already present, crowded but otherwise suitable areas being avoided.

C. FEEDING

More congeneric species of birds are segregated by feeding than in any other way, but the proportion is greater the broader the genera used (pp. 13–16). In particular, the paucity of such species in continental passerine birds is largely due to their genera differing less morphologically than those of most other birds, and if related genera are combined, the proportion of species separated by feeding is probably as high in passerines as in other groups. It would appear that subsequent to the differentiation of species separated by feeding, more continental passerine than other birds have evolved into further species separated by habitat. One possible reason is that the nature of the habitat, for instance whether the trees have broad leaves or needles, make a bigger difference to the way of life of a small than a large bird, and most passerine birds are small, most other land birds large. Perhaps, also, many small non-passerine land birds became extinct following the adaptive radiation of the passerines.

It is clear from previous chapters that many congeneric species segregated by feeding differ markedly from each other in size of body or beak, whereas most of those separated by range or habitat do not do so, but there are exceptions both ways, and size is not a reliable guide to the means of ecological isolation. Indeed, a few species are separated by habitat in one part of their range and by feeding in another, without a noticeable difference in morphology, for instance the Chiffchaff and Willow Warbler *Phylloscopus collybita* and *P. trochilus,* and also the Song Thrush and Blackbird *Turdus philomelos* and *T. merula,* in Scandinavia and England respectively (pp. 103–4, 114).

Coexisting passerine congeners differ more from each other in size on islands than continents (Lack 1944,

Schoener 1965). On islands fairly near the mainland, like the West Indies, this is largely because the few congeners established there are those which, on the mainland, differ most in size from each other (Grant 1969). On remote islands, the difference is largely due to local evolution, which presumably took place before other mainland genera arrived, notably in the Hawaiian sicklebills, Galapagos finches and certain white-eyes. Bigger size-differences are presumably needed on islands than the mainland because each island species takes a greater variety of foods.

In a few island species, the sexes differ greatly in beak-size and feeding (p. 239), notably in the extinct Huia *Heteralocha acutirostris,* shown in fig. 58. This presumably enables the species concerned to take a wider range of foods than would otherwise be available. On the mainland,

Fig 58. Beak difference between male and female Huia *Heteralocha acutirostris* (from Lack 1947).

Male excavates with short beak, female probes with long beak. Half natural size.

similar differences are found in a few waders, such as the Longbilled Dowitcher *Limnodromus scolopaceus* (Pitelka 1950, who was the first to suggest an ecological function), in pelicans and the larger grebes (Selander 1966), and in pursuing raptors, in the last of which there are corresponding differences in size of prey (pp. 133–4). But similar size differences in Grackles *Quiscalus,* the Capercaillie *Tetrao urogallus* and the Great Bustard *Otis tarda,* among others, have probably been evolved through sexual selection with promiscuous mating; one need not suppose that all size differences between the sexes have the same cause.

Adaptations for feeding

Hutchinson (1959), the first to discuss quantitative relationships between two coexisting congeneric bird species, pointed out that the one usually has a beak between 1·2 and 1·4 times as long as the other. But the figures then available were unrepresentative, and in particular included many from islands, where such differences tend to be unusually great, as already noted. Even on islands, however, the difference in beak-length between coexisting congeners is not always large (*pace* Grant 1968). In white-eyes for instance, it varies between a negligible amount and 1·7 times as long (pp. 218–9).

This is because beak-length is not the sole adaptation to a difference in feeding. In noddy terns *Anous* for instance, a longer (but thicker) beak is adapted for smaller prey than a shorter (but thicker) beak (p. 171), and the same applies to the Coal Tit *P.ater* compared with the Blue Tit *P.caeruleus* (pp. 24–25). Various other species which are segregated by feeding do not differ appreciably in beak. Some of them differ in overall size. In arboreal birds, this is often associated with a difference in the height in the trees at which, and the width of the branches on which, they seek for food (e.g. pp. 24, 187–8). In raptors it is often associated with a difference in the size of prey (p. 133).

Other species differ in the proportionate length of the tarsus or wing, which is often associated with a difference in hunting methods, for instance in finches (p. 83) and tanagers (p. 156). Differences in hunting methods and feeding stations are at least as important as differences in size of prey in separating congeners.

Although each species is adapted to its main foods, it may at times take a great variety, and may differ markedly in diet at different seasons of the year, and sometimes in different places. The Nutcracker has already been mentioned in the last connection (p. 246). Again, the Tawny Owl *Strix aluco* preys chiefly on mice in woodland but on small birds in London (Beven 1965), but the difference is not absolute, for woodland owls eat a few birds and London ones a few mice. Local differences in food are also found in various finches (Newton 1967a). An unusual degree of individual variation, at times in the same locality, occurs in the Oystercatcher *Haematopus ostralagus,* in which different birds specialize on limpets, cockles, mussels and earthworms respectively, each of which requires a different hunting technique (Norton-Griffiths 1967, 1968, Safriel 1967).

Overlap in diet

Many species segregated by feeding will take the same type of food when it is temporarily superabundant, for instance flying termites in the tropics or fruits at high latitudes. But the tropical birds which eat both fruit and insects have a succession of abundant fruits during at least most of the year, on which many different species prey; they are segregated mainly by their insect foods (pp. 155-7). Lemmings, which when abundant provide the main food of various different species of raptors, owls, skuas, corvids, stoats and foxes, may continue abundant for more than one season, but when they are sparse or absent, each main predator turns to different prey, helped by the fact that those in any one area are not closely related (Lack 1946).

Even when food is sparse, two coexisting congeners sometimes take the same type of prey, and the extent to which they can overlap in diet and yet coexist is uncertain (cf. Ashmole 1968b). This raises the question of whether any coexisting bird species take the same main prey. But most apparent cases of big overlap, in lemming predators, tropical fruit-eaters, northern and tropical seabirds, have now been resolved by detailed study, so it is reasonably certain that there are none, confirming the theoretical argument against coexistence under such circumstances. The 'evolutionary strategies' determining whether two species in the same area are more likely to be segregated by habitat or feeding, and if by feeding, the extent of any size differences between them, have been considered mathematically by MacArthur (1965, also with Levins 1964, 1967, with Pianka 1966 and with Wilson 1967), and by Schoener (1965, 1968, 1969).

Acquirement of segregation by feeding

How do coexisting congeners acquire differences in feeding sufficient for segregation? Most bird species have such varied diets that innate factors could play, at most, a small part in prey-recognition. But in some species, the possibility that the young learn from their parents is also ruled out, for instance megapode chicks are independent on emerging from the nest mounds, and fledgling swifts when they leave the nest. Since these very different types of birds acquire their feeding habits without parental guidance, many other species might do the same. At the other extreme, young Oystercatchers depend on their parents for several months, and each comes to specialize on the same type of prey as its parents, presumably through conditioning. However, it learns the required feeding technique by trial and error, unless it is a mussel-feeder. Those Oystercatchers which depend on mussels open the shell by either hammering or stabbing, and the young use the same means as their parents, presumably through watching them (Norton-Griffiths 1967). The young of other species might also learn from their parents, especially those with a long period of dependence on them.

Both nidifugous chicks and nidicolous fledglings try at first to eat a variety of objects, and though this behaviour evidently has an innate basis, for instance with respect to the size of objects pecked by chicks of different species, in the main they learn by experience which items are edible and which inedible (Hinde 1959). Birds also learn by experience to avoid brightly coloured poisonous insects. In addition, however, experiments have shown that both chicks and young passerines may learn where to find food, and also what to avoid, through observing others of their species (Klopfer 1957, 1959, Turner 1964, Dawson and Foss 1965, Alcock 1969). This holds not only while the young are with their parents, but also much later in flock-feeding species, in which inexperienced birds join others of their kind seen feeding.

The evidence assembled by Hinde (1959) suggests, however, that trial and error is the most important factor involved in segregation by feeding, and that each species learns through its own experience the foods that it can obtain most economically, which depends on its morphological characters of beak, wing, legs, musculature, and innate motor patterns. Young Chaffinches *F.coelebs,* for instance, learn by trial and experience the size and hardness of the seeds that they can tackle (Kear 1962). So do young Linnets *C.cannabina* and Greenfinches *C.chloris,* the Linnet coming to prefer smaller and the Greenfinch larger seeds, hence their segregation by feeding is acquired through experience (Newton 1967a). Probably the same applies to specific differences in feeding habits. For instance, the small Blue Tit *P.caeruleus* can hang upside down and hammer food, but the larger Great Tit *P.major* cannot feed adequately in this position (Hinde 1959, interpreting J. Gibb). In these two species the body-weight, length of tarsus and strength of leg muscles help to determine that the Blue Tit feeds more economically on smaller, the Great Tit on larger, twigs. Hinde therefore suggested that the morphological differences between coexisting congeneric species are not merely adapted to their respective diets and feeding methods, but largely determine them; probably this is why the differences in question are usually great. In a few species segregated by feeding, however, the morphological differences are small, and these would repay further study. The most economic feeding methods also depend on the abundance of different types of prey, and the capacity of birds to modify their feeding methods in relation to the available prey is responsible for the seasonal and local variations in diet found in various species, mentioned earlier.

In mixed flocks, a larger species may sometimes displace a smaller from a food item, and if food is sparse, the smaller may thereafter seek elsewhere, or for smaller items than those which attract the larger. But in general food-fighting seems quite unimportant in determining specific differences in food or feeding stations. An unusual example is found in the South American woodcreeper *Dendrocincla fuliginosa,* which commonly hunts low over columns of army ants for the insects which they disturb, provided that there are no ant-thrushes *Phaenostictus*. But the latter, when present, drive it away and it is forced to feed higher up where insects are sparser; hence it feeds low over army ants much more often in some than other parts of its range (Willis 1966b).

D. SOME CONSEQUENCES

The origin of species and adaptive radiation

As mentioned in chapter 1, ecological isolation does not in itself lead to the formation of new bird species, but is evolved later, being essential if new species, formed in geographical isolation, are to persist where they later meet. It was also long supposed that the origin of new species had nothing to do with the major trends of evolution, and that the differences between species are trivial and of no adaptive significance. In fact, the genetic discontinuities which separate species are the means by which they can become adapted to differences in their respective environments; and the differences between related species, by means of which they coexist, are an essential step in an adaptive radiation. In Darwin's finches (Lack 1947), I pointed out that every stage is present between incipient subspecies and an adaptive radiation; and their evolution is not peculiar, but similar to that producing the major adaptive radiations on the continents, where the early stages are hard to trace because so many forms have become extinct. Competitive exclusion is the precursor of adaptive radiation.

Species diversity

The theory of species diversity is due largely to MacArthur and his co-workers (see all references under his name, including joint authorship, in the bibliography). The number of bird species in an area depends essentially on the extent to which they can divide the available habitats and food resources. In particular, the greater number of species in the tropics than at high latitudes is associated with a greater diversity of habitats, a much greater tendency for a species to be restricted to only one type of habitat, a greater number of vegetational layers, a greater number of foods of different types available throughout the year, and hence a greater number of food specialists which can be separated by feeding (pp. 142–3).

Conversely, on islands, whether tropical or not, the fewer species present tend to have broader habitats and feeding stations than on the nearest mainland, probably because, with a reduced diversity of ecological resources, fewer species with broader requirements tend to displace more specialists (Lack 1969). Depauperate avifaunas are by no means restricted to islands. To mention only three

examples on continents, the number of species of acrocephaline warblers is greater in Austria than in similar
marshland habitats further west (p. 100), some montane
forests in eastern Africa support more species than others
(Moreau 1966), and more bird species occur in the alpine
zone of the Caucasus than on mountains further west
(Gaston 1968). Admittedly marshes, montane forests and
alpine zones are discontinuous habitats, so might be considered 'terrestrial islands', but there is no reason to think
that the smaller numbers of bird species on some of them
than others are due to difficulties of dispersal. Islands, I
suggest represent merely a striking example of a general
phenomenon, that bird species diversity is correlated with
ecological diversity.

Avifaunas

Sclater (1858) divided the world into six faunal regions. A
few birds, such as the Barn Owl *Tyto alba*, occur in all six,
but many in only one, and though some bird families are
widespread, others are restricted to one. In the latter case,
there are often unrelated ecological counterparts in another
region. Thus on different continents the seed-eating finch-
like birds, or the nectar-feeding birds, tend to be in
different families, which resemble each other through
convergent evolution.

It might have been thought that similar habitats in
different regions would normally be occupied by ecological counterparts, but this is only partly true. For
instance, the woods of mid-eastern USA and mid-Europe
respectively have a few bird species in common, and some
others are related counterparts, but others are so different
that it seems clear that the potential feeding stations for
forest birds are differently divided in the two regions.
Moreover differences of this type are not restricted to the
great biogeographical regions. Within Africa, for instance,
nearly all the species of lowland and highland evergreen

forest respectively are different (p. 140), and so are those on the adjoining Guinea islands of Principe and Sao Tomé (Snow 1950), and while some are congeners, others are in different genera or families, and are so different that the feeding stations must be differently divided.

Since birds sometimes occur far beyond their normal range, it may be wondered how the species in the main faunal regions, or for that matter on separate islands, come to be different. The main reason, I suggest, is that advanced earlier in this chapter, that birds are often excluded by a complex of other species present. Changes occur, as recorded on Pantapui (p. 255), and Mayr (1964) has also studied the mixing of North and South American birds which produced the present neotropical avifauna. But once a group of species has evolved together in a particular area and habitat (and such evolution might be much more rapid than formerly supposed), it becomes increasingly difficult for further species to find a place. Hence changes in natural habitats tend to be slow competitive exclusion being critical.

Other animals

Ecological isolation is a logical consequence of the potential competition between differently adapted species, so is presumably found in all animals, though it might well take different forms in other animals than birds. Bird numbers are commonly regulated by food, so the competition between species is largely for food, but the situation might well be different in animals limited in numbers by predators or parasites. It would, however, take me far beyond the limits of one book to enlarge on the problems of (i) ecological isolation in other animals, (ii) the origin of species, (iii) adaptive radiation, (iv) species diversity, and (v) the composition of faunas. My aim in this final section is to indicate that ecological isolation is not just a problem for the ornithological specialist, but occupies a central position with respect to principles of animal evolution and ecology.

Appendixes

1. Tits *Parus* on Formosa and the Philippines

Species	Subgenus	Habitat
P.ater (Coal)	*Periparus*	Subalpine fir forest, sparsely in pines at lower altitudes
P.holsti (Formosa Yellow)	*Machlolophus*	Very local in mountains
P.monticolus (Green-backed)	*Major*	Temperate broadleaved forest
P.varius (Varied)	*Sittiparus*	Below the montane forest (? broadleaved)

P.amabilis	*Pardaliparus*	Replaces next on eastern islands of Palawan and Balabac
P.elegans (Elegant)	*Pardaliparus*	
P.semilarvatus (White-fronted)	*Sittiparus*	Feeds in tops of trees

Notes

Based on Hachisuka and Udagwa (1951) and Delacour and Mayr (1946) respectively. All four Formosan species are in different subgenera, which in itself suggests that they differ in feeding stations, and some of them also differ in habitat. The large *P.holsti*, closely related to the mainland *P.spilonotus,* is much larger with a larger and thicker beak than the Formosan form of *P.monticolus,* which in turn is larger than the other two species. Of the latter, *P.varius* has a thick and *P.ater* a thin

271

beak. The two species in the subgenus *Pardaliparus* in the Philippines replace each other geographically and have thinner beaks than *P.semilarvatus*, which is closely related to *P.varius* on Formosa and in Japan.

2. The species of *Parus* in the mountains of central Asia

Species	Subgenus	Main range in central Asia
P.ater (Coal)	*Periparus*	Tian Shan *(P.a.rufipectus)*, E. Himalayas to W. China *(P.a. aemodius)*
P.cyanus (Azure)	*Cyanistes*	Tian Shan, Kunlun to Kokonor
P.davidi (Pére David's)	*Poecile*	E. Tibet, W. Szechwan and Kanau
P.dichrous (Brown Crested)	*(Lophophanes?)*	Himalayas, Burma, W. Szechwan and W. Yunnan
P.major (Great)	*Major*	widespread
P.melanolophus (Crested Black)	*Periparus*	W. Himalayas
P.montanus (Willow) *(songarus)*	*Poecile*	Tian Shan, and in west from Kansu to extreme N. Yunnan
P.monticolus (Green-backed)	*Major*	Himalayas, N. Burma, W. China
P.palustris (Marsh) *(hypermelaena)*	*Poecile*	Burma south to Mt. Victoria, Kansu to extreme N. Yunnan
P.rubidiventris (Black)	*Periparus*	widespread
P.superciliosus (White-browed)	*Poecile*	E. Tibet, W. Szechwan and Kansu
P.spilonotus (Black-spotted Yellow)	*Machlolophus*	E. Himalayas, N. Burma, Yunnan
P.venustulus (Yellow-bellied)	*Pardaliparus*	W. Szechwan and Kansu
P.xanthogenys (Yellow-cheeked)	*Machlolophus*	W. Himalayas

Notes

Nomenclature from Snow (1967). Some other workers have treated *P.ater aemodius, P.montanus songarus, P.palustris hypermeleana* (or *hypermelas*) and *P.rubidiventris rufonuchalis* as full species, or *P.spilono-tus* and *P.xanthogenys,* separated by Snow, as races of *P.xanthogenys,* while *P.melanolophus* is a geographical form of *P.ater,* though distinct enough in range and colouring to be treated as a full species. The attribution of *P.dichrous* to the subgenus *Lophophanes* is doubtful, as its resemblance to *P.cristatus* might be due partly to convergence.

The other species in mainland Asia not included above are *P.cinctus* (north), *P.nuchalis* (India) and *P.varius* (Korea, Japan). In addition, the Great Tit *P.major* is divided into the semispecies *minor* in eastern Asia, *cinereus* in southern Asia and *bokharensis,* treated as a full species by Snow (1967), in part of central Asia.

3. Main means of ecological segregation of *Parus* species in mountains of central Asia

F probably by food, G by geographical range, used solely for closely related species, H by habitat, i.e. forest type and altitude, and — no contact)

		1	2	3	4	5	6	7	8	9	10	11	12	13	14
1	*P.ater*	1													
2	*P.cyanus*	H	2												
3	*P.davidi*	H	—	3											
4	*P.dichrous*	F	—	H	4										
5	*P.major*	H	F	H	H	5									
6	*P.melanolophus*	G	—	—	F	H	6								
7	*P.montanus*	F	HF	H	HF	H	—	7							
8	*P.monticolus*	H	—	F	H	H	H	H	8						
9	*P.palustris*	H	—	H	H	F	—	H	H	9					
10	*P.rubidiventris*	F	H	H	F	H	F	HF	H	H	10				
11	*P.spilonotus*	H	—	—	H	F?	H	—	H	H	H	11			
12	*P.superciliosus*	H	—	H	H	F	—	HF	H	H	H	H	12		
13	*P.venustulus*	H	—	H	H	F	—	H	F	H	H	—	H	13	
14	*P.xanthogenys*	H	—	—	H	F?	H	—	H	—	H	G	—	—	

Note

In comparing lowland with highland species, it is not always easy to know whether H or — is the more appropriate category. Species which occur in the same habitat usually differ in size and size of beak and have been assumed to differ in feeding, but no critical studies have been made. Nearly all the differences in habitat, i.e. forest type, are associated with a difference, usually large, in altitude.

4. The *Parus* species of Africa south of the Sahara

Species	Main range	Habitat
AFER SUPERSPECIES		
P.afer (Grey)	southern Africa *(afer, cinerascens)*, Somalia *(thruppi)*	dry bush with scattered trees, acacia steppe
P.griseiventris (part of Grey)	Angola to W. Tanzania, Malawi, Rhodesia	open brachystegia woodland
P.fasciiventer (Stripe-breasted)	Mountains E. Congo, W. Uganda	montane gallery and bamboo forest
NIGER SUPERSPECIES		
P.niger (Southern Black)	W. Angola to Mozambique and south to E. Cape Province	mopane and other dry woodland
P.leucomelas (Black)	most of Africa from 15°N to 15°S except forest	acacia savanna, also clearings and wood edge; solely lowland where overlaps *P.albiventris*
P.albiventris (White-breasted)	S. Sudan to E. and C. Tanzania, E. Nigeria and Cameroons	acacia savanna, also clearings and wood edge solely highlands where overlaps *P.leucomelas*
P.leuconotus (White-breasted Black)	Abyssinia, Eritrea	woodland above 2000 m

APPENDIXES

Species	Main range	Habitat

RUFIVENTRIS SUPERSPECIES

P.*rufiventris*
(Cinnamon-
breasted)

S. Congo and Angola
east to C. Tanzania,
Mozambique and
Rhodesia

brachystegia woodland

P.*fringillinus*
(Red-throated)

S. Kenya, N. E.
Tanzania

acacia savanna

FUNEREUS SUPERSPECIES

P.*funereus*
(Dusky)

W. Africa, Congo,
N. Angola to W. E.
Uganda and
W. Kenya

lowland evergreen
forest feeding in
canopy

Notes

From Hall and Moreau (in prep., supplemented by Hall pers comm.), and following Snow (1967) for nomenclature, with superspecies from Hall and Moreau. Some other workers have treated two well marked subspecies of *P.afer* as separate species, namely *P.a.cinerascens* in southern Africa north of *P.a.afer* and *P.a.thruppi* in Somalia. The habitats in ascending order of richness are dry scrub with scattered trees, acacia steppe (annual grasses), acacia savanna (perennial grasses), mopane woodland, brachystegia woodland, gallery forest, evergreen forest.

All these species are of *P.major* type. Most have a wing-length around 80 mm, but *P.thruppi* and *P.fringillinus* are smaller. The beak is longest and heaviest in *P.afer*, a little less so in *P.albiventris*, *P.fringillinus* and *P.leucomelas*, a little smaller still in *P.funereus* (in which it is unusually broad) and *P.niger*, the last being close to European *P.major*. It is a little smaller in *P.rufiventris* and slenderest in *P.fasciiventer*, *P.griseiventris* and *P.leuconotus*. The difference between these last three species and *P.afer* is large. The adaptive significance of the beak differences has not been studied.

5. Ecological segregation of African species of *Parus*

(F by feeding, G by geographical range, H by habitat, Ha by habitat including altitude, — no contact. Species in same superspecies are bracketed.)

1	*P.afer*	1								
2	*P.griseiventris*	GH	2							
3	*P.fasciiventer*	GH	GH	3						
4	*P.niger*	H	—	—	4					
5	*P.leucomelas*	—	H	H	G(H)	5				
6	*P.albiventris*	—	—H	—	G(H)	G(Ha)	6			
7	*P.leuconotus*	—	—	—	GH	GHa	GH	7		
8	*P.rufiventris*	—	F	—	—	—	—	—	8	
9	*P.fringillinus*	—	—	—	—	HF?	HF	—	G	9
10	*P.funereus*	—H	—H	—H	—H	—H	H	—H	H	—H

Notes

Based on Hall and Moreau (in prep.), with difficult points clarified by Hall (pers. comm.). G is used solely for species in the same superspecies, and — for species in different superspecies that have separate ranges. The tendency in Africa for the main habitats, such as lowland forest, brachystegia woodland and acacia savanna, to occur over huge stretches of country means that many species are separated by both habitat and range, and it is hard to know which is primary. Most such cases are symbolized by —H (and scored as $\frac{1}{2}$ each in table 11). The addition of *P.afer cinerascens* and *P.a.thruppi*, which some workers consider to be full species, would further increase the number of species separated by range.

6. The North American species of *Parus*

SUBGENUS BAEOLOPHUS TUFTED TITS

SUPERSPECIES PARUS BICOLOR
P.(b.)atricristatus   Black-crested Tit   most of Texas, northern Mexico

P.bicolor	Tufted Tit	eastern half of USA (from central Iowa and Massachusetts south to Gulf coast and west to central Nebraska and Texas)
P.inornatus	Plain Tit	western USA (from southern Oregon and Idaho south to Baja California and western Texas)

SUPERSPECIES PARUS WOLLWEBERI

P.wollweberi	Bridled Tit	Highlands of Mexico, part of Arizona and New Mexico

SUBGENUS POECILE

SUPERSPECIES PARUS ATRICAPILLUS BLACK-CAPPED CHICKADEES

P.atricapillus	Black-capped Chickadee	from Alaska and Canada to middle of USA, and from Atlantic coast west to Rocky Mountain area and in places to west coast
P.carolinensis	Carolina Chickadee	southeastern USA (from southern Kansas and Pennsylvania south to Gulf of Mexico)
P.gambeli	Mountain Chickadee	Rocky Mountains and Pacific coast ranges (from British Columbia to Baja California and Texas)
P.sclateri	Mexican Chickadee	Mountains of Mexico and small part of Arizona and New Mexico

SUPERSPECIES PARUS CINCTUS BROWN-CAPPED CHICKADEES

P.cinctus	Siberian Tit or Gray-headed Chickadee	northern and western Alaska and Yukon
P.hudsonicus	Boreal or Hudsonian Chickadee	from Alaska and Canada south to northernmost parts of USA

10

| *P.rufescens* | Chestnut-backed Chickadee | western seaboard from southern Alaska to central California, locally inland to western Alberta and Montana |

Notes

Nomenclature from Snow (1967) except that the current AOU Check-List is followed in treaty *P.atricristatus* as a full species instead of as a subspecies of *P.bicolor*. These two forms interbreed where they adjoin, rather than intergrading, so are on the border between subspecies and full species, and it is convenient, for a discussion of ecology, to treat them as separate units. Snow and Dixon (1961) considered *P.wollweberi* too distinctive to be included in the subgenus *Baeolophus*, but while it is more different from the other three species than they are from each other, it seems much nearer to them than to any other species of *Parus*. Hence I have here placed it in the same subgenus but a different superspecies.

7. Main means of segregation of North American species of *Parus*

(F by feeding station, G by geographical range, H by habitat, and — no contact)

		1	2	3	4	5	6	7	8	9
1	*P.bicolor*	1								
2	*P.inornatus*	G	2							
3	*P.wollweberi*	G(F?)	G(F)	3						
4	*P.atricapillus*	F	—	—	4					
5	*P.carolinensis*	F	—	—	G	5				
6	*P.sclateri*	—	—	H(F)	G	G	6			
7	*P.gambeli*	—	H(F)	—	G	G	G	7		
8	*P.cinctus*	—	—	—	—	—	—	—	8	
9	*P.hudsonicus*	—	—	—	H	—	—	G(H)	H	9
10	*P.rufescens*	—	H(F)	—	G(H)	—	—	G(H)	—	G

Note

P.atricristatus was treated as a full species in earlier tables, but since the main object of table 20 is to provide a comparison with other regions, it seems fairest to adhere completely to the nomenclature of Snow (1967), in which it was treated as a subspecies of *P.bicolor*. Had *P. atricristatus* been included here, it would have been separated by G from *P.bicolor* and *P.inornatus*, by G(F?) from *P.wollweberi* and by F from *P.carolinensis*, with — for all other species, thus increasing the number of instances of separation by range.

8. Ecological segregation of the European and North American species of *Sitta*

(F by size, and presumably feeding station, G by range, H by habitat, and — no contact)

A. EUROPE

1	*S.europaea*	1		
2	*S.neumayer*	H	2	
3	*S.whiteheadi*	GH	—	3
4	*S.krüperi*	H	H	G

B. NORTH AMERICA

1	*S.canadensis*	1		
2	*S.carolinensis*	H	2	
3	*S.pusilla*	—	F	3
4	*S.pygmaea*	H	F	G

Notes

European forms of *S.europaea* and *S.neumayer* have a wing-length around 86 mm and *S.whiteheadi* and *S.krüperi* of around 73 mm, the second of these pairs being in both cases a little larger. The Corsican *S.whiteheadi* was for a long time treated as a relict subspecies of the North American *S.canadensis*, but this cannot be sustained (Löhrl 1960–61), and their similar small size, narrow beaks and white breasts are clearly due to convergent adaptation to coniferous forest. *S.white-headi* is not very close to the other white-breasted species in coniferous forest in Europe, *S.krüperi*.

The mean measurements of the wing-length and culmen of each of the four American species as given by Ridgway (1904) are as follows: (where two figures are given, they are the means for the smallest and largest subspecies measured).

Red-breasted Nuthatch *S.canadensis*   68 mm, 14·6 mm
White-breasted Nuthatch *S.carolinensis*   87–92 mm, 17·8–19·8 mm
Brown-headed Nuthatch *S.pusilla*   64 mm, culmen 14·5 mm
Pygmy Nuthatch *S.pygmaea*   65–67 mm, 14·1–16·3 mm

In a narrow zone of overlap, the small size of *S.canadensis* probably separates it by feeding from *S.carolinensis*. In southern Arizona, *S.carolinensis* is in spruce-fir and *S.pygmaea* in pine-oak forest, so they differ in habitat as well as size and presumably feeding (Marshall 1957).

9. Nuthatches *Sitta* of the mountains of central and southeast Asia

1   *S.castanea* Chestnut-bellied Nuthatch. India, Burma, Indochina. Lowlands and foothills to 1200–1400 m in tropical and subtropical races, but two races (in Kashmir and Laos-Tonkin) in montane temperate woodland. Extensive overlap with 4 in lowlands.

2   *S.europaea* European Nuthatch. (i) Tian Shan (presumably temperate forest), (ii) Kansu and Szechwan, subtropical. (Also widespread in northern Asia and Europe.)

3   *S.formosa* Beautiful Nuthatch. E. Himalayas from Sikkim to Assam, N. Burma, N. Laos, N. Tonkin. Mainly above 1200 m. Rare. Large

4   *S.frontalis* Velvet-fronted Nuthatch. India, Burma, S. Yunnan, Indochina. Lowlands and foothills to 1200–1400 m. Tropical and subtropical forest. Extensive overlap with 1. (Also Indonesia and Philippines.)

5   *S.himalayensis* White-tailed Nuthatch. Himalayas (except Kashmir), N. and E. Burma, W. Yunnan, Laos, N. Tonkin. Above 1500 m. Temperate forest, with preference for oaks. Overlaps 8 in N. Burma, with difference in habitat.

6   *S.leucopsis* White-cheeked Nuthatch. (i) Afghanistan, W. Himalayas, (ii) E. Tibet, W. China. High alpine spruce and fir forest.

7 *S.magna* Giant Nuthatch. E. Burma, N.W. Yunnan, N. Thailand. Above c. 1500 m. Dense evergreen forest or open pine. V. large.

8 *S.nagaensis* Naga Hills Nuthatch. E. Tibet and S. Kansu south through extreme N. E. India, N. Burma and Yunnan to central Burma (Mount Victoria), also N. Thailand, S. Annam. Above 1200–1400 m in light broadleaved forest and/or open pines. Overlaps 5 in N. Burma and 12 in E. Tibet, with difference in habitat. (Also mountains of Fokien.)

9 *S.tephronota* Eastern Rock Nuthatch. Tian Shan and W. Himalayas. 900–2500 m. Rocky hill slopes. (Also west to Iran.)

10 *S.victoriae* Chin Hills Nuthatch. Mt. Victoria (Burma). Above 2600 m. Humid evergreen broadleaved forest.

11 *S.villosa* Chinese Nuthatch. Kokonor and N. Kansu. Montane coniferous forest. (Also much of northern China.)

12 *S.yunnanensis* Yunnan Nuthatch. E. Tibet, W. Szechwan, Yunnan. Above 2700 m. Dry pine forest. Overlaps 8 in Tibet with difference in habitat.

Notes

Peters' 'Check-List' has been followed for nomenclature, *Sitta* being prepared by Greenway (1967a), who differed considerably in regard to the limits of *S.europaea* from Voous and Van Marle (1953), Vaurie (1957, 1959), and Ripley (1961), who in turn differed from each other. After examining the material in the British Museum (Natural History), I fully support Greenway. The critical points are as follows:

(i) *S.nagaensis* is a separate species from *S.europaea*, which it replaces at higher altitudes in two areas, first the border of eastern Tibet with China and secondly N. W. Fokien. Altitudinal replacement is characteristic of different species, not subspecies of the same species. The grey underparts of *S.nagaensis* contrast with the cinnamon-buff of *S.europaea*.

(ii) *S.castanea* is a separate species from both *S.europaea* and *S. nagaensis*. It does not come in contact with *S.europaea*. *S.nagaensis* replaces it at higher altitudes in Burma and parts of Indochina, without intergradation or interbreeding, in a manner typical of separate species and not of subspecies of the same species.

(iii) Given that *S.castanea* is a separate species, the form *cashmirensis* in Kashmir is a race of *S.castanea*, not *S.europaea*. It is close to *S.castanea* in its appearance, including its (reduced) sexual dimorphism and shape of beak, and while it is less bright below than the lowland forms of *S.castanea*, this is probably an adaptation for life in temperate forest and does not provide good evidence that it is a linking form between

S.castanea and *S.europaea*. It replaces *S.castanea* geographically, and is not near *S.europaea*.

(iv) Greenway followed most other recent workers in treating *S. victoriae* as a separate species from *S.himalayensis*. It is endemic to Mount Victoria and looks very like *S.himalayensis* but has a slender instead of thick beak. The latter might be enough in itself to justify specific separation. Whether it replaces *S.himalayensis* geographically is uncertain, because one specimen of the latter, now in the British Museum (Natural History), has been collected on Mount Victoria. However, it was not found there by Stresemann and Heinrich (1940), so is certainly rare and might not be resident.

Of the species listed in the table, *S.magna* is the largest with much the largest beak (wing c. 117 mm, culmen c. 27 mm), while *S.formosa* has a very slightly shorter wing and a much shorter beak (culmen 17 mm) and *S.tephronota* has a much shorter wing (c. 83 mm) but a relatively long beak (c. 22 mm). Next come four species with a wing-length around 75 to 77 mm, these, in descending order of size, being *S.leucopsis*, *S.europaea*, *S.castanea* and *S.nagaensis*, and finally come *S.himalayensis*, *S.frontalis*, *S.villosa*, *S.yunnanensis* and *S.victoriae*, with a wing-length around 70–73 mm.

10. Probable means of ecological segregation of *Sitta* species in the mountains of central and southeast Asia

	1	2	3	4	5	6	7	8	9	10	11	12
1 *S.castanea*	1											
2 *S.europaea*	G	2										
3 *S.formosa*	—?	—	3									
4 *S.frontalis*	F?	—	—?	4								
5 *S.himalayensis*	Ha	—	F?	Ha	5							
6 *S.leucopsis*	Ha	—	—	Ha	Ha(or —)	6						
7 *S.magna*	Ha	—	—?	Ha	—?	—	7					
8 *S.nagaensis*	Ha	G	F?	Ha	G(H)	H	F?	8				
9 *S.tephronota*	H	H	—	—	H	H	—	—	9			
10 *S.victoriae*	Ha	—	—	Ha	H(or G?)	—	—	Ha	—	10		
11 *S.villosa*	—	Ha	—	—	—	G	—	—	—	—	11	
12 *S.yunnanensis*	—	Ha	—	—	H	H	—	HF?	—	—	G	

(F probably by feeding stations, G by geographical range, used for closely related or very similar species, H by habitat, Ha where this is closely linked with altitude, and — no contact)

Notes

It is hard to know how to allocate some cases where two species differ markedly in altitude and therefore in range. On the above tentative assessment, the number of instances of each sort is: by geographical range 5, by habitat, mainly associated with a difference in altitude, $22\frac{1}{2}$, presumably by feeding station $4\frac{1}{2}$, and no contact 34. So as to restrict the table to full species, the two montane races of *S.castanea* were omitted. They differ by G(H) from the lowland forms of their species, *S.c.cashmirensis* by G and *S.c.tonkinensis* by H from *S.himalayensis*, by G from *S.nagaensis*, and by H from *S.frontalis*, *S.c.cashmirensis* by — and *S.c.tonkinensis* by F from *S.magna*, and they have no contact with the other species.

11. The European finches *Fringillidae*

Genus and species	Vernacular name	European breeding range
CARDUELINAE		
CARDUELIS		
C.cannabina	Linnet	widespread except much of north
C.carduelis	Goldfinch	widespread except much of north
C.chloris	Greenfinch	widespread except the most northern parts
C.flammea	Redpoll	all north, British Isles, Alps
C.flavirostris	Twite	western Norway, N.W. Britain and Ireland
C.hornemanni	Arctic Redpoll	extreme north
C.spinus	Siskin	omitting far north, widespread in north and east, also some areas in centre, local in south
CARPODACUS		
C.erythrina	Scarlet Rosefinch	primarily east of Baltic

Genus and species	Vernacular name	European breeding range
COCCOTHRAUSTES		
C.coccothraustes	Hawfinch	widespread in centre and south
LOXIA		
L.curvirostra	Crossbill	widespread in east and north, scattered areas elsewhere
L.leucoptera	Two-barred Crossbill	northeast
L.pityopsittacus	Parrot Crossbill	northeast and part of north
PINICOLA		
P.enucleator	Pine Grosbeak	far north
PYRRHULA		
P.pyrrhula	Bullfinch	widespread except extreme north and most of Spain, Greece and Mediterranean littoral
SERINUS		
S.citrinella	Citril	Alps and adjoining mountains, some mountains in northern Spain, Corsica, Sardinia
S.serinus	Serin	widespread in south and centre

FRINGILLINAE

FRINGILLA		
F.coelebs	Chaffinch	all except far north
F.montifringilla	Brambling	far north

Notes

The Arctic Redpoll *C.hornemanni* is by some authors treated merely as a subspecies of *C.flammea*, but it differs in colour and habitat, and though it interbreeds in some areas, in others it keeps separate. Until recently the Citril *S.citrinella* was put in *Carduelis,* not *Serinus,* and the Serin *S.serinus* was regarded as a subspecies of the Canary *S.canaria,* but current usage seems well founded.

12. Main means of ecological segregation in congeneric European finches

(F by food, G by geographical range, H by habitat, and — no contact)

(i) CARDUELIS

1	C.cannabina	1					
2	C.carduelis	F	2				
3	C.chloris	F	F	3			
4	C.flammea	F(H)	F(H)	F(H)	4		
5	C.flavirostris	G(H)	F	F	F(H)	5	
6	C.hornemanni	—	—	—	H	H	6
7	C.spinus	F(H)	F(H)	F(H)	F(H)	F(H)	—

(ii) LOXIA

1	L.curvirostra	1	
2	L.leucoptera	F(H)	2
3	L.pityopsittacus	F(H)	F(H)

(iii) SERINUS

S.citrinella from S.serinus by habitat

(iv) FRINGILLA

F.coelebs from F.montifringilla by geographical range

Note

In many cases in both *Carduelis* and *Loxia*, a difference in the preferred type of seed also involves a difference in habitat, but the difference in food is primary, so these have been scored as F(H).

13. Summer and winter ecology
of congeneric European passerine migrants to Africa

(Based on Voous 1960 for Europe and Moreau in prep. for Africa unless otherwise stated)

A. EUROPEAN SWALLOWS HIRUNDO

BREEDING SEASON, EUROPE (nesting portrayed in fig. 19)

Red-rumped Swallow *H.daurica*: discontinuously in extreme south, typically on low limestone hills, building tunnel-shaped nest on ceiling of cave or building.

Crag Martin *H.(Ptyonoprogne)rupestris*: southern Europe, chiefly in sunny mountain areas, nesting in small niche on steep rock face, usually on lower part.

(Barn) Swallow *H.rustica*: widespread, mainly in lowlying rural areas, especially round farms, nesting on ledges under cover, usually inside buildings with open access, but also in large caves, presumably the ancestral habit (e.g. Peus 1954).

WINTER IN AFRICA

Red-rumped Swallow: probably across northern tropical Africa, but uncertain owing to very small numbers involved and presence of native subspecies.

Crag Martin: mainly Sudan and Abyssinian highlands, one record from west (Senegal).

(Barn) Swallow: from S.Nigeria, N.E.Congo and tropical East Africa south to South Africa, but not normally in midwinter north of where stated.

Note:

Two martins in different genera are widespread in Europe, the House Martin *Delichon urbica*, which nests on vertical cliffs under an overhang, usually nowadays under eaves of houses, and the Sand Martin *R.riparia*, which tunnels a hole in sandy cliffs. These evidently differ from the Swallow and each other in diet, as the Sand Martin arrives, breeds and stops breeding before the other two, and the House Martin is the latest in these respects; but their foods have not been studied. There is considerable geographical overlap with the Swallow in winter quarters, where the House Martin feeds high and is rarely seen.

APPENDIXES

B. EUROPEAN PIPITS ANTHUS

BREEDING SEASON, EUROPE

Tawny Pipit *A.campestris*: south and middle Europe, on open, dry, sandy or limestone grassland or heaths with sparse vegetation, locally on arable land.

Red-throated Pipit *A.cervinus*: extreme north, in shrub or mossy tundra, usually in marshy places, often beside lakes or the sea. Here breeds alongside Meadow Pipit, but partly separated by preference for bushes or nearby water.

Meadow Pipit *A.pratensis*: north and middle Europe, in rather dense grassland, Calluna heaths and moors, locally (though not in the Alps) on alpine meadows.

Alpine Pipit *A.spinoletta spinoletta*: central and south European mountains, on alpine meadows and other low vegetation above the tree-line, often near streams.

Rock Pipit *A.spinoletta petrosus*: rocky seacoasts of northwest, feeding on grass near cliffs and below tideline, but on some western British islands penetrating a short way inland on sprayblown turf (where Meadow Pipit absent) and in Baltic, which is less salt, only on outermost islands, the Meadow Pipit breeding on the rest (P. Palmgren pers. comm.). Taxonomically the Alpine and Rock Pipits are normally treated as one species, but ecologically they are distinct, so they are treated separately here.

Tree Pipit *A.trivialis*: widespread except Mediterranean littoral, on drier grassland and heaths with scattered trees (the only species which normally breeds among trees, though the Tawny Pipit sometimes breeds where trees are present).

WINTER IN EUROPE

The Meadow, Alpine and Rock Pipits remain in Europe, but northern and alpine populations move south or to lower altitudes, to similar habitats to those in which they breed, the Alpine Pipit beside shallow inland waters.

WINTER IN AFRICA

Tawny Pipit: chiefly in acacia belt south of Sahara, but in east south to N.E. Kenya, in even drier habitats than in summer, often near grazing cattle.

Red-throated Pipit: some in Egypt, most in tropical west and east south to Nigeria, N. Congo and N. Tanzania, nearly always in short grass in damp spots, at times with cattle.

Tree Pipit: across northern tropical Africa south of Tawny Pipit, in east to Zambia, in grassland with scattered trees, savanna, open woodland and even brachystegia woodland.

C. EUROPEAN WAGTAILS MOTACILLA

BREEDING SEASON, EUROPE

White Wagtail *M.alba*: widespread, mainly on grassy or muddy borders of ponds or slow-moving rivers, or round rural dwellings and farms, and in other open areas with short grass, locally on seashore. In far north, where *M.cinerea* absent, also by rocky and fast-moving rivers, and in towns.

Grey Wagtail *M.cinerea*: southern and most of middle Europe, beside rocky streams and fast-flowing rivers with rocky or stony edges, locally in towns.

Yellow Wagtail *M.flava*: widespread in rather thick grassland or other heathy or marshy vegetation, in meadows, drier freshwater marshes, saltmarshes and tundra, often feeding alongside cattle where present.

WINTER IN AFRICA

White Wagtail: only a partial migrant, but many in the dry belt across northern tropical Africa, especially in villages and near water.

Grey Wagtail: only a partial migrant, but in East Africa it crosses the equator. Almost unknown in West Africa. In East Africa on clear rapid streams, elsewhere by ornamental waters and on sea-shore.

Yellow Wagtail: widespread (the various palaerctic subspecies mingling) in both dry and swampy grassland, typically feeding around large herbivorous mammals.

D. EUROPEAN SHRIKES LANIUS
(Summer and winter ranges in figs 20 and 21)

BREEDING SEASON, EUROPE

Redbacked Shrike *L.collurio*: widespread, except for most of Fenno-Scandia, Britain and Iberia, on heaths, grassland or cultivated land, at times in open forest, provided that there is dense ground cover, usually grass, and also many bushes or branches at middle heights; but where Woodchat *L.senator* is absent, it may frequent the habitat preferred by the latter (Durango 1954). Smaller than Great and Lesser Grey, similar in size to other two.

Great Grey Shrike *L.excubitor*: widespread except Britain, the Balkans and most of Italy, in shrub tundra, open birch forest, heaths, steppe, grassland and open cultivated land, provided that there are scattered tall bushes or trees. The largest species.

Lesser Grey Shrike *L.minor*: central, eastern and southeastern Europe, in generally similar habitat to Great Grey, but with rather more trees; common in orchards. Not quite so large as Great Grey, but with proportionately longer wing.

Masked Shrike *L.nubicus*: extreme southeast, in light wooded land. It keeps much more in cover under trees than the other species, so partly differs in habitat, and also feeds more like a flycatcher than they do and its beak is narrower, indeed Voous (1960) thought that it should not be put in *Lanius*.

Woodchat *L.senator*: central and south Europe, on heaths, grassland, cultivated land with bushes and open woodland, in generally similar habitats to Redbacked, but with sparse low ground vegetation or bare patches, and with open space above the ground between thinly growing trees; where Redbacked is absent, also breeds in bushy plains, but not if there is dense ground vegetation (Durango 1954).

WINTER IN AFRICA

Redbacked: from Kenya south to eastern Cape Province and Southwest Africa, in rather open country with thorny trees and bare soil.

Great Grey: The European races stay in Europe and Africa north of the Sahara, in similar habitats to summer.

Lesser Grey: in northern S.W. Africa and Botswana, overlapping Redbacked but mainly utilizing taller vegetation.

Masked: northern tropics east of Lake Chad, in low, hot acacia country.

Woodchat: across northern tropics in acacia savanna.

E. EUROPEAN FLYCATCHERS MUSCICAPA (INCLUDING FICEDULA)

In most modern treatments, only one European species, the Spotted Flycatcher *M.striata*, is included in *Muscicapa*, and the others in *Ficedula*, but following a usual practice in this book, I have used broad genera. One species, *M.(F.)semitorquata*, which lacks an English name, is usually treated as a subspecies, but I have followed Curio (1959) in accepting it as a species.

BREEDING SEASON, EUROPE

Three closely similar species with similar ecology replace each other geographically, the Pied Flycatcher *M.hypoleuca* in most of the north, west and centre, the Collared Flycatcher *M.albicollis* in east-central Europe and most of Germany, and *M.semitorquata* in the Balkans. They breed in fairly open oak forest, also montane beech woods, orchards, and in the north (Pied) in open pine woods and town parks. They feed especially by picking caterpillars and other insects off the leaves in the canopy, also by dropping from a perch to take insects off the ground, and take insects in the air much less often than the other species. The Pied and Collared have a narrow zone of overlap in the same type of habitat in part of Germany and on Gotland, where they sometimes interbreed. They also coexist in the primaeval forest of Bialowies in Poland, the Pied in dry pine and the Collared in damp alder-birch-spruce forest (Tischler 1942), which suggests the possibility that they might have had a wider area of coexistence in primaeval times. (The difference in habitat suggested by Dementiev and Gladkov, 1968, is not, I think, reliable, at least as translated).

Redbreasted Flycatcher *M.(F.)parva*: in eastern Europe, in tall dense beech, pine or other forest, feeding on small insects in the air by flying up from the canopy.

Spotted Flycatcher *M.striata*: widespread, at the wood edge and in very open, usually broadleaved, forest, also parks and gardens, feeding by flying out from a perch well below the level of the canopy for flying insects. In more open woods than Pied. The insects taken are presumably larger, on average, than those taken by the smaller Redbreasted Flycatcher.

WINTER

Pied Flycatcher: mainly West Africa and Congo, south to about Equator, associated with evergreen forest.

Collared Flycatcher: mainly south of Pied, to Malawi and Zambia, but overlapping in Congo; same habitat as Pied. Status in West Africa uncertain.

M.semitorquata: not adequately known, but seen in Uganda and Tanzania.

Redbreasted Flycatcher: mainly N.W. India.

Spotted Flycatcher: mainly south of the Equator (though some to the north), extending to South Africa, in open wooded and thorn country. Overlaps Pied and Collared in West Africa and Congo, where separated by habitat.

F. EUROPEAN REED WARBLERS ACROCEPHALUS

Following Parker and Harrison (1963), the Moustached Warbler is here included in *Acrocephalus* (formerly in *Lusciniola*).

BREEDING RANGE, EUROPE

Great Reed Warbler *A.arundinaceus*: widespread in middle and south.

Blyth's Reed Warbler *A.dumetorum*: east of Baltic, replacing Marsh Warbler.

Moustached Warbler *A.melanopogon*: much of south.

Aquatic Warbler *A.paludicola*: local in east-central Europe, also Italy.

Marsh Warbler *A.palustris*: middle Europe, especially eastern half.

Sedge Warbler *A.schoenobaenus*: widespread in north and middle.

Reed Warbler *A.scirpaceus*: widespread in middle and south.

APPENDIXES

Great Reed: tall, strong and not too dense reeds *Phragmites* over water, especially near edges of reedbeds or in dykes; much larger than Reed, with which it usually coexists.

Blyth's Reed: like Marsh (see below) (separated by range).

Moustached: over shallow water, chiefly in reedmace *Typha*; where in *Phragmites* with Reed Warbler, feeds lower and Reed Warbler higher (Zahavi 1957).

Aquatic: at lake margins, not over water, primarily in pure sedge *Carex*, which Sedge Warbler avoids (Koenig 1952).

Marsh: in willows and tall bushes, often but by no means always near water, with tall herbaceous vegetation, especially nettles *Urtica dioeca* and meadowsweet *Filipendula ulmaria*, locally in corn.

Sedge: on damp ground and at lake and river margins (not nesting though often feeding over water), in mixture of reeds *Phragmites*, sedge *Carex*, grass and scrub; in England also in habitat of Marsh Warbler where latter absent, and in Scotland in habitat of Reed Warbler where latter absent (Bell 1969).

Reed: throughout extensive and dense reed beds over water, including thin-stemmed reeds, so differs somewhat in habitat from Great Reed, but main difference from latter is much smaller size, and presumed size of prey. Often feeds on dry land (C. K. Catchpole pas. comm.)

WINTER

Blyth's Reed Warbler in India, Ceylon and Burma, and Moustached Warbler in Africa north of Sahara and part of Arabia, are separated by range from each other and the other five species, which winter south of the Sahara. The one record of the Aquatic Warbler is from West Africa. The Marsh Warbler is found mainly in East Africa from Kenya south to Natal, mainly in bushes. The other three are widespread from west to east in tropical and southern Africa. Their habitats require critical study and perhaps differ in different areas. Reed and Great Reed are recorded in reed-beds and tall grass, on both dry and wet ground, also in cultivation, and in Uganda the Reed is primarily in thickets in *Euphorbia-Capparis* savanna (M. Fogden, pers. comm.). The Great Reed is presumably separated from rest by its much greater size and hence size of prey. Further details in Moreau *(in prep.)*.

G. EUROPEAN HIPPOLAIS WARBLERS

BREEDING RANGE, EUROPE (mapped in fig. 22)

Icterine Warbler *H.icterina*: from southern Fenno-Scandia south to Alps, and from eastern France east to eastern Europe, replacing Olive-tree and Melodious geographically.

Olive-tree Warbler *H.olivetorum*: southern Balkans (larger than other three).

Olivaceous Warbler *H.pallida*: southeast Spain, southern Balkans, overlapping the others.

Melodious Warbler *H.polyglotta*: Iberia, most of France, Italy and small part of Dalmatian coast, replacing Icterine and Olive-tree geographically.

HABITATS

Icterine Warbler: open broadleaved woods, especially oaks, at times with few or no bushes, including city parks and gardens.

Olive-tree Warbler: open broadleaved woods, also olive groves and gardens.

Olivaceous Warbler: tall bushes of Mediterranean or semidesert type, often with scattered trees, sometimes in stream valleys, typically in drier climate than the other three. (A. Valverde pers. comm. for Spain, Niethammer 1943 and Rucner 1967 for Balkans).

Melodious Warbler: open broadleaved woodland, especially oak, with thick bushes, also thick bushes with scattered oaks, much more dependent on bushes than Icterine or Olive-tree Warblers.

WINTER RANGE (mapped in fig. 23)

Icterine: mainly in Zambia, Rhodesia, Botswana and S.W. Africa, in acacias. (Records in Cameroons and Kenya probably of birds in transit.)

Olive-tree Warbler: from Kenya south to Transvaal and Botswana, in acacias, sparse. Overlaps with Icterine and Olivaceous.

Olivaceous Warbler: across northern tropical Africa south to northern Tanzania, mainly in acacia savanna, in tall scrub.

Melodious Warbler: In West Africa; south of Olivaceous Warbler owing to its preference for evergreen wooded country, broadleaved savanna, secondary growth, forest clearings and mangroves, and its avoidance of acacia habitats.

H. GRASSHOPPER WARBLERS LOCUSTELLA

BREEDING SEASON, EUROPE

River Warbler *L.fluviatilis*: east, in marshy woodland or lines of trees with tall grass, beside rivers or sometimes lakes.

Savi's Warbler *L.luscinioides*: middle and parts of southern Europe in extensive reedbeds of *Phragmites* over water.

Grasshopper Warbler *L.naevia*: widespread in middle of Europe, in marshes and on edges of lakes with mixture of *Phragmites*, tall grass, brambles and low bushes, locally in dry scrub.

APPENDIXES

AFRICA IN WINTER (very insufficiently recorded)

River Warbler: the few records are in east from Kenya to the Transvaal, in grass in Brachystegia woodland, in thickets and in *Phragmites*.

Savi's Warblers: across northern tropical Africa, in swamps.

Grasshopper Warbler: recorded on passage on western edge of Sahara, but winter range not known.

J. EUROPEAN LEAF WARBLERS PHYLLOSCOPUS

BREEDING RANGE, EUROPE

Bonelli's Warbler *P.bonelli*: south and west-central, most of its range being south of Willow Warbler's.

Arctic Warbler *P.borealis*: extreme north, largely north of Greenish and wholly north of Wood Warbler (extends much further south in Asia).

Chiffchaff *P.collybita*: nearly everywhere except parts of Scandinavia and of Mediterranean littoral.

Wood Warbler *P.sibilatrix*: through middle and part of north, also much of Italy.

Greenish Warbler *P.trochiloides*: east of Baltic, largely south of Artic Warbler.

Willow Warbler *P.trochilus*: throughout north and middle.

ECOLOGY IN EUROPEAN TAIGA (northern forest)

Arctic Warbler: in tundra in patches of luxuriant birch, with taller and denser trees than those preferred by Willow, often with other broadleaved trees as well (O. Hilden pers. comm.) (inhabits a wider range of habitats in Asiatic USSR, Dementiev and Gladkov 1968).

Chiffchaff: coniferous forest, especially pine.

Greenish Warbler: undisturbed spruce forest mixed with broadleaved trees on rich damp soils (Suomalainen 1936) (a much wider range of habitats in USSR, Dementiev and Gladkov 1968).

(Wood Warbler: extends north to this zone in USSR but precise habitat not given by Dementiev and Gladkov 1968; presumably similar to that set out in next section.)

ECOLOGY IN WEST AND CENTRAL EUROPE

Bonelli's Warbler: low beech and oak woods, occasionally pine, normally with many bushes, and occasionally in bushes without trees, hence presumably, like Willow Warbler, feeds low down. Common on sunny submontane slopes in Alps, where Willow is solely in lowlands, and replaces Willow in lowland broadleaved woods in southern Europe.

Chiffchaff: broadleaved woods, rarely conifers, with thick bushes, feeding high in trees or taller shrubs.

Wood Warbler: shady broadleaved trees of large beeches or spreading oaks with extremely few or no shrubs, especially in lowlands, but also montane beech and conifers. It sometimes breeds alongside Willow Warbler or Chiffchaff, but usually where shrubless and bushy areas adjoin.

Willow Warbler: broadleaved woods, occasionally conifers, with thick bushes, feeding primarily low in bushes; at times in thick bushes without trees. Though common in far north, is entirely in lowlands, not montane, in middle Europe.

ECOLOGY IN SOUTHERN EUROPE

In southern Spain, Bonelli's Warbler and Chiffchaff coexist in lowland woods of evergreen and deciduous oak with rich bush layer. How they might be separated is not known, but since in mid-Europe Bonelli's requires bushes and the Chiffchaff feeds high, it is probably by feeding stations.

WINTER
ASIA

Arctic Warbler: Indonesia and Indochina.

Greenish Warbler: India, Burma and Indochina.

AFRICA

Bonelli's Warbler: across northern tropics south to about 10°N, in acacia steppe.

Chiffchaff: mainly north of Sahara in Mediterranean basin and Maghreb, but some south of Sahara from Senegal to Eritrea, especially in irrigated or relatively lush steppe. Rare south of 5°N and not found south of equator.

Wood Warbler: from about 9°N to a little south of equator, from 10°W to 35°E, mainly in evergreen growth, forest clearings and humid savanna.

Willow Warbler: from about 10°N south to South Africa, in all types of wooded habitats except evergreen forest.

K. EUROPEAN SYLVIA WARBLERS

BREEDING RANGE, EUROPE

(a) species which reach central Europe

Blackcap *S.atricapilla*: widespread except far north.

Garden Warbler *S.borin*: widespread except extreme south.

Whitethroat *S.communis*: widespread except far north.

APPENDIXES

Lesser Whitethroat *S.curruca,* widespread except far north and most of
southwest.
Barred Warbler *S.nisoria*: widespread in east.
Dartford Warbler *S.undata*: southern England, western and southern
France, Iberia, Italy.

(b) species restricted to southern Europe,
especially bordering Mediterranean
Subalpine Warbler *S.cantillans.*
Spectacled Warbler *S.conspicillata.*
Orphean Warbler *S.hortensis.*
Sardinian Warbler *S.melanocephala.*
Ruppell's Warbler *S.rüppelli*: local in eastern Meditterranean.
Marmora's Warbler *S.sarda*: islands of western Mediterranean.

ECOLOGY IN NORTHERN AND CENTRAL EUROPE
Blackcap: Broadleaved, rarely coniferous, woods with many broadleaved bushes, feeding mainly in trees and generally higher than
Garden Warbler.
Garden Warbler: Habitat as Blackcap, but feeds mainly in bushes, also
present in boreal birch forest where Blackcap absent.
Whitethroat: In fairly low, usually thorny, scrub, with or without
scattered trees, also at wood edge and in wide, open, woodland
clearings.
Lesser Whitethroat: in much taller and often denser bushes and secondary growth than Whitethroat, including at wood edge, chiefly in
broadleaved vegetation at low altitudes in western Europe, but in
uppermost forest zone of low dense conifers in mountains of central
Europe, also in pine plantations lower down.
Barred Warbler: thickets and bushes in light, including riverine, woods.
The Whitethroat is primarily outside woods, but it is not quite
clear how these two are separated.
Dartford Warbler: In England and France in thick *Calluna* heath and
low gorse *Ulex* (for Mediterranean habitats see below).

ECOLOGY IN SOUTHERN EUROPE
Blackcap: broadleaved woods of evergreen and deciduous oaks with
good bush layer.
Whitethroat: roadside hedges in cultivated land in S. Spain (personal
observation), local in broadleaved bushes of northern rather than
Mediterranean type in S. France (R. Leveque pers. comm.).
Most of the other seven species live in dry scrub (macchia etc.) and
several of them appear to overlap greatly in habitat, but their segregation might have been clearer when this region was well forested.

Dartford Warbler: chiely in gorse, broom and tree heather, fairly often with scattered pines, but not broadleaved trees. Not in such tall scrub as that often frequented by Sardinian, and much more resistant than Sardinian to cold winters (though killed by them in north of range), so in Spain tends to be in hills and mountains, with Sardinian in similar habitats lower down, where Dartford is scarce (which suggests possible direct exclusion by Sardinian). Apparently separated by habitat from Subalpine Warbler, reasons not clear, but partly linked with preference of Subalpine for broadleaved trees and taller and richer scrub.

Subalpine Warbler: in southern Spain (and the Maghreb) sparse in open broadleaved woods with much scrub, not in pure scrub where Sardinian Warbler is so common. In S. France especially in scrub, both with and without scattered trees, on sunny submontane slopes, whereas Sardinian is usually at low altitudes. This may be because Sardinian is readily killed in hard winters, but Subalpine migrates so escapes them. Where, as often, the two are found together in S. France, the Subalpine usually hunts at higher levels and the Sardinian in the lowest vegetation (J. Blondel pers. comm.).

Spectacled Warbler: clearly separated by habitat, as restricted to poor low xerophytic scrub in saltmarshes, especially *Salicornia*, and also in semi-desert inland, at times in mountains (e.g. Niethammer 1957). Overlaps with Dartford and Sardinian only where poorer scrub is giving place to richer.

Orphean Warbler: primarily in open oak and pine woods, often with very few bushes, in dry, sunny, lowlying areas, also in town gardens and olive groves; very little overlap with Blackcap, which prefers richer woods with more bushes (personal observations in S. Spain).

Sardinian Warbler: in S. France and Mediterranean islands is said to prefer scrub from one to two or more metres high, hence higher than that preferred by Dartford Warbler in same areas. But Sardinian occurs in all types of scrub, including low scrub and pure gorse, where it is commonest, as in southern Spain (and the Maghreb). It is a permanent resident, and its summer range may well be restricted by where it can survive cold winters, hence it is commonest at low altitudes; for which see under Dartford and Subalpine Warblers, already discussed. Tends to feed low, which helps to separate it from Subalpine.

Rüppell's Warbler: on arid ground with low xerophytic bushes, including low *Quercus coccifera* and *Calycotome villosa* (Watson 1964 and personal observations, modifying Meiklejohn 1934, 1935, Wettstein 1938, Stresemann 1943, Watson 1964). Separated geographically from Dartford and Marmora's Warblers. Coexists with Sardinian, and how separated is not known.

Marmora's Warbler: Appears to be simply a dark form of Dartford Warbler and one might have called it a geographical replacement of it on the islands of the western Mediterranean, did not the Dartford also occur there, though it is very scarce on both Corsica and Sardinia and absent from the Balearics. Habitat as Dartford, with gorse, broom, tree heath and, often, scattered pines, the scrub usually below one metre high, which there separates it from Sardinian, and separated from Subalpine and Spectacled in the same ways as the Dartford already noted. How separated from Dartford not known, but partly by range.

WINTER

Four species, the Dartford, Spectacled, Sardinian and Marmora's Warblers, stay in part on their breeding grounds, though some individuals cross the Mediterranean. They tend to be in the same habitats as in summer.

Blackcap: some in woods of Mediterranean basin and Maghreb, but most south of Sahara, extending south to Malawi; relatively sparse in West Africa, and in East Africa mainly montane above 1000 m. Chiefly in humid vegetation, evergreen scrub, at times forest, but locally in dry acacia steppe in West Africa. Whether separated from Garden Warbler by height of feeding stations, as in summer, is not known.

Garden Warbler: numerous only south of 8°N, hence in part separate in range from Blackcap, but much overlap; extends south to Transvaal and S. W. Africa. Chiefly in humid and (in East Africa) montane vegetation, including the edges of evergreen forest, dense secondary growth, more humid types of savanna, brachystegia woodland and acacia, chiefly along water courses.

Whitethroat: widespread south of Sahara, extending to 11°N in Nigeria and to Tanzania in east. In dry acacia steppe and savanna, bushes being essential, and avoiding dense growth of trees.

Lesser Whitethroat: mainly in northeastern Africa south of Sahara, from Sudan west of Lake Chad and just into north Nigeria; in thick and thorny bushes.

Barred Warbler: S. Sudan, Uganda and Kenya, from 12°N to 5°S, in acacia.

Subalpine Warbler: typically in acacia belt from Senegal to about 25°E, overlapping Lesser Whitethroat in range, but usually where there are trees.

Orphean Warbler: across northern Africa from Senegal to Eritrea, in dry acacia scrub.

Rüppell's Warbler: Sudan, in scrub.

L. EUROPEAN NIGHTINGALES AND BLUETHROATS LUSCINIA

BREEDING SEASON, EUROPE

Thrush Nightingale *L.luscinia* in east, and Nightingale *L.megarhychos* in west and south, replace each other geographically, with an overlap in north and east Germany, as shown in fig. 26. Both frequent lowlying broadleaved, including riverine, forest with dense low bushes, brambles and nettles. Where they overlap, Nightingale is in drier and Thrush Nightingale in damper habitats of this type, but outside this area their habitats are generally similar. The Nightingale also occurs in thickets and shrubberies without trees, especially in south.

Bluethroat *L.svecica*: far north, much of middle, and small part of southern Europe. Normally in thickets with few or no trees, in north in shrub tundra, in central and east Europe in bushes mixed with *Phragmites* at edges of shallow lowland lakes, in W. France in coastal saltmarshes with *Suaeda* and tamarisk (Mayaud 1958), in N. and C. Spanish mountains in dry *Cytisus* scrub. Has recently colonized treeless cultivated land in parts of middle Europe (Blaszyk 1963, Ern 1966).

WINTER

Thrush Nightingale: Kenya south to Transvaal and Botswana, in dense thickets.

Nightingale: in belt across tropical Africa, most numerous in west, in dense thickets. European forms of these two species replace each other geographically (but Asian *L.megarhynchos* overlap in East Africa around Equator) (see fig. 24).

Bluethroat: across Africa from Atlantic to Red Sea, north of 10°N, in edges of swamps or in bushes by water, in Abyssinia up to 2000 m.

M. EUROPEAN ROCK-THRUSHES MONTICOLA

BREEDING SEASON, EUROPE

Rock Thrush: *M.saxatilis*: southern and south-central Europe, on dry rocky slopes, in mountains above 1000 m.

Blue Rock Thrush *M.solitarius*: south, on maritime and barren inland cliffs up to subalpine zone, also in some towns. But in Crete, where Rock Thrush absent, Blue Rock Thrush extends to 2200 m and the same holds in W. Turkey, Himalayas and High Atlas (Watson 1964, Voous 1960).

WINTER

Rock Thrush: Eastward from Sierra Leone, from 16°N south to 8°N in west, but in east south through Kenya and Tanzania; chiefly montane, on bare rocky ground, also in savanna woodland. (Many of the eastern birds presumably breed in Asia.)

Blue Rock Thrush: many remain on breeding grounds, others winter in Saharan mountains, eastern Tschad and Darfur, with a few in Mauretania, Senegal and N.Nigeria. Hence nearly all are north of winter range of Rock Thrush with a small overlap in northern tropics. In similar habitats to Rock Thrush but a wider altitudinal range including down to sea level.

N. EUROPEAN WHEATEARS (OR DESERT CHATS) OENANTHE

BREEDING SEASON, EUROPE

Black-eared Wheatear *Oe.hispanica*: sandy or stony plains in southern Europe, at times with bushes.

Pied Wheatear *Oe.leucomela*: replaces *Oe.hispanica* in similar habitats in extreme east of Europe often with bushes or scattered trees.

Black Wheatear *Oe.leucura*: Spain and S.France, on cliffs, rock faces and scree, both on coast and inland.

Common Wheatear *Oe.oenanthe*: widespread on open, barren, stony, rocky or sandy ground, but in southern Europe almost entirely montane, so here separated by altitude from *Oe.hispanica*. Situation in Aegean complex, as both species occur on some large islands, separated by altitude, but other large and most small islands have only one or the other (Wettstein 1938, Stresemann 1950, Watson 1964). Competitive exclusion is evidently in operation, though the factors determining its outcome need clarification. My examination of the full information by Watson suggests that *Oe.oenanthe* may be primarily on limestone islands and *Oe.hispanica* on those with granite or gneiss, which suggest a difference in habitat.

(Isabelline Wheatear *Oe.isabellina*: recently found by Watson [1961, 1964] breeding in eastern Greece, but too rare in Europe for discussion here; barren plains.)

WINTER

Black-eared: across northern tropical Africa in dry acacia country, less barren than that where Isabelline and resident *Oe.deserti* occur.

Pied: solely in east, from 17°N to a little south of Equator in Tanzania, especially in highlands, often in bushy areas or with scattered trees.

Black: remains in breeding range.

Common: from about 16°N south to 8°N in west, but 15°S in east, chiefly in acacia steppe and savanna, from sea-level to 2500 m in east, commonest on less arid ground than that preferred by Black-eared.

APPENDIXES

Isabelline: from 17°N south to 14°N in west and to just south of Equator
in east, mainly in desert alongside *Oe.deserti*.

Of the three common winter visitors to Africa, the Isabelline is in the
driest country, the Common Wheatear is in the least dry, and the Black-
eared is intermediate, but they overlap extensively with each other.
Hartley (1949) found interspecific territorial exclusion in the species
which winter in North Africa, and Simmons (1951) reported the same
for species on autumn passage.

P. EUROPEAN REDSTARTS PHOENICURUS

Black Redstart: *Ph.ochruros*: central and southern Europe, originally
  on rocky slopes and cliffs in the mountains, but starting in middle of
  19th century, has spread to villages and towns, with big resulting
  extension of range across lowlands of north-central Europe. Many
  are resident, some move south and west within Europe, others (from
  Asia) to Abyssinia, Somalia and at times Sudan, chiefly in open or
  rocky montane forest.

Common Redstart *Ph.phoenicurus*: almost throughout Europe in open
  broadleaved and pine forest, also parks and gardens. Winters
  across northern tropical Africa from about 15°N, in west to about
  9°30′ and in Congo to 3°N in fairly dry habitats with trees.

Q. EUROPEAN CHATS IN THE GENUS SAXICOLA

BREEDING SEASON, EUROPE

Whinchat *S.rubetra*: northern and central Europe, mainly on damper
  ground than next species, with tall ground vegetation of grass,
  umbellifers or bracken, feeding mainly from flowers and other tall
  herbaceous vegetation.

Stonechat *S.torquata*: central and southern Europe, mainly on drier
  ground than last species, with short grass, *Calluna* heath or gorse,
  feeding mainly on ground.

Hence these two species differ primarily in feeding stations and their
difference in habitat reflects this.

WINTER

Whinchat: across northern tropical Africa from 14°N, in east (presumably Asian populations) to 15°S in Malawi, in open country
  including cultivated land.

Stonechat: in part completely resident, but various northern populations move south in Europe within breeding range, and palaearctic
  birds have been recognized in Abyssinia.

300

14. Main means of ecological segregation of congeneric European transequatorial migrant passerine birds in their summer and winter homes (symbols as in earlier tables)

SUMMER   WINTER

Hirundo swallows

	SUMMER			WINTER	
1 H.daurica	1		1	1	
2 H.rupestris	H	2	2	GH?	2
3 H.rustica	H	H	3	G	G

Anthus pipits

	SUMMER						WINTER				
1 A.campestris	1					1	1				
2 A.cervinus	—	2				2	H	2			
3 A.pratensis	H	H?	3			3	G(H)	G	3		
4 A.s.spinoletta	H	H	H	4		4	G(H)	G	H	4	
5 A.s.petrosus	H	H	H	H	5	5	G(H)	G(H)	H	H	5
6 A.trivialis	H	H	H	H	H	6	H	H	G	G	G

Motacilla wagtails

	SUMMER			WINTER	
1 M.alba	1		1	1	
2 M.cinerea	H	2	2	H	2
3 M.flava	H	H	3	H	H

Lanius shrikes

	SUMMER					WINTER			
1 L.collurio	1				1	1			
2 L.excubitor	HF	2			2	G	2		
3 L.minor	H	?	3		3	H	G	3	
4 L.nubicus	F	—	F	4	4	G	(G)F	G	4
5 L.senator	H	F	?	HF	5	G	G	G	G

APPENDIXES

Muscicapa flycatchers

SUMMER

1	M.albicollis	1			
2	M.hypoleuca	G	2		
3	M.semitorquata	G	G	3	
4	M.parva	F	F	F	4
5	M.striata	F	F	F	F

WINTER

1				
2	G	2		
3	G	G	3	
4	G	G	G	4
5	F?	F?	F?	G

Acrocephalus reed warblers

SUMMER

1	A.arundinaceus	1					
2	A.melanopogon	H	2				
3	A.paludicola	H	H	3			
4	A.palustris	H	H	H	4		
5	A.dumetorum	—	—	H	G	5	
6	A.schoenobaenus	H	H	H?	H	H	6
7	A.scirpaceus	F	HF	H	H	—	H

WINTER

1					
2	G	2			
3	?	G	3		
4	?	G	G?	4	
5	G	G	G	G	5
6	?	G	?	?	G
7	F	G	?	?	G

Hippolais warblers

SUMMER

1	H.icterina	1		
2	H.olivetorum	G	2	
3	H.polyglotta	G	G	3
4	H.pallida	H	H	H

WINTER

1			
2	?	2	
3	G	G	3
4	G(H)	G(H)	G(H)

Locustella grasshopper warblers

SUMMER

1	L.fluviatilis	1	
2	L.luscinioides	H	2
3	L.naevia	H	H

WINTER

1		
2	H	2
3	H?	G?

Phylloscopus leaf-warblers

SUMMER

1	P.bonelli	1				
2	P.borealis	—	2			
3	P.collybita	F?	H	3		
4	P.sibilatrix	H	—	H	4	
5	P.trochiloides	—	H	H	H?	5
6	P.trochilus	GH	H	HF	H	H

WINTER

1					
2	G	2			
3	H	G	3		
4	G(H)	G	H(G)	4	
5	G	G	G	G	5
6	G	G	G	H	G

SUMMER   *Sylvia warblers*

	1	2	3	4	5	6	7	8	9	10	11	12
1 S.atricapilla	1											
2 S.borin	F	2										
3 S.cantillans	H	—	3									
4 S.communis	H	H	H	4								
5 S.conspicillata	H	—	H	H	5							
6 S.curruca	H	H	—	H	—	6						
7 S.hortensis	H	—	H	H	H	—	7					
8 S.melano-cephala	H	—	FH	H	H	—	H	8				
9 S.nisoria	H	H	—	H	—	H	—	—	9			
10 S.rüppelli	H	—	H	H	—	—	H	?	—	10		
11 S.sarda	H	—	H	—	H	—	H	H	—	G	11	
12 S.undata	H	H	H	H	H	H	H	GH	—	G	G?	

WINTER

	1	2	3	4	5	6	7	8	9	10	11
	1										
2	$\frac{1}{2}$G$\frac{1}{2}$F?	2									
3	G(H)	H	3								
4	H	H	H?	4							
5	H	—	—	H	5						
6	H	H	G	?	—	6					
7	H	—	?	?	—	?	7				
8	H	—	—	—	H	—	—	8			
9	H	H	—	?	—	?	?	—	9		
10	H	—	—	—	—	—	—	H	—	10	
11	H	—	—	—	H	—	—	H	—	G	11
12	H	—	—	—	H	—	—	H	—	G	G?

SUMMER   WINTER

Luscinia nightingales

	1	2			1	2
1 L.luscinia	1					
2 L.megarhyncha	G	2		2	G	2
3 L.svecica	H	H		3	G(H)	G(H)

SUMMER   WINTER

Monticola rock-thrushes

SUMMER

1	M.saxatilis	1
2	M.solitarius	H

WINTER

1		1
2	G	

Oenanthe wheatears

SUMMER

1	Oe.hispanica	1		
2	Oe.leucomela	G	2	
3	Oe.oenanthe	GH	GH	3
4	Oe.leucura	H	—	H

WINTER

1		1		
2	$\frac{1}{2}$G$\frac{1}{2}$?	2	
3	$\frac{1}{2}$G$\frac{1}{2}$?	?	3
4	GH	GH	GH

Phoenicurus redstarts

SUMMER

1	P.ochuros	1
2	P.phoenicurus	H

WINTER

1		1
2	G(H)	

Saxicola chats

SUMMER

1	S.rubetra	1
2	S.torquata	F(H)

WINTER

1		1
2	G	

15. Ecological isolation in the other congeneric European passerine birds

Based on Voous 1960 unless otherwise stated.

LARKS ALAUDA (SENS. LAT.)

Following Harrison (1966), the genera *Galerida* and *Lullula* are merged in *Alauda*.

Woodlark *A.(Lullula)arborea*: widespread except much of north, but in south chiefly in hills and low mountains; on dry grassland and heathland with widely scattered trees, especially pine or juniper, and locally in cultivated land of similar appearance.

Skylark *A.arvensis*: widespread, in open treeless grassland, grassy moorland, sand dunes, drier marshes and cultivated fields, usually on more fertile ground than the other three species, in south primarily montane.

Crested Lark *A.(Galerida)cristata*: widespread in middle and south, in sandy areas with coarse grass, often along sandy roads and edges of fields.

Thekla Lark *A.(Galerida)thekla*: southern Spain, on barren hillsides with bare stony patches among *Cistus* and broom (Abs 1963), also sand dunes on Coto Donana where *A.cristata* absent (A. Valdeverde pers. comm.); the only species in Majorca. (All four species separated by habitat, so scored as 6H.)

SHORT-TOED LARKS CALENDRELLA

Short-toed Lark *C.brachydactyla*: widespread in south, on sandy plains and steppe.

Lesser Short-toed Lark *C.rufescens*: south and east Spain, on arid salty flats near sea (and in Asia in more arid semi-desert than *C.brachydactyla*). (H).

ACCENTORS PRUNELLA

Dunnock or Hedge Sparrow *P.modularis*: widespread, in broadleaved and coniferous scrub in and outside woods in lowlands, and in central Europe in stunted montane conifer forest below the alpine zone.

Alpine Accentor *P.collaris*: in central and southern mountains, in rocky parts of alpine zone. (H).

GOLDCRESTS OR KINGLETS REGULUS

Firecrest *R.ignicapillus*: mainly in centre and south, hence partly replaces Goldcrest geographically. In big area of overlap, Firecrest typically at lower and Goldcrest at higher altitudes, e.g. Switzerland (Glutz 1962) and Corsica (Jourdain 1911–12), and Firecrest often in broad-leaved woods, Goldcrest normally in conifers, but Firecrest's habitat has a few conifers and at times it breeds in pure conifers.

Goldcrest *R.regulus*: mainly in centre and north, hence partly separated by range from Firecrest. Normally in spruce and fir, sometimes pine, but also in oak woods in British Isles (where Firecrest absent until last few years), usually with Yew *Taxus baccata* in understorey. The two species are not normally found in same wood. ($\frac{1}{2}$G, $\frac{1}{2}$H).

THRUSHES TURDUS (see also Willgohs 1951 for Scandinavia)

Blackbird *T.merula*: widespread, typically in oak or other heavy broad-leaved woods rich in bushes, avoids pure conifers. In far north dependant on rural farmsteads. In middle and southern Europe also in towns. In dry weather and in winter, often searches by tossing fallen leaves aside.

Redwing *T.iliacus*: widespread in north and east, in light open forest, especially in birch and willow, or mixed woodland, usually with bushes, and often at wood edge; avoids dark woods of pure conifers (Tyrvainen 1969). In winter migrates south to grassy fields, hedgerows and open woods.

Song Thrush *T.philomelos*: widespread; in Fenno-Scandia prefers dark spruce woods with bushes and avoids open woods of birch and pine, in middle Europe breeds in both spruce woods and pure broadleaved woods with bushes. In the latter coexists with Blackbird, but is smaller and takes snails in times of drought and frost, which Blackbird does not normally do.

Fieldfare *T.pilaris*: widespread in north and east, nesting in isolated clumps of broadleaved trees, especially birch, or at the wood edge, or in open woodland bare of bushes and near boggy ground. Migrates south to fields and hedgerows rich in berries.

Ring Ousel *T.torquatus*: widespread in mountain areas. The subspecies in Fenno-Scandia and the British Isles breeds above the treeline in rocky and often treeless moorland, often in a stream valley with some bushes. The central European subspecies breeds primarily in montane coniferous forest, mainly on damper soils. Migrates to North Africa, especially the Saharan Atlas, where it is away from the rest.

Mistle Thrush *T.viscivorus*: widespread, in northern and montane central Europe primarily in open dry pine woods, but in west-central Europe in open broadleaved woodland and parkland, feeding especially on dry grassland.

Hence in the breeding season all these species are separated from each other primarily by habitat, except that the Blackbird and Song Thrush coexist in part of their ranges and are separated by feeding stations, and the Redwing and Fieldfare at times breed in the same woods and are presumably separated by feeding stations (linked with a difference in size). (14H, 1F). The situation in winter, notably in the British Isles, requires further study.

TREE-CREEPERS CERTHIA (ranges shown in fig. 25).

Short-toed Tree-Creeper *C.brachydactyla*: widespread in south and middle, also in part of east, chiefly in broadleaved woods at low altitude; where *C.familiaris* is absent in southern Europe, also in conifers and at higher altitudes, notably in central Spain and parts of Mediterranean littoral (also the Maghreb).

Tree-creeper *C.familiaris*: widespread in north and east, also in parts of middle and south, primarily in coniferous woods; mainly at higher altitudes in area of overlap with *C.brachydactyla*, but elsewhere also at low altitudes. Also in broadleaved woods in British Isles and Corsica (Jourdain 1911–12), where *C.brachydactyla* absent. Hence both species have more restricted habitats in their area of overlap than outside it ($\frac{1}{2}$G, $\frac{1}{2}$H).

BUNTINGS EMBERIZA

Yellowbreasted Bunting *E.aureola*: solely northeast, in willow and birch with rich herbaceous vegetation at edge of taiga.

Cretzschmar's Bunting *E.caesia*: extreme southeast, on dry, stony, treeless hill slopes with xerophilous bushes; the similar-looking *E.hortulana* there occurs only at higher altitudes (Niethammer 1943, Makatsch 1950, Peus 1954, Stresemann and Portenko 1960).

Corn Bunting *E.calandra*: widespread in middle and south, in grass steppe and corn fields with tall herbs or sparse low bushes. Partly overlaps in habitat, especially in farmland, with *E.citrinella* or *E.hortulana*, but has a much larger and broader beak, so probably takes mainly different foods.

Rock Bunting *E.cia*: widespread in south and local in mid-Europe, on dry, sunny and often steep rocky slopes in subalpine and montane zones, in open forest or open bushy areas, also vineyards; the Ortolan prefers barer, flatter and more open country (Scharnke and Wolf 1938).

Cirl Bunting *E.cirlus*: widespread in south, also in western part of mid-Europe, in light open woodland (especially in south), and heathland or farmland with scattered trees and bushes. Habitat closely similar to that of Yellowhammer, which replaces it to north and east, and at higher altitudes in area of overlap. In southern England almost restricted to wooded farmland on chalk or limestone, and it shows the same preference in New Zealand, where it was introduced (Falla et al. 1966).

Yellowhammer *E.citrinella*: widespread except Spain and Mediterranean littoral, in open woodland, wooded heaths and farmland with trees like Cirl Bunting, occurring at higher altitudes where their ranges overlap.

Ortolan *E.hortulana*: widespread except parts of west and northwest, in rather arid steppe and wide dry fields, in south at higher altitudes than Cretzschmar's. Habitat more open than that of Yellowhammer, for small overlap with Corn Bunting see latter.

Blackheaded Bunting *E.melanocephala*: solely southeast, in tall, xerophilous macchia, orchards, and trees beside fields, chiefly at low altitudes. Separated by habitat from Cretzschmar's, Ortolan, Rock, Corn (here primarily in cornfields) and Cirl (here in more wooded country at higher altitudes) (references as for Cretzschmar's).

Little Bunting *E.pusilla*: solely northeast, in open birch forest with grass meadows or willow scrub (Andersson et al. 1968). Is also smaller than other buntings in this region, so is provisionally classed as differing from them in both habitat and feeding.

Rustic Bunting *E.rustica*: solely northeast, in swamps with willow or birth in coniferous taiga, in more closed and wooded habitats than Yellowbreasted or Reed.

Reed Bunting *E.schoeniclus*: widespread in marshy ground with grass, reeds and bushes.

Nearly all these species differ from each other in habitat, while many have no contact with each other, as summarized below ($\frac{1}{2}$G, 17—, 33H, $4\frac{1}{2}$F).

1	*E.aureola*	1									
2	*E.caesia*	—	2								
3	*E.calandra*	—	H	3							
4	*E.cia*	—	H	H	4						
5	*E.cirlus*	—	H	(H)F	H	5					
6	*E.citrinella*	H	—	(H)F	H	GH	6				
7	*E.hortulana*	H	H	(H)F	H	H	H	7			
8	*E.melanocephala*	—	H	H	H	H	—	H	8		
9	*E.pusilla*	HF?	—	—	—	—	H	H	H	9		
10	*E.rustica*	H	—	—	—	—	H	H	—	HF?	10	
11	*E.schoeniclus*	H	H	H	H	H	H	H	H	HF?	H

(Watson [1964] found the Cinereous Bunting *E.cineracea* breeding on arid ground in Greece, but it is too rare to be considered here.)

SPARROWS PASSER

House Sparrow *P.domesticus*: widespread, in towns, villages and farmland, breeding on houses (rarely in trees).

Spanish Sparrow *P.hispaniolensis*: southern Europe. In Sardinia (also Canaries) replaces House Sparrow geographically in similar habitats. In Spain and Macedonia (also parts of Maghreb and Asia Minor) separated by habitat from *P.domesticus*, frequenting thickets away from cultivation, especially along rivers, and often breeding in nests of storks or raptors (as does House Sparrow at times). In most of rest of southern Europe (and rest of Maghreb) hybridizes freely with House Sparrow, in Italy resulting in stable population often treated as separate species *P.italiae*.

Tree Sparrow *P.montanus*: widespread, in lightly wooded country with old trees (for nesting holes) on edges of cultivated land, also in lines of trees and wooded gardens. Separated by habitat from other species (including Spanish Sparrow where rural in Spain and Sardinia, Steinbacher 1960). In east and southeast Asia, where House Sparrow absent, Tree Sparrow takes its place in the towns. ($2\frac{1}{2}$H, $\frac{1}{2}$ no segregation).

STARLINGS STURNUS

Rose-coloured Pastor or Starling *S.(Pastor)roseus*: in eastern Europe, in grass steppe with rocky hills, breeding irregularly where locust swarms have settled.

Spotless Starling *S.unicolor*: Spain and islands of west Mediterranean, replacing next species geographically in similar habitats, with apparently a small zone where neither is found.

Starling *S.vulgaris*: widespread, except where replaced by Spotless Starling, in farmland, pastures, forest edge, villages and towns. (1G, 1—, 1F).

CROWS CORVUS

Raven *C.corax*: widespread, in tundra, mountains above the treeline, very open (but not closed) forest, cliffs and deserts. Eats carrion, small mammals, seeds and many other foods, but is so much larger, with a larger beak, than all the other species that it presumably differs from them in its main natural foods.

Hooded Crow *C.cornix*: northern and eastern Europe, Ireland and Italy, with narrow zone of interbreeding with *C.corone*, on heaths, moorland, sea cliffs and rather open woodland. Foods similar to those noted for Raven, but presumably different on average.

Carrion Crow *C.corone*: replaces Hooded Crow in western and southern Europe, in similar habitats and eats similar foods; locally in towns.

Rook *C.frugilegus*: middle Europe, in grassy country, parkland, riverine woods, and cultivated land with clumps of trees, including villages, but not large towns. Probes in soil for invertebrates to much greater extent than other species, and has relatively long pointed beak with bare patch at the base, so presumably differs in its basic diet, though it takes a wide variety of other foods at times, including seeds. Also differs in habitat from the two crows.

Jackdaw *C.monedula*: widespread, in grassy country, parkland, very open woods and sea cliffs, locally in towns, but decidedly smaller than all other species, and in contrast to Rook, takes almost all its animal and plant food from the surface (Lockie 1956). (1G, $\frac{1}{2}$H, $7\frac{1}{2}$F).

These presumed differences in food refer primarily to natural habitats; in cultivated land all the species may depend on grain (Holyoak 1968).

CHOUGHS PYRRHOCORAX

Alpine Chough *P.graculus*: south and central Europe, in alpine zone above treeline. Nowadays eats much village garbage in winter.

(Redbilled) Chough *P.pyrrhocorax*: in south on lower mountains and rocky hills, in west on sea cliffs, usually feeding in grass. Partly differs in altitude from Alpine, but much overlap. As it has a longer and thinner beak, it presumably probes more deeply in soil or takes smaller insects, which doubtless separates it where the two coexist on montane pastures ($\frac{1}{2}$H$\frac{1}{2}$F).

SUMMARY

The combined totals for the different types of interspecific segregation in the families included here are $3\frac{1}{2}$ separated by geographical range, 18 with no contact, $59\frac{1}{2}$ by habitat, $14\frac{1}{2}$ by feeding and $\frac{1}{2}$ with no segregation (two sparrows in part of their range). Of the $14\frac{1}{2}$ instances of separation by feeding, $12\frac{1}{2}$ (those in buntings and crows) are associated with a big difference in size or size of beak.

16. Ecological isolation in other (non-passerine) European birds

A. LAND BIRDS

Based on Voous 1960 unless otherwise stated.

Note

Only those differences which are probably critical for isolation are set out, for range and other ecology see general works, notably Voous (1960), from whom the main information has been derived.

RAPTORS FALCONIFORMES

Accipiter: The large Goshawk *A.gentilis* takes larger birds than the small Sparrowhawk *A.nisus*, and the latter differs in habitat from the Levant Sparrowhawk *A.brevipes*.

Aquila: The northern and montane Golden Eagle *A.chysaetos* differs in range and habitat from Imperial *A.heliaca* of the southern plains. Both much larger than the other two species, which differ at least partly in habitat, Spotted *A.clanga* in open forest and lakes, Lesser Spotted *A.pomarina* in dense forest.

Buteo: Common Buzzard *B.buteo* is replaced in far north and Scandinavian mountains by Rough-Legged *B.lagopus*. In Greece, Long-legged *B.rufinus* is in open and *B.butes* in wooded country, but *B.butes* is in wooded and open country elsewhere.

Circus: Marsh Harrier *C.aeroginosus* breeds in marshes, so differs in habitat from Hen *C.cyanus*, Pallid *C.macrourus* and Montagu's *C.pygargus*, but how three latter might be segregated is not known. Pallid is solely in extreme east, where other two, which are widespread, also breed. Montagu's is smaller than Hen, so probably takes smaller prey, Pallid frequents drier country than other two.

Hieraaëtus: Bonelli's Eagle *H.fasciatus* of open montane country differs in habitat from Booted *H.pennatus* of woodland.

Milvus: Black Kite *M.migrans* differs partly in habitat and partly in food from the Red Kite *M.milvus*, since in Europe it usually breeds in woods near water, eats much fish and scavenges by water, whereas Red Kite breeds in open hill country with woods, and eats small mammals and mammal carrion. Both also eat a variety of other foods.

Falco: 10 species breed in Europe. The four large species presumably take mainly larger prey than the rest. Three of these, the northern Gyr *F.rusticolus*, eastern Saker *F.cherrug* and southern Lanner *F.biarmicus*, have separate ranges, but the Peregrine *F.peregrinus* overlap with all three. It takes more birds in flight and many fewer mammals and gamebirds off the ground than the Gyr or Saker. Of the others, the Kestrel *F.tinnunculus* specializes on small rodents, the Hobby *F.subbuteo* on hirundines and dragonflies (see fig. 26), the Merlin *F.columbarius*, a northern species, on small passerines such as larks and pipits on moorland, the Lesser Kestrel *F.naumanni*, a southern species, on grasshoppers, beetles and small land vertebrates, the Redfooted Falcon *F.vespertinus*, an eastern species, on large insects, caught especially at dusk, and Eleanora's Falcon *F.eleonorae* of the Mediterranean islands on small passerine autumn migrants and large insects.

GAMEBIRDS GALLIFORMES

Alectoris: Four species replace each other geographically in Europe, (as shown in fig. 27), the Redlegged Partridge *A.rufa* in the west, the Barbary *A.barbara* in Sardinia and Gibraltar (introduced from North Africa), the Rock Partridge *A.graeca* in the mountains of southern and eastern Europe and the Chukor *A.chukar* in part of Balkans. Where *A.graeca* adjoins *A.rufa* in west and *A.chukar* in east respectively, it occurs at a higher altitude, but these two latter species occur at high altitudes elsewhere (Watson 1962ab).

Lagopus: The (Rock) Ptarmigan *L.mutus* of the arctic-alpine zone is separated by habitat from both the Willow Ptarmigan *L.lagopus* of subarctic willow and birch forest and the Red Grouse *L.scoticus* of British heather moors. *L.scoticus* replaces *L.lagopus* geographically in the British Isles (and is by some workers considered merely a subspecies).

BUSTARDS OTIDIDAE

Otis: The Great Bustard *O.tarda* is so much larger than the Little Bustard *O.tetrax* that they presumably differ in diet.

SANDGROUSE PTEROCLIDIDAE

Pterocles: The way in which the Pintailed Sandgrouse *Pterocles alchata* might be separated from the Black-bellied Sandgrouse *P.orientalis* is not known.

PIGEONS COLUMBIDAE

Columba: The Wood Pigeon *C.palumbus*, naturally a woodland bird, there feeds on buds, flowers and seeds of trees and of various herbs; it nests among the twigs, or in a fork, of a tree. In cultivated fields, it depends primarily on grain in late summer and clover in winter, supplemented by acorns and beechmast in season, and cultivated brassicas in snow. The smaller Stock Dove *C.oenas*, which nests in holes in trees, rabbit burrows and sandy cliffs, does not depend on trees for food, and eats primarily weed seeds, including those of *Sinapsis, Brassica, Stellaria, Polygonum* and *Chenopodium*, hence it depends mainly on different seeds from *C.palumbus*, though it also takes grain. The Rock Dove *C.livia* is of similar size to *C.oenas*. It breeds in rocky cliffs by the sea and inland, with a feral form in towns nesting inside buildings and in protected sites outside them. In towns it depends on grain by the docks or bread provided by man, and in the countryside on the same weed seeds as *C.oenas*, supplemented by grain. Hence these two species are in potential competition, but as they feed in the general neighbourhood of their nests, they largely differ in habitat (Murton and Isaacson 1964, Murton and Westwood 1966, Murton and Clarke 1968, Murton 1968).

Streptopelia: The Turtle Dove *S.turtur* breeds in farmland hedges and feeds on the seeds of *Fumaria, Stellaria* and grasses (Murton and Isaacson 1964), so differs in habitat from the Collared Dove *S. decaocto*, which breeds in evergreen shrubs or trees in suburbs and feeds mainly on grain put out for poultry.

OWLS STRIGIDAE

Asio: The Short-eared Owl *A.flammeus* of open moors, grassland and marshes differs in habitat from the Long-eared Owl *A.otus* of light woodland.

Strix: The large Great Grey Owl *S.nebulosa* and medium-sized Ural Owl *S.uralensis* occupy mainly separate ranges in Scandinavia (Svärdson 1949), but coexist in northern USSR (Voous 1960); they differ fairly markedly in size and hence presumably in size of prey. The Tawny Owl *S.aluco* is more southern than either, but overlaps with them, notably in the USSR; it differs in its habitat, light woodland, and also presumably in size of prey.

Eight other owls, each in a different genus, breed in Europe, ranging from the huge Eagle Owl *B.bubo* to the tiny Pygmy Owl *Glaucidium passerinum*, but though at least some of them differ in their prey, critical study has not been made.

NIGHTJARS CAPRIMULGIDAE

Caprimulgus: It is not known how the Nightjar *C.europaeus* differs from the Red-necked Nightjar *C.ruficollis* where they coexist in Spain.

SWIFTS APODIDAE

Apus: The Alpine Swift *A.melba* is so much larger than the other species that it presumably differs either in where it feeds (more in the open) or in the size of insects taken. It is not known how the widespread Common Swift *A.apus* differs in its food from the similar-sized Pallid Swift *A.pallidus* of southern Europe, but since *A.apus* raises one brood in a shorter season and *A.pallidus* two broods in a more extended season, they evidently differ. The somewhat smaller White-rumped Swift *A.caffer* coexists with them in south-western Spain. These 4 species differ in nesting sites, *A.melba* under roofs, many pairs entering by the same hole and nesting alongside each other, *A.apus* in small holes in trees, cliffs or buildings, each pair normally having a separate entrance and no contact inside with other pairs, *A.pallidus* on the top of ledges under eaves, at times in old nests of Swallow *H.rustica*, and *A.caffer* in closed tunnel nests of Redrumped Swallow *H.daurica* on ceilings of caves or culverts. As in the *hirundinidae* (see p. 95), this presumably means that there has been potential competition between them for nesting sites, though probably not at the present time, when most of them nest mainly in human buildings. Because *H.daurica* nests in limestone caves, *A.caffer* is separated by habitat from the other species of *Apus* in Spain. The situation in winter quarters in Africa is inadequately known (Moreau in prep.).

APPENDIXES

WOODPECKERS PICIDAE

Dendrocopos: Three species differ markedly in size (measurements from Niethammer 1938), the Great Spotted *D.major* (wing 135, culmen 29 mm), Middle Spotted *D.medius* (wing 125, culmen 26·5 mm) and Lesser Spotted *D.minor* (wing 89, culmen 16·5 mm). Correlated with this, *D.major* feeds mainly on trunks and large branches and *D.minor* on small and high branches and twigs; in addition *D.major* extracts pine seeds from cones, which *D.minor* does not do, but *D.medius* has not been studied. The Syrian Woodpecker *D.syriacus* replaces *D.major* geographically in the southeast, and where they overlap, inhabits dry lowlying cultivated land with scattered trees, whereas *D.major* is in hill forest. The White-backed Woodpecker *D.leucotos* has not been studied critically, but is larger with a proportionately larger beak, than *D.major* (wing 142, culmen 37 mm), so presumably differs in food, at least from *D.medius* and *D.minor*, and in habitat from *D.syriacus*

Picus: The Green Woodpecker *P.viridis* of southern and western Europe is replaced geographically in eastern Europe (and Asia) by the Grey Woodpecker *P.canus*, but they overlap in much of central Europe, where *P.canus* inhabits denser woods with a higher proportion of conifers and feeds much less on the ground in the open than *P.viridis*, though outside their area of overlap they do not differ in these respects (Voous 1960).

B. THE WADING BIRDS

HERONS AND STORKS CICONIIFORMES

Ardea: The Grey Heron *A.cinerea* feeds beside almost all types of open waters, but rarely in reedbeds or other cover like the Purple Heron *A.purpurea*, which also has a much narrower beak and breeds much later in the summer.

Egretta: The Great White Heron *E.alba* is much larger, so presumably takes larger prey than, the Little Egret *E.garzetta*. (*E.alba* is sometimes put in *Ardea*.)

Ciconia: The White Stork *C.ciconia* breeds in open country and often feeds on drier ground than the Black Stork *C.nigra*, which breeds mainly in forest.

RAILS RALLIDAE

Porzana: The Spotted Crake *P.porzana* is decidedly larger than the Little *P.parva* and Baillon's *P.pusilla*, but how the two latter might differ is not known.

314

PLOVERS CHARADRIIDAE

Charadrius: On the sea shore, the southern Kentish Plover *C.alexan-drinus* is largely replaced geographically by the northern Ringed *C.hiaticula,* but they overlap on the Danish, German and Dutch North Sea coasts, where the Kentish is on sandy and salty flats by the sea, estuaries and brackish lagoons, and the Ringed on shingle or stones with sand on open beaches. They are separated by habitat from the freshwater Little Ringed *C.dubius,* which breeds on sandy shores of, or banks in, rivers and shallow lakes, at times in estuaries. In the Baltic, the Ringed is on the outer and the Little Ringed on the inner islands (cf. the Rock and Meadow Pipits p. 287). On the mainland in Scandinavia the Ringed breeds on the tundra north of, and at a higher altitude than, the Little Ringed (Voipio 1956). Hence these three species are separated from each other by different means in different parts of Europe.

Pluvialis: In Europe the Golden Plover *P.apricaria* is largely separated geographically from the Grey *P.squatarola* (solely arctic USSR); and the Golden is on wet moors and tundra, the Grey on arid tundra. In winter they move south, and the Golden feeds on grassland and grassy mudflats, the Grey on the sea shore.

SANDPIPERS SCOLOPACIDAE

Gallinago (Capella): The widespread Common Snipe *G.gallinago* breeds in marshes, the northern Great Snipe *G.media* in less swampy and more wooded ground. Many Common Snipe winter further north than any Great Snipe.

Limosa: The Bar-tailed Godwit *L.lapponica* breeds further north than the Black-tailed *L.limosa,* with a large gap between. In winter, the Bar-tailed feeds on the sea shore and estuaries near the sea, the Black-tailed higher up in estuaries, and by freshwater rivers and lakes.

Numenius: The Curlew *N.arquata* breeds mainly south of the Whimbrel *N.phaeopus.* The boundary between their ranges has been shifting north, probably correlated with the amelioration of the northern summer (cf. *Fringilla* species p. 90). Where these two species meet, the Curlew is solely in cultivated land, and man is cultivating more northerly valleys than formerly (Voipio 1956). Many Curlew winter north of the normal winter range of the Whimbrel, and they feed both on the sea shore and on grassy mudflats, whereas Whimbrel feed primarily on the sea shore.

Tringa: Of the seven European species, six breed in Lapland. The Spotted Redshank *T.erythropus* and Greenshank *T.nebularia* have so much longer legs and beaks than the rest that they presumably take different foods from them. The Spotted Redshank has a long narrow beak and usually wades in deep water, the Greenshank has a broader and uptilted beak, and usually feeds on bare mud or in shallow water, so they presumably take mainly different prey. Further, the Spotted Redshank nests on dry ridges in open forest and the Greenshank in small bogs or beside small lochs in open forest or on moors. The Redshank *T.totanus,* smaller than these two, is larger than the rest, so presumably takes mainly different prey and it breeds beside slow-flowing rivers, in damp grassland and on coastal salt marshes. Next in descending order of size, and similar to each other, are the Wood Sandpiper *T.glareola,* which breeds in large open swamps, and the Green Sandpiper *T.ochrophus,* which differs in habitat as it breeds in boggy ground in forest, in the old nest of a thrush *Turdus* in a tree. The Marsh Sandpiper *T.stagnatilis* is also similar in size to the last two, and how it might be segregated from them is not known; in Europe it is solely in the east at middle latitudes. The smallest species, the Common Sandpiper *T.hypoleucos,* also differs in habitat, as it breeds beside rocky streams and stony lake shores (but, in the absence of the Redshank, the North American subspecies breeds in damp grassland). Hence some of these 7 species differ from each other completely, and others partly, in their breeding habitats. Their differences in beak and leg, being adapted to different ways of feeding, probably provide the main means of segregation outside the breeding season, but this needs checking, and especially in view of the similarity in the summer foods of artic American *Calidris* sandpipers which differ in size (see pp. 134–6), the possibility that these *Tringa* species differ in their main foods in summer should not be taken for granted. For pictures, see p. 15.

Stints (peeps) Calidris: One high arctic species, the Knot *C.canutus,* is much larger than the rest, so presumably differs in its preferred prey. Three species, similar in size to each other, are the Sanderling *C.alba,* restricted in Europe to stony tundra in Spitsbergen, the Purple Sandpiper *C.maritima,* on damp tundra or near lakes, in lowlands in Spitsbergen but solely in mountains in Iceland and Fenno-Scandia, and the Dunlin *C.alpina,* widespread in northern Europe in moss tundra, damp highland moors and coastal marshes, at lower altitudes than the Purple where their ranges overlap. Smaller than these three and with relatively short beaks are the Little Stint *C.minutus* and Temminck's *C.temminckii,* which breed in moss and shrub tundra respectively.

In autumn and winter, after they have moved south, the Knot, Dunlin and Little Stint, which differ markedly from each other in size, feed primarily on coastal mudflats, while the Sanderling feeds on firm sand, especially where waves are breaking, the Purple Sandpiper on rocks near the tideline and Temminck's Stint beside freshwaters. Hence these latter three are separated by habitat. On the north German coast in winter, the Knot takes mainly molluscs, the Dunlin nereid worms and the Sanderling insects (crustacea elsewhere) (Ehlert 1964). Note that various arctic American species which differ in size overlap in their foods in summer, but are segregated in winter (see pp. 134–6).

Phalaropus: The Grey (Red) Phalarope *P.fulicarius* breeds almost entirely north of the Rednecked *P.lobatus*, and beside barer pools. In winter both are oceanic, and perhaps differ in range.

C. THE FRESHWATER SWIMMING BIRDS
Including the marine species in families predominantly consisting of freshwater species

DIVERS (LOONS) GAVIIDAE
Gavia: The White-billed Diver *G.adamsii* and Great Northern *G.immer* occupy separate ranges in the extreme north of Europe, so does the Black-throated *G.arctica,* but the last overlaps with them in Asia and North America, so is probably separated primarily by its smaller size, and presumed smaller prey, also by its preference for smaller and richer lakes in forest, whereas the other two frequent large barren lakes in tundra. (But the Great Northern breeds in forest lakes in North America.) The Redthroated *G.stellata,* which overlaps in range with the others, is smaller, so probably takes mainly different prey, and breeds on small shallow lakes in tundra, often near the sea, in which it usually feeds. In winter, all four move out to sea, and their ecology needs study, since Madsen (1957) reported their diets to be very similar.

GREBES PODICIPITIDAE
Podiceps: Grebes are of three sizes, which presumably differ from each other in their main diet (i) the large Great Crested *P.cristatus* and Rednecked *P.griseigena,* (ii) the northern Horned (Slavonian) *P.auritus* and southern Blacknecked *P.nigricollis,* which replace each other geographically and (iii) the small Dabchick *P.ruficollis.* How the first two species are segregated is not known, and in Denmark their diet is similar, but different from that of the other three (Madsen 1957). Species in all three size-groups often coexist on larger waters, but smaller waters have only the smaller species. Their winter ecology needs study.

APPENDIXES

Pelecanus: How the White Pelican *P.onocrotalus* might be separated from the Dalmation *P.crispus* is not known.

Phalacrocorax: Of the two species in freshwater, the Cormorant *P.carbo* is so much larger than the Pygmy Cormorant *P.pygmaeus* that they presumably take different prey. The Cormorant also feeds in estuaries and coastal salt water, where it takes mainly flatfish and prawns off the bottom. Hence it differs from the Shag *P.aristotelis,* which is exclusively marine and feeds well offshore, on free-swimming, not bottom-living, fish; these two species also differ in nesting sites (Lack 1945a, Pearson 1968).

WATERFOWL ANATIDAE

Anas: 8 species of surface-feeding ducks breed in Europe, as many as 7 of them on the south side of the Baltic and 6 together in several other areas, but the Marbled Teal *A.angustirostris* is restricted to the extreme south of Europe. The 6 breeding in the Gulf of Bothnia differ in nesting sites and habitat preferences (Hilden 1964), but with so much overlap that these do not result in ecological segregation. Probably the main differences are in feeding. In southern England in winter, as portrayed in fig. 28, the Pintail *A.acuta,* which has a long neck, up-ends for plants on the bottom, the Shoveller *A.clypeata* strains micro-organisms through the fringing filters of its beak from the surface of the water, the Wigeon *A.penelope* grazes on grass on land and on estuarine *Zostera,* and the relatively unspecialized large Mallard *A.platyrhynchos,* and small Teal *A.crecca* also take mainly different foods from each other and from the rest (Olney 1963). But each at other times feeds like one or more of the others, as several of them up-end and several 'dabble' water through their beaks. Further the Gadwall *A.strepera* looks very like a Mallard and the Garganey *A.querquedula* like a Teal, and how they might be separated is not known. Many ducks have been extending their breeding ranges in the last fifty years, so the present position might be unstable.

Aythya: Of the diving ducks in the genus *Aythya,* the Scaup *A.marila* eats mainly animal food, the Pochard *A.ferina* and Ferruginous Duck *A.nyroca* mainly vegetable matter and the Tufted *A.fuligula* both, its animal food being smaller than that of the Scaup. In winter the Scaup goes to the sea, the rest stay on fresh waters.

Bucephala: The Goldeneye *B.clangula* and Barrow's Goldeneye *B. islandica* (solely in Iceland) differ in range.

Melanitta: How the Common Scoter *M.fusca* and Velvet Scoter *M.nigra* might be separated is not known.

APPENDIXES

Somateria: How the Common Eider *S.mollissima* and King Eider *S. spectabilis* might be separated is not known.

Mergus: The Goosander (American Merganser) *M.merganser* breeds in forest and feeds in rocky, fast-flowing streams, the Redbreasted Merganser *M.serrator* breeds in open country and feeds in lakes, slower rivers and estuaries, and the Smew *M.albellus* breeds in forest and feeds in smaller lakes and rivers than the other two. The large Goosander and small Smew differ greatly in the size of fish and invertebrates taken (Madsen 1957), and the Redbreasted Merganser eats smaller fish than the Goosander (Hilden 1964). In winter, the Redbreasted Merganser is in salt water, the other two in fresh waters.

Tadorna: The Shelduck *T.tadorna* is in Europe a coastal species. It overlaps on the Black Sea with the Ruddy Shelduck *T.casarca,* which breeds inland (but the Shelduck breeds inland in Asia and how they might be separated there is not known).

Anser: The Greylag Goose *A.anser* differs in breeding range from the Whitefront *A.albifrons,* and in its habitat, lowlying open ground, from the Bean Goose *A.fabalis* in conifer forest and the Lesser Whitefront *A.erythropus* in montane willow tundra in Fenno-Scandia, and from the Pinkfoot *A.brachyrhynchus* in montane areas in Iceland. The last differs in breeding range from the Bean (which is by some authors treated as conspecific). In winter, these 5 species are commonly found in different places.

Branta: The Brent Goose *B.bernicla* breeds on moss tundra and the Barnacle Goose *B.leucopsis* on rocky cliffs. In winter, the Brent feeds largely on *Zostera* on the sea shore and in estuaries, and the Barnacle on sprayblown grassland above the tideline.

Cygnus: Bewick's Swan *C.columbianus,* the Whooper *C.cygnus* and the Mute *C.olor* have different breeding ranges. Their winter ecology needs study.

COOTS FULICINAE

Fulica: In the world as a whole, the Coot *F.atra* is largely separated by range from the Crested Coot *F.cristata,* but in the only part of Europe where the latter occurs, namely southern Spain, both are found, and how they might be separated is not known.

FRESHWATER TERNS CHLIDONIAS

Chlidonias: The Whiskered Tern *C.hybrida* feeds on small animals, including insects, caught on the surface of the water, and the White-winged Black Tern *C.leucoptera* on insects, especially dragonflies, caught on the wing (Voous 1960), but the much more widespread

Black Tern *C.niger* feeds in both ways, and how it might be separated from the other two, especially where they all coexist, is not known.

D. THE SEABIRDS

PETRELS PROCELLARIIFORMES

Puffinus (Procellaria): The North Atlantic Shearwater *P.diomedea* is much larger than the Manx *P.puffinus,* but whether they differ in size of prey is not known.

SKUAS (JAEGERS) STERCORARIIDAE

Stercorarius: The Longtailed Skua *S.longicaudus* breeds inland on tundra and preys mainly on lemmings, the rather larger Arctic Skua *S.parasiticus* breeds near the coast, and mainly robs seabirds for their prey, though it takes a variety of other foods, locally including lemmings when they are numerous. The yet larger Pomarine Skua *S.pomarinus,* which breeds in arctic USSR, feeds primarily on lemmings, but breeds chiefly on marshy coastal plains, so differs in habitat from the Longtailed. All three species are circumpolar, and though they can be found nesting alongside each other, there is nowhere where both the Arctic and Pomarine are common, and the Longtailed is commonest in a different habitat from the others (Pitelka *et al.* 1955). All three winter at sea. The Great Skua *Catharacta skua,* sometimes put in the same genus, feeds like the Arctic, but is much larger and robs larger species of seabirds of their prey.

GULLS AND SEA TERNS LARIDAE

Larus: 10 species of gulls breed in Europe, 7 in the north, 5 in the south, with 2 common to both. They are in five main size-groups, and it has been assumed here that those in each of these groups differ in their main natural prey, though nowadays many feed together on fish waste and town garbage. The Little Gull *L.minutus,* the smallest species, breeds inland and feeds primarily on insects. In Britain, the Herring Gull *L.argentatus* feeds on the coast and nests mainly on cliffs, and the similar Lesser Blackback *L.fuscus* feeds further out to sea and nests on flat areas in from the cliffs; Herring Gulls remain for the winter, whereas most Lesser Blackbacks migrate south. In the Mediterranean, the Herring Gull again feeds on the coast and the similar-sized Audouin's Gull *L.audouinii* far out to sea. For the rest, too little is known for discussion here.

Sterna: The Caspian Tern *S.caspia* and Little Tern *S.albifrons* are so much larger and smaller respectively than the other four species that it may be assumed that they feed on mainly different prey. The

northern Arctic Tern *S.paradisaea* and more southern Common
Tern *S.hirundo* differ partly in range, and in their main area of
overlap in northern Europe, the Arctic breeds on the coast and the
Common chiefly inland by fresh waters. But in Iceland, where the
Common is absent, the Arctic also breeds and feeds inland, and in
the south of Europe, where the Arctic is absent, the Common breeds
chiefly on the coast. The two species breed beside each other in
some coastal areas, notably in N.E. England, where they bring
similar prey species of similar size to their young (Pearson 1968),
but this is exceptional. The Roseate Tern *S.dougallii* is of similar size
and often breeds with them, but is thought to feed further; it also
snatches food from other terns. The Sandwich Tern *S.sandvicensis*
is rather larger than the other three. It breeds in larger and
more widely spaced colonies, suggesting that it feeds further from
the shore than the Common or Arctic, which definitely holds in the
Black Sea (Gause 1934, quoting Formosov). In N.E. England the
Sandwich feeds its young on the same species as the Common and
Arctic Terns, but its prey is, on average, larger (Pearson 1968).
Especially in view of the situation in various Pacific terns (see pp.
166–171), it looks as though the European marine terns take the
same species of fish, but at different distances from the shore and
of a different range of sizes.

AUKS ALCIDAE

Uria: The Common Guillemot *U.aalge* and far northern Brünnich's
Guillemot *U.lomvia* replace each other geographically, with some
overlap. Where they overlap in northern USSR, they apparently
bring the same species of prey to their young (Belopolskii 1957,
Kartaschew 1960), but in Alaska, the proportion of fish to invertebrates was 63:34 in Brünnich's, 95:6 in the Common, so their diets
partly differ (Bedard 1969, citing Swartz 1966). Where they overlap
on Bear Island, the Common Guillemot nests on the tops of stacks
and on flatter areas above the vertical cliffs, while Brünnich's uses
ledges on vertical cliffs; but each occupies both types of nesting
site in the rest of its range (personal observation). This segregation
in the area of overlap implies that there is competition for nesting
sites, so it is further suggestive that each of the European auks in
other genera likewise has a distinctive site, the Razorbill *Alca torda*
in a covered niche, the Black Guillemot *Cepphus grylle* in piles of
boulders near the foot of cliffs, the Puffin *Fratercula arctica* in a
burrow and the Little Auk *Plautus alle* in a small hole. The Black
Guillemot differs from all the rest in feeding inshore, and the Little
Auk, which is much the smallest, in feeding mainly on crustacea.

But at least in northern USSR, not only both guillemots, but also the Razorbill and Puffin, bring the same main species of fish to their young (Belopolskii 1957, Kartaschew 1960), and the same was found for the Guillemot and Puffin in N.E. England (Pearson 1968). However, Fisher and Lockley (1954) reported that the prey brought to their young by the large Guillemot, medium Razorbill and small Puffin were respectively larger, intermediate and smaller; Harris (1963 and pers. comm.) confirmed that in Wales the sand-eels *Ammodytes tobianus* and sprats *Clupeus sprattus* brought to their young by the Guillemot were on average twice as long as those brought by the Puffin, with those brought by the Razorbill intermediate, and Pearson (1968) found that the fish brought by the Guillemot were on average four times as heavy as those brought by the Puffin. Hence these species differ in the mean size of their prey, though with some overlap.

17. Ecological isolation of congeneric species in selected families in Usambara, Tanzania

		Number of instances of isolation by			
Family and genus	range	habitat	feeding station	beak- or body-size	unknown (same habitat)
PASSERINES					
Pycnonotidae					
Andropadus	–	8	1	1	–
Phyllastrephus	2	3	2	8	–
Laniidae					
Dryoscopus	–	1	–	–	–
Laniarius	–	3	–	–	–
Lanius	–	1	–	–	–
Malaconotus	–	6	–	–	–
Tschagra	–	1	–	–	–
Prionopidae					
Prionops	–	3	–	–	–
Muscicapinae					
Batis	–	3	–	–	–
Bradornis	–	1	–	–	–
Muscicapa	–	1	–	–	–
Tchitraea	–	1	–	–	–
Trochocercus	–	1	–	–	–

APPENDIXES

Family and genus	range	habitat	Number of instances of isolation by		
			feeding station	beak- or body-size	unknown (same habitat)
Sylviinae					
Acrocephalus	–	–	–	–	1
Apalis	–	12	–	–	–
Bradypterus	–	3	–	3	–
Camaroptera	–	1	–	–	–
Cisticola	–	10	3	1	–
Eremomela	–	1	–	–	–
Prinia	–	–	–	–	1
Sylvietta	–	1	–	–	–
Turdinae					
Alethe	–	–	–	1	–
Cercotrichas	–	1	–	–	–
Cichladusa	–	1	–	–	–
Cossypha	–	3	–	–	–
Turdus	–	6	–	–	–
Nectariniidae					
Anthreptes	–	13	–	–	2
Nectarinia	–	49	–	5	1
Estrildidae					
Estrilda	–	13	–	–	2
Hypargos	–	–	–	1	–
Lagonosticta	–	1	–	–	–
Lonchura	–	2	–	1	–
Viduinae					
Vidua	–	1	–	–	–
Ploceidae					
Euplectes	–	6	–	6	9
Ploceus	–	45	–	–	10
Quelea	–	–	–	–	1
NON-PASSERINES					
Capitonidae					
Buccanodon	–	–	–	–	1
Lybius	–	3	–	–	–
Pogoniulus	2	4	–	–	–
Trachyphonus	1	2	–	–	–
Musophagidae					
Tauraco	–	1	–	–	–

323

APPENDIXES

Notes

Derived from tables which I drew up for each genus in the same way as in previous chapters, based on the ecological differences recorded by Moreau (1948) in his appendix. Since he wrote, various small genera have been merged, and the larger modern genera have been used for the present analysis, which was checked by R. E. Moreau prior to publication. The specific nomenclature is that used by Moreau. The generic and other changes from Moreau's list are as follows:

Pycnonotidae: *Andropadus* includes *Arizelocichla masukuensis, A. milanjensis, A.tephrolaema* and *Eurillas virens*.

Laniidae: *Malaconotus* includes *Chlorophoneus nigrifrons, C.sulphureopectus* and *Telephorus quadricola. Tschagra* includes *Antichromus minutus*.

Prionopidae: *Prionops* includes *Sigmodus retzii* and *S.scopifrons*.

Muscicapinae: *Muscicapa* includes *Alseonax cinereus* and *A.minimus*.

Sylviinae: *Acrocephalus* includes *Calamoecetor leptorhyncha. Bradypterus* includes *Sathrocercus cinnamomeus* and *S.mariae. Camaroptera* includes *Calamonastes simplex. Prinia* includes *Heliolais erythroptera*.

Turdinae: *Cercotrichas* includes *Erythropygia leucophrys* and *E.leucoptera. Turdus Geokichla gurneyi*.

Nectariniidae: *Nectarinia* includes four species of *Chalcomitra,* three of *Cinnyris,* one of *Cyanomitra* and one of *Drepanorhynchus*.

Estrildidae: *Estrilda* includes *Coccopygia melanotis, Uraeginthus bengalensis* and *U.cyanocephalus. Hypargos* includes *Mandingoa nitidula. Lonchura* includes *Amauresthes fringilloides, Spermestes cucullatus* and *S.nigriceps*.

Ploceidae: *Euplectes* includes *Coliuspasser albonotatus, C.ardens* and *Urobrachys axillaris. Ploceus* includes *Anaplectes melanotis*.

Other birds

Capitonidae: *Lybius* includes *Tricholaema melanocephalum,* and *Pogoniulus* includes *Viridibucco simplex* and *V.leucomystax* (a full species); *P.bilineatus* is separated by range from *P.(V.)simplex*. Moreau added (pers. comm.) one further family, the turacos Musophagidae, of which *Tauraco fischeri* is in lowland forest and *T.hartlaubi* in highland forest.

18. Habitat restriction of forest and woodland birds in selected families in Usambara, Tanzania

(derived from Moreau 1948, Appendix)

The habitats selected for analysis, followed by Moreau's numbers for them in his appendix, are lowland rain forest 1a (1–3), edge of lowland rain forest 1a (4), highland rain forest 3a (1–3), edge of highland rain forest 3a (4), riverine forest 1b (1, 2), wooded grassland 1c (2) and semi-desert with thorn trees 1d (1). Intermediate rain forest 2a (1–4), which has a mixture of the lowland and highland species, was omitted here, as were all types of cultivated ground and the habitats without trees. The nomenclature is that of Moreau, but with the broader genera used in appendix 17. The families analysed by Moreau were, in the sequence in the following lists, the bulbuls Pycnonotidae, flycatchers Muscicapidae, thrushes and chats Turdidae, warblers Sylviidae, shrikes Prionopidae and Laniidae, sunbirds Nectariniidae, weavers Ploceidae, estrildines Estrildidae, barbets Capitonidae and turacos Musophagidae. Where a species is noted as 'restricted' to a particular habitat, it means that it is not found in any of the other habitats analysed here (but it may also occur in the man-modified environments also studied by Moreau).

Species restricted to interior of lowland rain forest (17):

> *Phyllastrephus flavostriatus, Muscicapa cinerea, Batis mixta, Trochocercus bivitatus, Neocossyphus rufus, Cossypha natalensis, Apalis caniceps, Prionops scopifrons, Nicator chloris, Anthreptes pallidigaster, A.neglectus, Ploceus bicolor, Spermophaga ruficapilla, Buccanodon olivaceum, Pogoniulus simplex, P.bilineatus, Turaco fischeri.*

Species restricted to edge of lowland rain forest (13):

> *Andropadus virens, Suaheliornis kretschmeri, Parisoma plumbeum, Chloropetella holochlorus, Hyliota australis, Tchitrea perspicillata, Camaroptera brevicaudata, Dryoscopus cubla, Malaconotus quadricolor, Anthreptes collaris, A.reichenowi, Hypargos niveoguttatus, H. nitidulus.*

Species restricted to interior of highland forest (17):

> *Andropadus tephrolaema, A.masukuensis, Trochocercus albonotatus, Turdus olivaceus, T.gurneyi, Sheppardia sharpei, Alethe fülleborni, A. montana, Pogonocichla stellata* (rare in lowlands), *Seicercus ruficapilla, Bradypterus mariae, Laniarius fülleborni, Malaconotus nigrifrons, Ploceus nicolli, Cryptospiza reichenowi, Pogoniulus leucomystax, Turaco hartlaubi.*

Species restricted to edge of highland forest (17):

> *Andropadus milanjensis, Muscicapa minima, Bradypterus cinnamomeus, Apalis murina, Artisornis metopias, Laniarius ferrugineus, Nectarinia mediocris.*

Species common to interior of lowland and highland rain forest (5):
Phyllastrephus flavostriatus (in different subspecies), *P.fischeri* (in different subspecies), *Apalis melanocephala* (in different subspecies), *Nectarinia olivacea, Buccanodon leucotis.*
Species common to edge and interior of lowland rain forest (1):
Chloropeta natalensis.
Species common to lowland rain forest and lowland forest edge (1):
Chloropetella holochlorus
Rare rain forest species of uncertain range (1):
Apalis moreaui (found at intermediate altitude).
Species restricted to riverine forest (5):
Phyllastrephus terrestris, P.strepitans, P.cerviniventris, Platysteira peltata, Lybius melanopterus.
Species restricted to wooded grassland (11):
Bradornis griseus, Batis molitor, Eremomela scotops, Prionops retzii, Malaconotus sulphureopectus, Nectarinia bifasciatus, N.senegalensis, N.veroxii, Anaplectes melanotis, Lybius torquatus, Pogoniulus pusillus.
Species restricted to semi-desert with thorn trees (13):
Bradornis microrhynchus, Batis minor, Apalis flavida, Sylvietta brachyura, Eremomela griseoflava, Prionops poliocephala, Eurocephalus rüppellii, Nilaus minor, Lanius cabanisi, Dryoscopus pringlii, Nectarinia hunteri, Anthreptes orientalis, Ploceus intermedius.

19. Habitat restriction in European forest species

While some species are widespread in both broadleaved and coniferous woods, and others are virtually restricted to one or the other, some species are somewhat and others much commoner in one than the other, and it is hard to allocate such borderline cases. In the following lists, '(C)' after a species found primarily in broadleaved woods means that locally it is regular in conifers (in two warblers only where there are broadleaved bushes), and '(B)' after a species found primarily in conifers means that locally it is regular in broadleaved woods. Birds found primarily at the wood edge or in extremely open woodland are omitted:
Species regular in both broadleaved and coniferous woods (29):
Passerines: Jay *Garrulus glandarius,* Great Tit *Parus major,* Willow Tit *P.montanus,* Longtailed Tit *Aegithalos caudatus,* Wren *T.troglodytes,* Mistle Thrush *Turdus viscivorous,* Song Thrush *T.philomelos,*

Redstart *P.phoenicurus,* Robin *Erithacus rubecula,* Orphean Warbler *Sylvia hortensis,* Lesser Whitethroat *S.curruca,* Willow Warbler *Phylloscopus trochilus,* Greenish Warbler *P.trochiloides,* Chiffchaff *P.collybita,* Firecrest *Regulus ignicapillus,* Spotted Flycatcher *Muscicapa striata,* Pied Flycatcher *M.(F.)hypoleuca,* Dunnock *Prunella modularis,* Redpoll *Carduelis flammea,* Bullfinch *P.pyrrhula,* Chaffinch *Fringilla coelebs,* Brambling *F.montifringilla.* Others: Great Spotted Woodpecker *Dendrocopos major* (esp. C), Three-toed Woodpecker *Picoides tridactylus,* Black Woodpecker *Dryocopus martius,* Wryneck *Jynx torquilla,* Cuckoo *Cuculus canorus,* Wood Pigeon *Columba palumbus.*

Species primarily in broadleaved woods (30):

Golden Oriole *O.oriolus,* Blue Tit *Parus caeruleus,* Azure Tit *P.cyanus,* Marsh Tit *P.palustris,* Sombre Tit *P.lugubris,* Nuthatch *Sitta europaea* (C), Short-toed Treecreeper *Certhia brachydactyla* (C), Fieldfare *Turdus pilaris,* Redwing *T.musicus,* Blackbird *T.merula,* Nightingale *Luscinia megarhynchos,* Thrush Nightingale *L.luscinia,* Melodious Warbler *Hippolais polyglotta,* Icterine Warbler *H.icterina,* Olive-tree Warbler *H.olivetorum,* Blackcap *Sylvia atricapilla* (C), Garden Warbler *S.borin* (C), Wood Warbler *Phylloscopus sibilatrix* (C), Bonelli's Warbler *P.bonelli* (C), Arctic Warbler *P.borealis,* Collared Flycatcher *Muscicapa(F.)albicollis,* Redbreasted Flycatcher *M.(F.) parva,* Hawfinch *C.coccothraustes.* Others: Grey Woodpecker *Picus canus* (C), Green Woodpecker *P.viridis,* Whitebacked Woodpecker *Dendrocopos leucotos,* Syrian Woodpecker *D.syriacus,* Middle Spotted Woodpecker *D.medius,* Little Spotted Woodpecker *D.minor,* Stock Dove *Columba oenas.*

Species primarily in conifers (15):

Nutcracker *Nucifraga caryocatactes,* Siberian Jay *Cractes infaustus,* Coal Tit *Parus ater* (B), Crested Tit *P.cristatus* (B), Siberian Tit *P.cinctus,* Corsican Nuthatch *Sitta whiteheadi,* Treecreeper *Certhia familiaris* (B), Ring Ousel *Turdus torquatus* (also moors), Goldcrest *R.regulus* (B), Siskin *Carduelis spinus,* Serin *Serinus serinus,* Pine Grosbeak *Pinicola enucleator,* Crossbill *Loxia curvirostra,* Parrot Crossbill *L.pytyopsittacus,* Two-barred Crossbill *L.leucoptera.* Others: none.

Note

My aim was to include all forest birds in the families from pigeons to passerines inclusive (in the standard sequence), but omitting owls, as no predatory birds were included for Amani. Main reference Voous (1960).

20. Geographical and ecological isolation in congeneric turacos Musophagidae

(from Moreau 1958)

Crinifer: Both species live in wooded savanna. *C.piscator* in the west and *C.zonurus* further east replace each other geographically, and though the boundaries of their ranges overlap in a small area, they have not been found together, so may interdigitate rather than overlap.

Corythaixoides: The three species live in country dominated by various species of acacia. The more southern *C.concolor* does not overlap in range with *C.personata* and *C.leucogaster*. The two latter overlap, but differ partly in habitat, *C.personata* being associated with bigger flat-topped species of acacia and *C.leucogaster* living in dryer country; probably more important, *C.leucogaster* has a much smaller beak and eats the leaves and buds of acacias to a much greater extent than any other turaco except perhaps *C.concolor*. These two latter species also live in drier country than any other turacos.

Musophaga: Both species live in gallery forest, *M.violacea* in the west and *M.rossae* to the east and south of it, and are separated by range except for one possible narrow overlap.

Tauraco: As shown in fig. 33a, 9 of the 10 species replace each other geographically without overlap, except for *T.leucolophus* and *T. hartlaubi* in a relatively small area southwest of Lake Rudolph where they differ in habitat, as *T.leucolophus* lives in lowlying gallery forest, *T.hartlaubi* in montane forest. Of the others, *T.macrorhynchus* lives in lowland evergreen forest, *T.bannermani*, *T.johnstoni*, *T.leucotis* and *T.ruspolii* in highland evergreen forest of various types, and *T.erythrolophus* and *T.porphyreolophus* in gallery forest; these differences are partly responsible for their separate ranges. *T.bannermani* and *T.erythrolophus* form a superspecies, *T.leucotis* and *T.ruspolii* another.

There is one further species *T.corythaix,* which as shown in fig. 33b, overlaps greatly in range with several of the others. By Peters and most other workers, it was divided into 5 geographically replacing species, and if this is upheld, the number of species separated by range in this genus is correspondingly increased. But Moreau (1958) merged them in one large species, though he said that there was about as much reason to keep them separate, as they do not intergrade. Using the five species of Peters' Check-List, *T.corythaix sens. strict.* of evergreen forest in South Africa is separated by habitat from *T.porphyreolophus* in gallery forest, and by range from all the rest. *T.fischeri*, locally in lowland forest in East Africa, is separated by habitat from *T.hartlaubi* in montane forest, and by range from all the rest. *T.livingstonii*, in eastern Africa, in lowland and highland evergreen forest, is separated by habitat from *T.erythrolophlus* and *T.porphyreolophus* in gallery forest and by range from all the rest except for a relatively small overlap in one area with *T.hartlaubi*. *T.schütti*, in lowland and highland evergreen forest in central Africa, overlaps in range with *T.johnstoni* on Mount Ruwenzori, but the records by Chapin (1939) indicate that it occurs mainly at a lower altitude and *T.johnstoni* higher up, though with an overlap in altitudinal range. Finally *T.persa* found in West Africa in evergreen and gallery forest, overlaps with *T.macrorhynchus*, and how they might be separated is not known.

21. Geographical and ecological isolation in congeneric birds of paradise and bower birds

(from Gilliard 1969)

A. BIRDS OF PARADISE PARADISAEIDAE

Manucodia: Among the relatively unspecialized manucodes, *M.ater* is widespread in the lowlands of northern and southern New Guinea, *M.jobiensis* is restricted to the northern lowlands but there overlaps with *M.ater*, *M.chalybatus* is in the highlands from about 5–600 m, so is separated by altitude from them. There is only one species on each of the outlying islands, *M.ater* on the Louisiades, Aru and Western Papuan Islands, *M.jobiensis* on Japen, and a fourth species *M.comrii*, on the D'Entrecasteaux and Trobriand islands. (1G, 1Ha, and 1?, but G on islands.)

Ptiloris: Of the three rifle birds, *P.paradisaea* is in eastern Australia, *P.victoriae* further north in a small part of N. Queensland, and *P.magnificus* in Cape York and New Guinea. (3G.)

Paradigalla: The two paradigallas replace each other geographically in the mountains of New Guinea, *P.carunculata* in the west and *P. brevicauda* in the centre. (1G.)

Drepanornis: Of the two sickle-billed birds of paradise in this genus, *D.albertisii* is widespread in the mountains of New Guinea above 600 m and *D.brouijnii* is restricted to the northwestern lowlands. (Ha.)

Epimachus: Of the two sickle-billed birds of paradise in this genus, *M.fastosus* is restricted to the high central and eastern mountain ranges above about 1800 m and *M.meyeri* is more widespread in the mountains, at lower altitudes than *M.fastosus* where both species occur, but extending high where *M.fastosus* is absent (as on Mount Hagen). (Ha.)

Astrapia: The five astrapias replace each other geographically in the mountains of New Guinea, *A.rothschildi* in the Huon peninsula, and the other four in the main ranges from west to east as follows: *A.nigra, A.splendidissima, A.mayeri,* and *A.stephaniae.* Both of the two last occur in a narrow zone of overlap on Mts. Giluwe and Hagen, where *A.mayeri* has been taken at a rather higher altitude than *A.stephaniae.* (5G.)

Parotia: The four six-wired birds of paradise replace each other geographically, *P.wahnesi* in the mountains of the Huon peninsula and the other three in the mountains of the main ranges from west to east as follows: *P.sefilata, P.carolae* and *P.lawesi.* (4G.)

Diphyllodes: *D.magnificus* is widespread in New Guinea, (including Misol and Salawati in Western Papuan Islands), *D.respublica* is restricted to Waigeu and Batanta in Western Papuan Islands.

Paradisaea: Of the 7 species in this genus, two are restricted to smaller islands, *P.decora* in the D'Entrecasteaux archipelago and *P.rubra* in the Western Papuan Islands. Most of the others are separated geographically on the mainland of New Guinea, *P.apoda* in the south and on the Aru Islands, *P.raggiana* in the east, *P.minor* in the north except for the Huon Peninsula, where *P.guilielmi* is found. Finally *P.rudolphi* occurs in the eastern mountains of New Guinea above about 1350 m, so is separated by altitude from the others in this area (*P.raggiana and P.minor*). (See also fig. 32 p. 157) (6G, 1Ha.)

B. BOWER BIRDS PTILONORHYNCHIDAE

Ailuroedius: Two catbirds are widespread in New Guinea, *A.buccoides* from sea level to about 850 m, and *A.melanotis*, which has a discontinuous range, mainly above 850 m where it is in the same areas as *A.buccoides,* though elsewhere it is also at low altitudes. On the islands, *A.buccoides* is the only one on Japen and most of the Western Papuan Archipelago, and *A.melanotis* is the only one on the Aru

Islands and Misol (in the Western Papuan Archipelago), so here they replace each other by range. A third species, *A.crassirostris,* is in eastern and southeastern Australia, so is separated by range from the other two, including *A.melanotis,* which is in N.E. Australia. (2G, 1Ha [and partly G].)

Amblyornis: Of the three gardener bower birds in New Guinea, *A. macgregoriae* is widespread in the central and eastern mountains, *A.inornatus* is restricted to the western mountains and *A.subalaris* to the mountains of the southeast. (3G.) (A fourth species, *A.flavifrons,* is known solely from specimens of unknown origin; perhaps in western New Guinea or the Western Papuan Archipelago.)

Sericulus: Of the three regent bower birds, *S.aureus* is widespread in west and south New Guinea, *S.bakeri* is known only from a small area in N.E. New Guinea and *S.chrysocephalus* is in eastern Australia. (3G.)

Chlamydera: Of the four species of *Chlamydera,* the Spotted Bower Bird *C.maculata* is widespread in eastern and southern Australia, the Great Grey Bower Bird *C.nuchalis* is in northern Australia, the Fawn-breasted *C.cerviniventris* is in the coastal lowlands of the eastern half of New Guinea, chiefly in the south, and also in one area in the northwest, and *C.lauterbachi* is in three widely separated parts of the interior of central New Guinea (in different areas from *C.cerviniventris*). *C.cerviniventris* also reaches the tip of Cape York, where it lives in mangroves whereas *C.nuchalis* is in open forest inland. Hence the two latter species are largely separated by range, and differ in habitat in a small area of overlap. (4G, but 1H in a small area.)

22. Differences in feeding and morphology in tanagers in Trinidad

Species	Percent of time observed			Percent of time seen searching for insects on				
	on fruit	on flowers	hunting for insects	foliage	branches twigs	ground	flowers seedheads	aerial
Tanagra								
guttata	74	—	26	92	8	—	—	—
gyrola	70	—	30	8	90	—	1	2
mexicana	53	—	47	4	91	—	2	2
Thraupis								
palmarum	43	9	48	89	1	—	—	10
virens	53	10	37	56	17	—	11	16
Ramphocelus								
carbo	45	5	50	77	x	13	2	7
Tachyphonus								
luctuosa	29	—	71	94	3	—	—	3
rufus	60	10	30	32	3	51	—	14
Tanagra								
trinitatis	0	—	100	—	80+	—	—	—
violacea	97	—	3	—	—	—	—	—

Notes

From Snow and Snow (*in prep.*). The main difference between *Tanagra gyrola* and *T.mexicana* is that *T.gyrola* was found to spend 85% of time searching for insects on twigs $\frac{1}{2}$–2 inches in diameter and *T.mexicana* 55% of time on twigs less than $\frac{1}{2}$ inch in diameter. *Thraupis palmarum*, unlike the others, often feeds on large slippery leaves such as those of palms. *Ramphocelus carbo*, closely related to the *Tachyphonus* species, feeds in the leaf canopy of shrubs and low secondary growth, not high trees.*Tachyphonus luctuosa* usually feeds above 25 feet and *T.rufus* usually below 25 feet. Some of these species live in forest and others outside it, but no congeners occupy separate habitats except *T.guttata* in highland and *T.gyrola* in lowland forest. The winglength and beak-length respectively are around 70 and 8 mm in the species of *Tanagra*, 90 and 10·5 mm in *Thraupis*, 80 and 13 mm in *Ramphocelus*, 64–87 and 10–14 mm in *Tachyphonus*

23. Food species recovered from Ascension Island sea-birds

	Brown Noddy *Anous stolidus*	Black Noddy *Anous tenuirostris*	Fairy Tern *Gygis alba*	Sooty Tern *Sterna fuscata*	Redbilled Tropicbird *Phaethon aethereus*	Yellowbilled Tropicbird *Phaethon lepturus*	White Booby *Sula dactylatra*	Brown Booby *Sula leucogaster*
FISH								
Ophioblennius	+	+	+++	++			+	+
Centrolophus	+		++			+	+	++
Holocentrus		+++		+		+		+
Danichthys		+++					+	++
Selar	++	+	+	+			++	+
Oxyporamphus	+				+			
Hemiramphus	+							
Exocoetus		+	+	+	+	+	+	+
Cypsilurus and Hirundichthyes			+		+	+	++	++
Benthodesmus		+	+	+			++	++
Scomberesox				+				+
Trichiurus			++					
Decapterus							++	++
Engraulis		+						
Fistularia								
MOLLUSCS								
Hyaloteuthis				+	+	+	+	+

Notes

From Stonehouse (1962) p. 120. To simplify the table, only the generic names of the fish are given, and since every species represented is in a different genus, this should not cause confusion. Further species taken by *Gygis alba* but no other species were in the genera *Lampanyctus*, *Belone*, *Priacanthus*, *Ranzania*, *Sternoptyx*, *Diaphus* and *Nomeus*. The diet of the frigate bird *Fregata aquila* included *Exocoetus*, *Cypsilurus* and *Hirundichthyes*, normally obtained by robbing other species.

24. Measurements of congeneric species of Hawaiian sicklebills Drepanididae

Species (bracketed if in same superspecies)	Island or islands	Mean (range of means if on more than one island)		Main foods
		wing-length (mm)	culmen along its curvature from forehead to tip (mm)	
LOXOPS				
L.coccinea	widespread	63–65	10–11	Caterpillars in buds and on leaves
L.maculata	widespread	(61)65–70	12–15	Insects in crevices in bark
L.virens	Hawaii	65	14	Nectar, insects
	Kauai	68	19	Insects on trunks and branches
	other islands	65–67	15–16	Nectar, insects
L.parva	Kauai	60	13	Insects in leaves, nectar
L.sagittirostris	Hawaii	81	21	Insects in leaves, some nectar
HEMIGNATHUS				
H.obscurus	Hawaii, Lanai, Oahu	78(82)	43(53)	Nectar of lobelias, insects in crevices and decaying wood
H.procerus	Kauai	90	68	
H.lucidus	Maui, Cahu, Kauai	76–81	29–32	
H.wilsoni	Hawaii	84	27	Removes bark for weevils

PSITTIROSTRA				
P.psittacea	widespread	95–98	17–18	Fruits of *Freycinetia*, and lobelias
P.bailleui	Hawaii	92	13	Seeds of *Sophora*, etc.
P.flaviceps	Hawaii	c.98	19	(Beans of *Acacia koa*?)
P.kona	Hawaii	88	20	Seeds of *Myoporum*
P.palmeri	Hawaii	109	21	Beans of *Acacia koa*
P.cantans	Laysan	85	18	Insects, seeds, shoots, birds' eggs
DREPANIS				
D.pacifica	Hawaii	107	45 ⎫	Nectar of lobelias
D.funerea	Molokai	102	60 ⎭	

Notes

Based on Amadon (1950), for foods summarizing Perkins (1903).
The genera with only one species are *Pseudonestor xanthophrys* which eats primarily cerambycid beetles, *Ciridops anna* which ate the fruits of the palm *Pritchardia*, and three nectar-feeding species with beaks of different size, which presumably feed from at least partly different species of flowers, namely *Himatione sanguinea*, *Palmeria dolei* and *Vestiaria coccinea*.

335

25. Ecological isolation in congeneric Drepanididae in Hawaiian Islands

LOXOPS

1	*L.coccinea*	1				
2	*L.maculata*	F	2			
3	*L.virens*	F	F	3		
4	*L.parva*	F	F	F	4	
5	*L.sagittirostris*	F	F	F	—	

HEMIGNATHUS

1	*H.obscurus*	1			
2	*H.procerus*	G	2		
3	*H.lucidus*	F	F	3	
4	*H.wilsoni*	F	—	G	

PSITTIROSTRA

1	*P.psittacea*	1					
2	*P.bailleui*	F	2				
3	*P.flaviceps*	F?	F?	3			
4	*P.kona*	F	F	F?	4		
5	*P.palmeri*	F	F	F?	F	5	
6	*P.cantans*	G	G	G	F	F	

DREPANIS

1	*D.pacifica*	1	
2	*D.funerea*	G	

26. Ecological isolation in White-eyes Zosteropidae

A. AFRICA

Nomenclature is that of Moreau (1967) except as in note under the mainland forms, and measurements are from Moreau (1957), the first figure being the mean wing-length in mm and the second the mean culmen in mm. The measurements for the mainland species show the limits of the usual mean values for different populations, but not necessarily the full range, as Moreau did not put them under species, and in certain areas it is hard to be sure which of them refer to which species.

Mainland

Z.senegalensis (53–59, 11·8–13·5) occupies most of the African lowlands, but is replaced by *Z.abyssinica* (52–58, 11·5–13·4) in the dry lowlands of Abyssinia, Somalia and part of Kenya, and by *Z.pallida* (56–64, 12·4–14·0) in southern Africa. At a higher altitude in separated montane areas in Abyssinia, Kenya and N.Tanzania occurs *S.poliogastra* (60–64, 13·7–15·4), which intergrades with lowland *Z.senegalensis* in W.Kenya but elsewhere is distinct. On Mount Cameroon, *Speirops melanocephala* (63, 15·0) is in montane rainforest above *Z.senegalensis* (54, 12·1).

Contrary to Moreau (1967) I have here followed (a) Hall and Moreau (*in prep.*) in treating *Z.poliogastra* as a separate species and in merging *Z.virens* in *Z.pallida,* and (b) Amadon and Basilio (1957) and Eisentraut (1963) in treating *S.melanocephala* as a separate species from *S.lugubris* of Sao Tomé.

Gulf of Guinea

Fernando Po *Z.senegalensis* at lower altitudes (56, 13·4), *Speirops brunnea* in high montane rain-forest (65, 13·5) (Eisentraut 1968).
Principe *Z.ficedulina* (53, 13·5), *Speirops leucophaea* (68, 14·9).
Sao Tomé *Z.ficedulina* (55, 12·7), *Speirops lugubris* (74, 15·9).
Annobon *Z.griseovirescens* (62, 15·8).

Indian Ocean

Socotra *Z.abyssinica* (56, 13·5).
Pemba (Tanzania) *Z.vaughani* (53, 13·7).
Mahé (Seychelles) *Z.modesta* extinct (60, 14·6).
Marianne (Seychelles) *Z.mayottensis* (59, 14·7).
Aldabra *Z.madaraspatana* (52, 11·5).
Grand Comoro *Z.senegalensis* on lower ground (53, 12·8), *Z.mouronensis* in tree heath above 1700 m (62, 14·0)(Benson 1960).
Moheli (Comoros) *Z.madaraspatana* (53, 12·5).
Anjouan (Comoros) *Z.madaraspatana* (57, 14·4).
Mayotte (Comoros) *Z.mayottensis* (53, 14·0).
Madagascar *Z.maradaspatana* (below 1200 m 55, 13·5, above 1200 m 61, 14·1).
Reunion *Z.borbonica* widespread in wooded country (55, 13·5), *Z. olivacea* in forest above 500 m (58, 15·6) (Milon 1951).

Mauritius *Z.borbonica* widespread (55, 14·0), *Z.olivacea* in remote forest (52, 16·4); apparently these two do not meet (Rountree et al. 1952, Moreau 1957 p. 401).

B. MAINLAND OF ASIA AND AUSTRALIA
Nomenclature and measurements from Mees (1957, 1961, 1969)

Asia

Z.erythropleura breeds in Ussuria (61, 12·5), *Z.japonica* in nearly all of China and Japan (57–60, 12·4–14·3) and *Z.palpebrosa* almost throughout India, Burma and the countries of southeast Asia (52–57, 12·0–13·1), but both the two latter species have been recorded in Szechwan and Yunnan, where their ecology is not known, and this is also in the regular wintering area of *Z.erythropleura*. In the Malay Peninsula, *Z.everetti* (53, 11·9) occurs in heavy forest above 700 m, while *Z.palpebrosa* (53, 12·9) here breeds in mangroves, the forest edge and cultivated land at lower altitudes.

Australia

Z.lateralis occurs in the southwest, south, east and northeast (56–62, 12·4–13·3), just overlapping in the west with *Z.lutea* (55–57, 12·3) which extends along the coast from there north and then east to east side of Cape York. Where they meet, *Z.lutea* breeds in mangroves and *Z.lateralis* in other coastal habitats (Serventy and Whittell 1962), but apparently the separation may not be absolute (Mees 1969 p. 25). *Z.chloris* (60, 14.7 is on small islands off the north and east coasts of Cape York peninsula, but not on the mainland.

ISLANDS IN THE INDO-AUSTRALIAN REGION AND PACIFIC
Notes

Nomenclature is that of Mees (1957, 1961, 1969), but if a species accepted by him has been treated as a subspecies in Peters' Check-List by Mayr (1967), Mayr's specific name precedes that used by Mees in brackets, and the same for the one genus accepted by Mees but not Mayr, and I have followed the same procedure for the one species accepted by Mayr and treated as a subspecies by Mees. Mees is also followed for range, habitats and measurements. The two measurements given are the means in mm for the wing-length and whole culmen respectively, solely for males where the sexes were separated by Mees. Where two figures are given for each measurement, they refer to the means for the largest and smallest subspecies on the island or archipelago in question. All the main islands are included below, but minor islands in a group are omitted unless of special interest, while small offshore islands are listed under the nearest largest island. Subspecific names are omitted.

Laccadives, Andamans, Nicobars

Z.palpebrosa (53–56, 12·8–13.4).

Ceylon

Z.palpebrosa cultivated lowlands to about 1200 m (54, 11·9).
Z.ceylonensis montane forest, mainly above *Z.palpebrosa* (57, 14.3).

Sumatra

Z.palpebrosa cultivation and forest 200–1600 m, but not true lowlands (51, 11·8).
Z.atricapilla forest (not cultivation) 1000–2200 m in centre and south, but extending higher to alpine meadows on northern mountains where *Z.montana* absent (57–58, 13·0–13·2).
Z.montana local on southwest side, in uppermost montane zone of dwarf shrubs and open country of rhododendrons and Vaccinium, i.e. above *Z.atricapilla* (58, 13·4).
(*Z.salvadorii* endemic species on Engano Island to the south (58, 11·6).)

Java

Z.flava north coast of western part, in mangroves and cultivation (51, 11·4).
Z.palpebrosa hills and mountains at middle altitudes, chiefly at forest edge and in gardens (52, 12·1).
Z.montana mountains of west, chiefly in uppermost shrub zone and more or less open country, hence above *Z.palpebrosa* (56–59, 12·2–13·8).
Lophozosterops javanica mountains from 900 m to summits (64, 13·6–14·6).
(*Z.chloris* on offshore islets to north (56, 13·4), *Z.natalis* on Christmas I. to south (61, 15·1).)

Sumbawa and Flores (also Rensch 1931)

Z.chloris from sealevel to 600 m, occasionally to 1300 m, primarily in cultivated land, including bushes and light woods (56, 13·9).
Z.wallacei from sealevel to 800 m in light monsoon forest and shrub country (54–55, 13·5).
Z.palpebrosa middle altitudes chiefly at forest edge (800 m) (52, 11·8).

APPENDIXES

Z.montana 1000–2800 m, primarily in *Casuarina* forest, also sparsely in montane rainforest (56, 12·9).
Lophozosterops dohertyi 300–1100 m, in light rainforest (62, 13·7–14·6).
Lophozosterops superciliaris 1000–1500 m, in *Casuarina* forest and montane rainforest (68, 15·5).
Heleia crassirostris lowlands to 1050 m (69, 18·0).
Bali has *Z.palpebrosa, Z.montana* and *Lophozosterops javanica,* Lombok *Z.chloris* and *Z.montana,* Sumba *Z.chloris* and *Z.wallacei.*

Timor

Z.chloris coastal and lowlands, to 1200 m (57, 12·8).
Z.montana montane 1800–2600 m (59, 12·9).
Heleia muelleri lowlands of west up to 845 m (68, 16·3).

Borneo (also Smythies 1960)

Three geographically replacing coastal species:
  (i) *Z.chloris* islands off southwest coast (56, 13·4).
  (ii) *Z.flava* small part of southeast coast (51, 11·5).
  (iii) *Z.palpebrosa* Natuna I. and small part of west coast, mangroves and coastal (53, 12·9).
Z.everetti north, submontane forest (51, 11·9).
Z.atricapilla solely two mountains in north, above 1000 m (55, 13·5).
Oculocincta squamifrons hills and mountains from 200 to 1200 m, in west and north, especially in moss forest, feeding in dense scrub; very warbler-like (50,11·5).
Chlorocharis emiliae mountains of north and northwest above 1300 m, especially in highest zone of low trees, conifers, rhododendrons and heather, in moss forest (64–69, 15·6–17·8).

Celebes (also Stresemann 1940)

Three geographically replacing species of lowlands and mid-levels, especially in secondary forest and cultivated belts:
  (i) *Z.consobrinorum* S.E. peninsula (54, 12·5).
  (ii) *Z.atrifrons* N. peninsula, N. central, offshore eastern islands (55–59, 12·9–13·4).
  (iii) *Z.anomala* S.W. peninsula, primarily in secondary forest and deforested hills, not open cultivated areas (58, 14·0).
Z.chloris S.W. peninsula, N. central, and islets off S.E. and N. peninsulas. Coastal and lowlands, extending up to 1000 m. Hence coexists with *Z.anomala* in S.W. and to much smaller extent with *Z.atrifrons* in N.C. and comes close to *Z.consobrinorum* in S.E. In S.W. is in cultivated lowlands which *Z.anomala* avoids, and is absent from secondary forest and deforested hills where *Z.anomala* is common (53–55, 12·2–13·3).

340

Z.montana widespread in mountains, mainly above 1500 m, in closed forest and above the treeline (54–57, 11·9–13·6).

Lophozosterops squamiceps in mountains above 1000 m (62–65, 14·7–16·0).

Moluccas

Northern: *Z.atriceps* lowlands (58, 14·3), *Z.montana* mountains (56–58, 12·6–13·4).

Buru: *Z.buruensis* lower and mid-montane zone in forest (59, 13·6), *Z.montana* mountains (presumably in uppermost zone) (58, 13·0), *Madanga ruficollis* (endemic genus) mountains of northwest (72, 15·1).

Ceram: *Z.chloris* on small offshore islands but not Ceram itself (60, 14·7), *Z.atrifrons* midlevels of west and centre around 700 m. in secondary forest (55, 13·1), *Z.montana* montane, so above *Z.atrifrons*, chiefly at 1000–2000 m. (56, 13·3), *Tephrozosterops stalkeri* (endemic genus) mountains, often with *Z.atrifrons* (69, 16·6), *Lophozosterops pinaiae* central mountains at 1200–2100 m (74, 17·4). One record of *Z.kuehni* has been questioned.

Ambon: *Z.kuehni,* probably in hill forest (55, 13·0).

Kei and Aru Islands

Z.chloris solely on islets, not main Kei or Aru islands (60, 14·7).

Z.novaeguineae Aru islands, of necessity in lowlands (mid-levels in New Guinea) (54, 12·3).

Z.grayi in Great Kei (63, 15·0) and *Z.uropygialis* (61, 14·8) on Little Kei are closely related to each other.

New Guinea Mainland
(Also Rand and Gilliard 1967, and for adjoining islands)

Z.atrifrons 500–1800 m (also hilly islands, where occurs down to sea-level) (57–58, 13·0–13·5).

Z.fuscicapilla mountains at 1200–2200 m in western half New Guinea, forest and secondary (59, 13·0).

Z.novaeguineae mountains of eastern half of New Guinea, locally in lowlands. Hence no recorded overlap with *Z.fuscicapilla* except for one western population in Arfak Mts. (53–62, 12·1–14·2, montane forms much larger than lowland).

Schouten Islands (northern New Guinea)

Z.atrifrons on Japen (57, 13·0) *Z.mysorensis* on Soepiori (60, 14·5).

341

APPENDIXES

D'Entrecasteaux and Louisiades (southeastern New Guinea)

Z.atrifrons montane in D'Entrecasteaux including Goodenough (57, 13·5), Z.(atrifrons)meeki on Tagula in Louisiades (58, 13·7).
Z.fuscicapilla montane on Goodenough (D'Entrecasteaux) (61, 15·0).
Z.griseotincta most Louisiades (63, 16·1; 64, 17·6 on Rossel), not on Tagula.

Bismarks (east of New Guinea)

Z.(atrifrons)hypoxantha on large islands up to at least 1000 m (57, 13·2).
Z.griseotincta on very small islands (62, 15·8).

Philippines (also Delacour and Mayr 1946)

Two species replace each other geographically:
  (i) Z.nigrorum low and moderate altitudes in primary forest and secondary growth, from sea level to 1500 m, Luzon, Mindoro and west central islands south to Negros (52–55, 12·3–13·2).
  (ii) Z.everetti in similar habitat to last and replacing it geographically in Samar, Cebu and east central islands south to Mindanao (53–57, 12·5–13·3).
Z.japonica Batan (far north), Luzon and immediately adjoining islands. Hence overlaps with Z.nigrorum in Luzon, but frequents cultivated lowlands (54, 13·1).
Z.montana all main islands, montane from 1000 to 2500 m. Especially in shrub zone above forest, but in Luzon also in pine (56–57, 13·0–13·9).
Lophozosterops goodfellowi mountains of Mindanao above 1250 m (66–72, 15·3–16·2).

Palawan

West of the main Philippine islands, formerly thought to have no Zosterops, has Z.montana (Salomonsen 1962).

Solomons (also Mayr 1945)

Almost every island has only one species, which in nearly all cases occurs at all altitudes. Omitting for the moment the islands with two species, and taking first the outer chain from north to south and secondly the inner group of islands, the species concerned are: Z. griseotincta (62, 15·8) on Nissan, Z.metcalfii (59–63, 14·0–14·6) on Choiseul, Ysabel and Florida, Z.stresemanni (69, 17·4) on Malaita, Z.rendovae (Z.ugiensis in Peters) (67–68, 15·0–15·9) on Guadalcanal and San Cristobal, Z.vellalavella (63, 16·0) on Vellalavella and Bagga, Z. luteirostris (61, 15·6) on Gizo, Z.(luteirostris)splendida (61, 16·2) on Ganonga, and Z.(rendovae)kulambangrae (63–66, 16·3–17·0) on New Georgia, Rendova and Tetipari. Each of the other main islands has two species, Bougainville *Z.metcalfii* (60, 14·4) in the lowlands and *Z.rendovae* (*Z.ugiensis* in Peters) (*c.* 68, 16) in the mountains, Kulambangra *Z.(rendovae)kulambangrae* (63, 16·5) in the lowlands and *Z.murphyi* (66, 17·8) in the mountains, Rennell *Z.(griseotincta)rennelliana* (64, 16·4) and *Woodfordia superciliosa* (78, 21·2), and Santa Cruz (not always included in Solomons) *Z.sanctaecrucis* (68, 15·5) and *Woodfordia lacertosa* (83, 24·3) (the largest of all Zosteropidae).

New Hebrides and Banks

Z.flavifrons forest and secondary growth at all altitudes, also gardens (59–66, 14·6–17·8).
Z.lateralis cultivated land, often with *Z.flavifrons* but perhaps in more open habitats (64–69, 15·0–15·7).

Lifu (Loyalties)

Z.lateralis primarily in gardens and open country (63, 14·3).
Z.minuta gardens, scrub and forest (56, 13·8).
Z.inornata forest trees, also scrub (74, 20·7) (derived from *Z.lateralis*).

New Caledonia

Z.lateralis open country, brush and cultivated land (63, 13·4).
Z.xanthochroa primarily forest, but also gardens (60, 15·1).

Fiji Islands

Z.lateralis gardens and similar habitats (62, 13·7).
Z.explorator forest inland (61, 15·0).

Samoa

Z.samoensis habitat not specified, but recorded in forest at 1200 m. (60, 14·2).

Norfolk Island (amplified by Mees pers. comm.)

Z.lateralis, colonized in 1904, primarily in open or cultivated land with trees, but also in forest (62, 13·1).
Z.tenuirostris primarily in forest, also secondary woodland (68, 18·0).
Z.albogularis primaeval forest (78, 18·8).

Lord Howe

Z.lateralis (60, 15·5).
Z.strenua (71, 23·2), now extinct.

Marianas

Z.conspicillata forest edge and secondary growth (53–57, 12·4–13·1).

Caroline Islands

From east to west, Kusaie has only *Z.cinerea* (c. 64, c. 15), Ponape has *Z.conspicillata* (55, 13·7), *Z.cinerea* (61, 13·3) and *Rukia longirostra* (71, 23·3), Truk and Yap have *Z.conspicillata* (56–57, 12·7–13·9) and *Rukia ruki* (81, 20·4) and *R.oleaginea* (c. 73, 18·5) respectively, Palau has *Z. conspicillata* (56, 13·0), *Z.cinerea* (66, 16·8) and *Megazosterops (Rukia) palauensis* (82, 20·6).

Note

Mees (1957, 1961, 1969) recognized the genus *Megazosterops* for the largest species on Palau which Mayr (1967) merged in *Rukia*, used *Z.rendovae* for Mayr's *Z.ugiensis*, treated Mayr's *Z.rennelliana* as a subspecies of *Z.griseotincta* and made what Mayr treated as *Z.atrifrons meeki*, *Z.atrifrons hypoxantha*, *Z.luteirostris splendida* and *Z.rendovae kulambangrae* into full species. These differences are trivial and do not affect the ecological picture. The genus *Hypocryptadius*, endemic to the Philippines, was at one time included in the Zosteropidae but is now agreed not to belong here.

27. Ecological isolation in congeneric West Indian passerine birds

Notes

I have included here all the islands designated faunistically as the West Indies, e.g. in the map by Bond (1960), omitting Trinidad and Tobago (included for cricket teams but faunistically part of Venezuela), and the Dutch Lesser Antilles, Aruba, Curacao and Bonaire, which are likewise faunistically part of South America. I have used the term 'Greater Antilles' for the four large islands of Cuba, Jamaica, Hispaniola (Haiti and the Dominican Republic) and Puerto Rico. Ranges and ecology have been set out solely for the West Indies, and are taken from Bond (1960) unless otherwise stated. Bond (1960) has also been followed for scientific nomenclature, but changes made by later revisers, notably in the recent volumes of Peters' 'Check-List', have been noted. English names are also from Bond but are omitted if merely a direct translation of the Latin name or if simply the name of the island to which a species is restricted. Regretfully, there is no place for the often vivid vernacular names. Genera with only one species in the West Indies are omitted. The symbols for segregation are G by geographical range, H by habitat, Ha by habitat closely linked with altitude, F by feeding, Fs by a big size difference presumably linked with a difference in feeding, and ? for unknown. As in previous chapters, the segregation of each species has been scored against every other in the same genus, but when there are many geographically replacing species, this misleadingly inflates the total isolated by range, so I have also given the score counting each such species only once, this being placed in brackets after the overall score for G.

Kingbirds *Tyrannus:* The Grey Kingbird *T.dominicensis* is on all the islands in open settled country with some trees. The Loggerhead Kingbird *T.caudifasciatus,* found in the Greater Antilles, originally replaced the Grey in open forest (Wetmore 1927, Wetmore and Swales 1931), but has now spread locally into open country with trees, so partly overlaps with it in habitat, and here their possible separation has not been studied. The Giant Kingbird *T.cubensis,* restricted to Cuba and the southern Bahamas, lives in deep lowland forest, especially pines, (Barbour 1943), so differs from the other two in habitat. So does the Tropical Kingbird *T.melancholicus,* found in semi-arid country on Grenada. (Probable segregation 2G, 4H, but two of latter might involve Fs.)

Crested flycatchers *Myiarchus:* These are mapped in fig. 50 (p. 225). Three species occur in Jamaica, the large Rufous-tailed *M.validus* (wing 103, culmen 17·0 mm) in the highlands, the medium-sized *M.stolidus* (wing 86, culmen 14·3 mm) in the lowlands and the small Dusky-capped *M.tuberculifer* (= *M.barbirostris*) (wing 73, culmen 11·4 mm) in both lowlands and highlands. There is only one species on each of the other islands, on Bond's classification *M.stolidus* in the other Greater Antilles and also Martinique, and the Rusty-tailed *M.tyrranulus* in the Lesser Antilles except Martinique (3G, 1Ha, 2Fs). But in the revision by Lanyon (1967), from whom the foregoing measurements of the Jamaican forms have been taken, the *M.stolidus* superspecies is divided into *M.stolidus* on Jamaica and Hispaniola, *M.sagrae* on Cuba, Grand Cayman and the Bahamas, *M.antillarum* on Puerto Rico and the Virgins and *M.oberi* on the Lesser Antilles south to St Lucia inclusive. On St Vincent and Grenada, this superspecies is replaced by *M.nugator,* derived from *M.tyrranulus* of South America. *M.nugator* and *M.oberi* are of similar size, so are ecological equivalents though in different superspecies. *M.stolidus* is of similar size on Hispaniola and on Jamaica, though on the latter it occurs with a larger and a smaller species, whereas it is the sole species on Hispaniola. (18(5)G. 1Ha, 2Fs.)

Elaenia: The Caribbean *E.martinica* occurs in both open country and forest in the Caymans, Puerto Rico, the Virgins and northern Lesser Antilles, where it is the sole Elaenia present. But in the southern Lesser Antilles of Grenada, the Grenadines and St Vincent, it co-exists with the Yellow-bellied *E.flavogaster,* which frequents solely open country. On Grenada, presumably through competitive displacement, *E.martinica* is restricted to montane forest. On St Vincent, where *E.flavogaster* might be a relatively recent colonist, *E.martinica* is found primarily in forest, but also in small numbers in open country alongside *E.flavogaster,* with minor differences in feeding stations (Crowell 1968). The Greater Antillean Elaenia *E.fallax,* on Hispaniola and Jamaica, is separated by range from the other two. (2½G, ½Ha.) (The Jamaican Yellow-crowned Elaenia *Myiopagis cotta* is presumably separated by feeding from *E.fallax,* but is in a separate genus, so is not scored here.)

Crows Corvus: The Cuban *C.nasicus,* also on Caicos, the Whitenecked *C.leucognaphalus* on Hispaniola and formerly Puerto Rico, and *C.jamaicensis* on Jamaica, are separated by range. The Palm Crow *C.palmarum* coexists in pine forest with the larger *C.leucognaphalus* on Hispaniola (tentatively classified as Fs) and is very local on Cuba. (4(or 3)G, 1Fs, 1?.)

Mockingbirds *Mimus:* The St Andrew species *M.magnirostris* (treated as a subspecies of *M.gilvus* in Peters) is separated by range. The Tropical *M.gilvus* in the Lesser Antilles is separated by range from the Northern *M.polyglottos* in the Greater Antilles and Bahamas. The Bahama Mockingbird *M.gundlachii* coexists with *M.polyglottos* both on Jamaica, where it is restricted to scrub forest in lowland limestone and *M.polyglottos* to open country with trees (including just outside the limestone forest), and also in the Bahamas, where *M.gundlachii* frequents both scrub forest and open country with trees and its possible separation from *M.polyglottos* requires study. (5(or 3)G, 1H, reduced to 2G, 1H if *M.magnirostris* is treated as a subspecies.)

Thrashers *Margarops:* The Pearly-eyed *M.fuscatus* and the smaller Scaly-breasted *M.fuscus* occur in most of the Lesser Antilles, and how they are separated is not known, but the use for the latter of a separate genus *Allenia* in Peters, and the difference in size, including size of beak, suggest a difference in feeding. *M.fuscatus* is also in the Virgins, Puerto Rico and the southern Bahamas. (1Fs.)

Thrushes *Turdus:* In Jamaica, the relatively small White-eyed *T.jamaicensis* lives in high montane forest, and the much larger White-chinned *T.aurantius* in damp woodland in the hills and in highland forest (provisionally classified as $\frac{1}{2}$Ha, $\frac{1}{2}$Fs).

Two other species are on St Vincent and Grenada, the Cocoa Thrush *T.fumigatus* in montane forest and the Bare-eyed *T.nudigensis* (also on other islands) in lowland woods and gardens. The La Selle Thrush *T.swalesi* is in montane forest in part of Hispaniola. (8(3)G, 1$\frac{1}{2}$Ha, $\frac{1}{2}$Fs.)

Thrushes *Mimocichla*: The Grand Cayman Thrush *M.ravida* (now extinct, Johnston 1969) was separated geographically from the Redlegged *M.plumbea*, found in the Bahamas, Cuba, Cayman Brac, Hispaniola, Puerto Rico and Dominica. (1G.) *Mimocichla* has been merged with *Turdus* in Peters, and as *M.plumbea* replaces the species of *Turdus* geographically in the West Indies, this means that the score for *Turdus* becomes 18(7)G, 2$\frac{1}{2}$Ha, $\frac{1}{2}$Fs; the additional separation by habitat refers to *M.plumbea* and *T.swalesi* in Hispaniola.

Solitaires *Myadestes*: The Cuban *M.elisabeth* is separated geographically from the Rufous-throated Solitaire *M.genibarbis* of Jamaica, Hispaniola and some of the Lesser Antilles. (1G.)

Gnatcatchers *Polioptila*: The Cuban *P.lembeyi* is separated geographically from the Blue-grey Gnatcatcher *P.caerulea*, which breeds in the Bahamas (but the latter species winters in Cuba). (1G.)

Vireo: Six island species in the White-eyed Vireo superspecies replace each other geographically, *V.nanus* on Hispaniola, *V.modestus* on Jamaica, *V.caribaeus* on St Andrew, *V.crassirostris* on the Bahamas and Caymans, *V.gundlachii* on Cuba and *V.latimeri* on Puerto Rico. (Of these, *V.modestus*, *V.caribaeus*, *V.crassirostris* and *V.gundlachii* were treated as subspecies of the North American *V.griseus* by Hamilton 1958, but *V.nanus* and *V.latimeri* were kept separate.) These frequent trees and shrubs, though Bond noted that *V.latimeri* and *V.gundlachii* feed higher than the others. The larger and widespread Black-whiskered *V.altiloquus* feeds mainly in the trees. This reverses the usual trend with size in forest birds, but the shrub layer often consists of very fine twigs. The Yucatan Vireo *V. magister*, a pronounced geographical form of *V.altiloquus*, is found in the West Indies on Grand Cayman (it is treated as a subspecies by Hamilton 1958). No island has more than two species of vireos except Jamaica, where the Blue Mountain Vireo *V.osburni* coexists with the other two in highland forest, but is intermediate in size with an unusually deep and decurved beak (wing 72 mm, culmen 13 mm, cf. 57 and 8·9 mm in *V.modestus* and 83 and 15·8 mm respectively in *V.altiloquus*; Ridgeway 1904), so is presumably separated by feeding. (27(8)G, 8Fs, 1F?; but on Hamilton's classification 5(4)G, 4Fs, 1F?.)

Dendroica warblers: The Yellow Warbler *D.petechia* is widespread except on the southernmost Lesser Antilles, and is separated by habitat from the rest, as it breeds primarily in coastal mangroves, though it also spreads into lowland forest, especially on the smaller islands. Except for the Bahamas, each of the other main islands has only one further breeding species, and these are separated by range from each other: the Olive-capped *D.pityophila* on Cuba (in pines), *D.vitellina* on the Caymans and Swan I. (in thickets), the Arrow-headed *D.pharetra* on Jamaica (in montane forest), the Pine Warbler *D.pinus* on Hispaniola (in pines), *D.adelaidae* on Puerto Rico (in lowand limestone forest), St Lucia (in lowland and montane forest) and Barbuda, and *D.plumbea* on Guadeloupe and Dominica (in forest, especially montane). However, three species coexist in pine forest in the northern Bahamas, *D.pityophila* (also on Cuba), *D.pinus* (also on Hispaniola) and the Yellow-throated *D.dominica*. Of these, *D.dominica* is separated from the other two by its much longer beak and its habit of probing in crevices for food, while *D.pityophila* feeds lower down than *D.pinus* (Bond 1948). (18(7)G, 7H, 3F.)

The situation is complicated in winter by the arrival of various other species of *Dendroica* from North America, and. how they might be separated, if they are, is not known.

Ground warblers *Microligea*: The Green-tailed *M.palustris* and White-winged *M.montana* occur on Hispaniola, the former creeping close to the ground in dense thickets and the latter feeding higher up in the branches and creepers of more open thickets (Wetmore and Swales 1931). (1F.) *M.montana* is restricted to mountains, *M.palustris* is in both montane rain forest and arid lowland scrub-forest.

Teretistris warblers: The Yellow-headed *T.fernandinae* and the Oriente Warbler *T.fornsi* replace each other geographically in western and eastern Cuba respectively. (1G.)

Palm tanagers *Phaenicophilus*: The Grey-crowned *P.poliocephalus* of S.W. Haiti and Gonave is replaced in the rest of Hispaniola by the Black-crowned *P.palmarum*. (1G.)

Grackles *Quiscalus*: The Greater Antillean *Q.niger* and the Lesser Antillean *Q.lugubris* replace each other geographically. (1G.)

Orioles *Icterus*: Five species replace each other geographically, the Black-cowled *I.dominicensis* in the Bahamas, Cuba, Hispaniola and Puerto Rico, *I.leucopteryx* in Jamaica, Grand Cayman and St Andrew, *I.oberi* in Montserrat, *I.bonana* in Martinique and *I.laudabilis* in St Lucia. (10(5)G.) (The Troupial *I.icterus* was introduced to Puerto Rico, where it occurs in scrub-forest and cultivated land with trees alongside *I.dominicensis,* but it is much larger, so presumably differs in feeding.

Blackbirds *Agelaius:* The Yellow-shouldered *A.xanthomus* in Puerto Rico is separated by range from the other two. The Tawny-shouldered *A.humeralis* in Cuba and a small part of Haiti frequents rather open lowland country, preferring a drier habitat to the Redwing *A.phoeniceus,* which frequents marshes in Cuba and the Bahamas. (2G, 1H.)

West Indian bullfinches *Loxigilla:* The Puerto Rican *L.portoricensis,* Greater Antillean *L.violacea* on Hispaniola, Jamaica and the Bahamas, and the Lesser Antillean *L.noctis* replace each other geographically. Formerly *L.noctis* and the larger *L.portoricensis* co-existed on St Kitts, where the endemic form of *L.portoricensis,* much larger than that on Puerto Rico, is now extinct. (2½G, ½F.)

Grassquits *Tiaris:* Both the Yellow-faced *T.olivacea* and the Black-faced *T.bicolor* are widespread, and at least on Jamaica *T.bicolor* is restricted to very dry areas and feeds mainly on the ground or over 3 feet above it, while *T.olivacea* is less common in dry areas and usually feeds between 6 inches and 3 feet from the ground (Pulliam 1968). The Cuban *T.canora* there replaces *T.bicolor.* (1G, 1F, 1?.)

28. Ecological isolation in congeneric non-passerine West Indian land birds

The opening paragraph of appendix 27 applies equally here. Some pigeons and parrots are known to have become extinct through human disturbance and others are now so rare that they may not occupy their full range or habitats, hence the record is less complete than for passerine birds.

Chondrohierax kites: The Hook-billed *C.uncinatus* in Grenada and the Cuban *C.wilsonii* in eastern Cuba are separated by range. (1G.)

Accipiter hawks: The Sharp-shinned *A.striatus* breeds in Puerto Rico, Hispaniola and Cuba, the larger *A.gundlachi* coexists with it in Cuba and is presumably separated by food. (1Fs.)

Buteo hawks (or buzzards): The Red-tailed Hawk *B.jamaicensis* is widespread and much larger, presumably taking larger prey, than both the Broad-winged *B.platypterus,* found with it in Cuba, Puerto Rico and a few of the Lesser Antilles, and *B.ridgwayi,* found with it in Hispaniola. (1G, 2Fs.)

Pigeons *Columba:* The White-crowned *C.leucocephala* and Red-necked *C.squamosa* are widespread, the Plain *C.inornata* is in the Greater Antilles and the Ring-tailed *C.caribaea* on Jamaica. *C.leucocephala* occurs low down, *C.squamosa* mainly high, *C.caribaea* high (Bond, Kepler, see also Gosse 1851 pp. 166–167). (2Ha, 4?.)

APPENDIXES

Zenaida doves: The Zenaida Dove *Z.aurita* is widespread, the Mourning Dove *Z.macroura* and White-winged *Z.asiatica* are in the Greater Antilles, and the Violet-eared *Z.auriculata* is in some of the Lesser Antilles. It is not known how they are separated where they are on the same island. (2G, 4?.)

Quail doves *Geotrygon:* The Ruddy *G.montana* is in the Greater and Lesser Antilles, in both lowland and hill forest, the Key West *G.chrysia* in the Bahamas, Cuba, Hispaniola and part of Puerto Rico, in lowland forest, the Grey-headed *G.caniceps* in Cuba and Hispaniola, primarily in montane forest, the Crested *G.versicolor* in Jamaica in high montane forest, separated by altitude from *G. montana* on wooded lower slopes (Gosse 1851 pp. 166–167), and the Bridled *G.mystacea* in the Virgins and Lesser Antilles, partially separated from *G.montana* by its preference for drier woodland. (4G, 2½H, 3½?.)

Parrots *Amazona:* In the Lesser Antilles, a large species *A.guildingii* is on St Vincent, another *A.versicolor* on St Lucia, and two, the very large *A.imperialis* higher up and the somewhat smaller *A.arausiaca* lower down, on Dominica. One certain and one probable further species, on Guadeloupe and Martinique respectively, are extinct (Greenway 1967). The Greater Antilles have much smaller species, *A.leucocephala* on Cuba, *A.ventralis* on Hispaniola, *A.vittata* on Puerto Rico, and the Yellowbilled *A.collaria* and slightly smaller Blackbilled *A.agilis* on Jamaica, where their means of separation are not known. (34(8)G, 1Ha, 1?, but 53(10)G, 1Ha, 1? including the extinct forms.)

Parakeets *Aratinga:* *A.chloroptera* on Hispaniola and formerly Puerto Rico, *A.euops* on Cuba and *A.nana* on Jamaica replace each other geographically. (3G.) (A fourth species, *A.pertinax,* has been introduced to St Thomas.)

Cuckoos *Coccyzus:* The widespread Mangrove Cuckoo *C.minor* and the Yellow-billed Cuckoo *C.americanus* on the Greater Antilles, coexist in lowland scrub forest, and how they are separated there is not known, but the Mangrove Cuckoo is much the commoner and is also found in other types of forest including mangrove. (1?.)

Hyetornis cuckoos: The Chestnut-bellied *H.pluvialis* on Jamaica and the Bay-breasted *H.rufigularis* on Hispaniola have separate ranges. (1G.)

Lizard cuckoos *Saurothera:* *S.merlini* of the Bahamas, *S.longirostris* of Hispaniola, *S.vetula* of Jamaica and *S.vieilloti* of Puerto Rico, replace each other geographically. (10(4)G.)

Eared owls *Asio:* The Short-eared Owl *A.flammeus* breeds in open grassland in Hispaniola and Puerto Rico, the Stygian Owl *A.stygius* in woodland in Cuba and Hispaniola. (1H.)

Nightjars *Caprimulgus*: *C.cubanensis* on Cuba and Hispaniola, *C.nocti-therus* on Puerto Rico (formerly thought extinct; Reynard and Wetmore 1962), *C.rufus* on St Lucia, and *C.cayennensis* on Martinique, have separate ranges. (6(4)G.) The two last coexist on Trinidad.

Chimney swifts *Chaetura*: The Short-tailed *C.brachyura* on St Vincent, the Grey-rumped *C.cinereiventris* on Grenada (and possibly St Vincent) and the Lesser Antillean *C.martinica* on the other Lesser Antilles (possibly including St Vincent) evidently have separate breeding ranges. (3G.) The two former coexist on Trinidad.

Black swifts: On my, but not Bond's, classification, the Collared Swift *Streptoprocne collaris* is in the same genus as the Black Swift *Cypseloides niger*. The latter breeds in the Greater and Lesser Antilles; the Collared Swift, found on Cuba, Hispaniola and Jamaica, is separated by its much larger size. (1Fs.)

Emerald hummingbirds *Chlorostilbon*: *C.maugaeus* on Puerto Rico, *C.swainsonii* on Hispaniola and *C.ricordii* on Cuba replace each other geographically. (3G.)

Mango hummingbirds *Anthracothorax*: *A.mango* on Jamaica and *A.prevostii* on St Andrew and Old Providence are separated geographically from each other and from *A.dominicus* on Hispaniola and Puerto Rico and *A.viridis* on Puerto Rico. On Puerto Rico *A.viridis* is restricted to the highlands and *A.dominicus* to the lowlands, except that it has recently spread high up where there are roads or other clearings; but on Hispaniola *A.dominicus* occurs in both lowlands and highlands. (5½(3½)G, ½Ha.)

Bee hummingbirds *Mellisuga*: *M.helenae* on Cuba is separated geographically from *M.minima* on Hispaniola and Jamaica. (1G.)

Todies *Todus*: *T.multicolor* on Cuba, *T.mexicanus* on Puerto Rico and *T.todus* on Jamaica are separated geographically from each other and from the two species on Hispaniola, of which the Broadbilled *T.subulatus* occurs primarily in the semi-arid lowlands and the Narrowbilled *T.angustirostris* in dense, damp and usually montane, forest. The two latter species coexist only in extremely limited areas (Wetmore and Swales 1931, Wetmore and Lincoln 1933). Hence though the specific names 'Narrow-billed' and 'Broad-billed' suggest separation by feeding, they are primarily separated by habitat, including altitude. (9(4)G, 1H.)

Woodpeckers *Melanerpes* and *Centurus*: These two genera are usually merged. *M.portoricensis* on Puerto Rico, *M.herminieri* on Guadeloupe, *C.superciliaris* on the Bahamas and Cuba, *C.radiolatus* on Jamaica and *C.striatus* on Hispaniola, are separated from each other by range. (1G + 3G on Bond's classification, 10(5)G if the two genera are combined.)

29. Habitats of forest passerines in three West Indian islands

The habitats compared are lowland dry limestone scrub forest and highland rain forest in Jamaica (abbreviated to J) and Puerto Rico (P), and dry and moist lowland forest on St John, Virgin Islands (S), based on Bond (1969), Kepler (1969 and pers. comm.), and Robertson (1962), respectively.

A. PASSERINES

Tyrannus dominicensis low, entirely at edge, (P, S) (not really in forest J)
Tyrannus caudifasciatus both, primarily at edge (J)
Myiarchus stolidus low (J, P)
Myiarchus tuberculifer both (J)
Myiarchus validus high (J)
Contopus caribaeus both (J)
Elaenia martinica both (but commoner in dry) (S)
Elaenia fallax high (J)
Myiopagis cotta both (J)
Corvus leucognaphalus both, at least on Hispaniola (but now rare and only on high ground in Puerto Rico)
Mimus polyglottos low (P) (high only outside forest J)
Mimus gundlachii low (J)
Margarops fuscatus both (P, S)
Turdus jamaicensis high (J)
Turdus aurantius both (J, but only at edge in high forest)
Mimocichla plumbea both (P)
Myadestes genibarbis high (J, but also low in N.E.)
Vireo modestus both (J)
Vireo latimeri low (P)
Vireo osburni high (J)
Vireo altiloquus both (commoner low) (J, P, S)
Dendroica petechia low, dry (J, P, S)
Dendroica adelaidae low (P) (but also high on St Lucia)
Dendroica pharetra mainly high (J)
Coereba flaveola both (J, P, S)
Euneornis campestris high (J)
Tanagra musica both (P)
Pyrrhuphonia jamaica both (J)
Spindalis zena both (J, P)
Nesospingus speculiferus high (P)
Icterus dominicensis both (P) (but primarily on edge in montane forest)

Icterus leucopteryx both (J)
Nesopsar nigerrimus high (J)
Loxigilla portoricensis both (P)
Loxigilla violacea both (J)
Loxipasser anoxanthus high (J)

Note

Platypsarus niger (J) and *Corvus jamaicensis* (J) omitted because neither low nor high, but intermediately in the hills.

B. NON-PASSERINES FROM CUCKOOS TO WOODPECKERS

Coccyzus minor low (J, P), both dry and moist (S)
Coccyzus americanus low (J, P)
Hyetornis pluvialis high (J)
Saurothera vetula both (J)
Saurothera vieilloti mainly low (P)
Caprimulgus noctitherus low (P)
Chlorostilbon maugaeus both (P)
Anthracothorax mango low (J)
Anthracothorax dominicus low (P) (but both Hispaniola)
Anthracothorax viridis high (P)
Sericotes holosericeus low (P), mainly dry (S)
Orthorhyncus cristatus low (P), more in dry than moist (S)
Trochilus polytmus both (J)
Mellisuga minima both (J)
Todus todus both (J)
Todus mexicanus both (P)
Melanerpes portoricensis both (P)
Centurus radiolatus both (J)

Notes

The other land birds have not been included, pigeons Columbidae because the statements by different authors do not always agree and because present distributions might have been modified by human shooting or the sowing of grain, parrots *Amazona* spp. because though they are now confined to high ground, this might be due to the destruction of rich lowland forest or their capture in former times in the lowlands by man, owls Strigidae because they are nocturnal, so records of habitats might be incomplete, raptors because they are not attached to particular habitats in the same way as other species, and all extinct species because knowledge of their habitats might be incomplete.

APPENDIXES

30. Tentative survey of ecological isolation in congeneric passerine birds in other tropical archipelagoes

A. PHILIPPINES (based on Delacour and Mayr 1946)

The Philippines occupy a similar latitudinal range to the West Indies and are fairly isolated, Luzon being some 350 km south of Taiwan and Mindanao a similar distance northeast of Borneo, but with small islands in between. Roughly 30 per cent of the congeneric passerine birds are separated from each other by range, a much smaller proportion than in the West Indies, perhaps because the individual islands are closer. There are too many instances in which the means of segregation are not known to say whether differences in habitat or feeding are the more frequent, but both are important.

B. SOLOMON ISLANDS
(based on Mayr 1945, Cain and Galbraith 1956)

Of the genera with two species in the archipelago, separation is by range in the crows *Corvus* and flower-peckers *Dicaeum*, by habitat in the cuckoo-shrikes *Edilosoma* and the flycatchers *Pachycephala*, and by unknown means in the leaf-warblers *Phylloscopus*. The 3 species of cuckoo-shrikes in *Coracina* are separated by habitat, as is one species of *Monarcha* flycatchers from the other two, but how the other two might be separated is not known. Four of the 5 starlings in the genus *Aplonis* are separated by habitat, and the other is larger than the rest. Two species of honey-eater *Myzomela* are on one island, where they are separated by habitat, and each of the other 4 is on a different island from these two and from each other. Of the 7 species of *Rhipidura* flycatchers, one is separated from the rest by habitat, others are separated by range, habitat or feeding (highly tentatively 4G, 9H, 3F, 4?). Finally the separation of the 10 species of white-eyes *Zosterops* (following Mees) was set out on pp. 342–3. ($9\frac{1}{2}$G, $1\frac{1}{2}$H.)

Hence most congeneric species are separated from each other by habitat and somewhat fewer by range. Separation by range is frequent in white-eyes and honey-eaters. Separation by habitat is in most cases between closed forest and open woodland (and the forest edge), or between lowland and montane forest.

C. FIJI ISLANDS (based on Mayr 1945)

Of the genera with two species in the archipelago, separation is by range on different islands in the flycatchers *Rhipidura* and *Mayrornis*, by habitat in the flycatchers *Myiagra* and white-eyes *Zosterops*, and by size and presumably feeding in the flycatchers *Clytorhynchus*, honeyeaters *Myzomela* and parrot-finches *Erythrura*. No passerine genus has more than two species in the islands.

D. SAO TOMÉ, PRINCIPE AND ANNOBON IN GULF OF GUINEA
(based on Snow, 1950, Amadon, 1953a)

The fourth Guinea island, Fernando Po, is excluded as it is only 32 km from the mainland. The following summary concerns solely the native species, and introduced or probably introduced species are omitted. Of the genera with two species, separation is by range in the paradise flycatchers *Terpsiphone*, white-eyes *Zosterops* and white-eyes *Speirops*. Of the 3 native species of *Ploceus*, one is separated by range from the other two, which differ markedly in beak and presumably feeding. Of the four species of sunbirds *Nectarinia*, there is a large and a small on one island and a different large and different small species on another island. (In addition, two species of *Lamprotornis* starlings have been recorded from Principe, but there seems to be no good evidence that one of them, *L.ornatus*, breeds there.)

Hence most congeneric species are separated by range, the rest by size and presumably feeding, and none are separated by habitat, a situation recalling that of the passerine birds of the Galapagos and Hawaiian archipelagoes. In addition, many genera present on Sao Tomé are absent from Principe and conversely, so there is much geographical replacement by species in different genera, as in the West Indies. Moreover some of the species found on Principe are so different from any on Sao Tomé and conversely, that the ecological niches must to some extent be differently subdivided on the two islands.

References

Abs M. 1963. Vergleichende Untersuchungen an Haubenlerche (*Galerida cristata* [L.]) und Theklalerche (*Galerida theklae* A.E. Brehm). Bonn.zool.Beitr. 14:1–128.

Alcock J. 1969. Observational learning in three species of birds. Ibis 111:308–321.

Ali S. 1949. Indian Hill Birds. (Bombay).

Ali S. 1953. The Birds of Travancore and Cochin. (Bombay).

Amadon D. 1947. Ecology and the evolution of some Hawaiian birds. Evolution 1:63–68.

Amadon D. 1950. The Hawaiian honeycreepers (Aves, Drepaniidae). Bull.Amer.Mus.Nat.Hist. 95:151–262.

Amadon D. 1953a. Avian systematics and evolution in the Gulf of Guinea. Bull.Amer.Mus.Nat.Hist. 100:393–452.

Amadon D. 1953b. Migratory birds of relict distribution: some inferences. Auk 70:461–469.

Amadon D. and Basilio A. 1957. Notes on the birds of Fernando Poo Island, Spanish Equatorial Africa. Amer.Mus.Novit. no. 1846.

Amann F. 1954. Neuere Beobachtungen an Weiden- und Alpenmeisen, *Parus atricapillus*, mit vergliechenden Angaben über die Nonnenmeise, *Parus palustris*. Orn.Beob. 51:104–109.

Andersson G., Gerell R., Källander H. and Larsson T. 1968. (Notes on the habitat of the Little Bunting (*Emberiza pusilla* Pall.)). Fågelvärld 27:136–141

Andrewartha H.G. and Birch L.C. 1955. Distribution and Abundance of Animals. (Chicago).

Angwin J. 1968. A preliminary study of the ecology of birds in riverine acacia woodland. M.Sc. thesis. Univ.East Africa (Nairobi)

Ashmole N.P. 1968a. Competition and interspecific territorially in *Empidonax* flycatchers. Syst.Zool. 17:210–212.

Ashmole N.P. 1968b. Body size, prey size, and ecological segregation in five sympatric tropical terns (Aves: Laridae). Syst.Zool. 17:292–304.

REFERENCES

Ashmole N. P. and Ashmole M. J. 1967. Comparative feeding ecology of sea birds of a tropical oceanic island. Peabody Mus.Nat.Hist. Bull. 24.

Austin O. L. and Kuroda N. 1953. The birds of Japan. Their status and distribution. Bull.Mus.Comp.Zool. 109:278-637.

Bagenal T. B. 1951. A note on the papers of Elton and Williams on the generic relations of species in small ecological communities. J.Anim. Ecol. 20:242-245.

Baker E. C. S. 1922. The Fauna of British India. Birds. Vol. 1 (London).

Baker E. C. S. 1942. Cuckoo Problems. (London).

Baker J. R. 1938. The evolution of breeding seasons. Evolution. Essays presented to E. S. Goodrich (Oxford) pp. 161–177.

Balát F. 1962. Contribution to the knowledge of the avifauna of Bulgaria. Acta Acad.Sci Cechoslovenicae 34:445-492.

Balgooy M. M. J. Van. 1969. A study on the diversity of island floras. Blumea 17:139-178.

Bannerman D. A. 1963. A History of the Birds of the Canary Islands and of the Salvages. (Edinburgh and London).

Bannerman D. A. and Bannerman W. M. 1965. A History of the Birds of Madeira, the Desertas and the Porto Santo Islands. (Edinburgh and London).

Bannerman D. A. and Bannerman W. M. 1968. History of the Birds of the Cape Verde Islands. (Edinburgh).

Barbour T. 1943. Cuban Ornithology. Mem.Nuttall Orn.Club no. 9 (Cambridge, Mass.).

Bates R. S. P. and Lowther E. H. N. 1952. Breeding Birds of Kashmir. (London).

Bédard F. 1967. Ecological segregation among plankton-feeding Alcidae (*Aethia* and *Cyclorhynchus*). Ph.D. thesis. Univ. of British Columbia.

Bédard J. 1969. Adaptive radiation in Alcidae. Ibis 111:189-198.

Bell B. D. 1968. Population ecology of the Reed Bunting *Emberiza schoeniclus schoeniclus* (L.). Ph.D. thesis. Univ. Nottingham.

Bell B. D. 1969. Some thoughts on the apparent ecological expansion of the Reed Bunting. Brit.Birds, 62:209-218.

Belopolskii L. O. 1957. Ecology of Sea Colony Birds of the Barents Sea. (English transl. R. Ettinger and C. Salzmann, Jerusalem, 1961).

Benson C. W. 1960. The birds of the Comoro Islands. Ibis 103b: 1–106 (esp. pp. 88–91).

Betts M. M. 1955. The food of titmice in oak woodland. J.Anim.Ecol. 24:282-323.

Beven G. 1959. The feeding sites of birds in dense oakwood. London Nat. 38:64-73.

Beven G. 1965. The food of Tawny Owls in London. London Bird Rep. 29:56-72.

REFERENCES

Bezzel E. 1957. Beiträge zur Kenntnis der Vogelwelt Sardiniens. Anz. Orn.Ges.Bayern 4:589–707.

Blaszyk P. 1963. Das Weisssternige Blaukehlchen *Luscinia svecica cyanecula* als Kulturfolger in der gebüschlosen Ackermarsch. J.Orn. 104:168–181.

Bond J. 1948. Origin of the bird fauna on the West Indies. Wilson Bull. 60:207–229.

Bond J. 1960. Birds of the West Indies. (London).

Bowman R. I. 1961. Morphological differentiation and adaptation in the Galapagos finches. Univ.Calif.Publ.Zool. 58:1–302.

Bowman R. I. 1963. Evolutionary patterns in Darwin's finches. Occ. Pap.Calif.Acad.Sci. 44:107–140.

Bowman R. I. and Biller S. L. 1965. Blood-eating in a Galapagos finch. Living Bird 4:29–44.

Brewer R. 1963. Ecological and reproductive relationships of Black-capped and Carolina chickadees. Auk 80:9–47.

Brock S. E. 1914. Ecological relations of bird-distribution. Brit.Birds 8:29–44.

Brosset A. 1961. Écologie des Oiseaux du Maroc Oriental. Trav.Inst.Sci. Chérifien Ser.Zool. 22 pp. 155.

Brown W. L. and Wilson E. O. 1956. Character displacement. Syst.Zool. 5:49–64.

Burleigh T. D. 1958. Georgia Birds. (Oklahoma).

Cain A. J. 1969. Speciation in tropical environments: summing up. Biol. J.Linn.Soc. 1:232–236. Speciation in Tropical Environments *ed.* R. H. Lowe-McConnell.

Cain A. J. and Galbraith I. 1956. Field notes on birds of the eastern Solomon Islands. Ibis 98:100–134, 262–295.

Chapin J. P. 1939. The Birds of the Belgian Congo. Part II. Bull.Amer. Mus.Nat.Hist. 75.

Colquhoun M. K. 1941. Visual and auditory conspicuousness in a wood-land bird community: a quantitative analysis. Proc.Zool.Soc.Lond. A 110:129–148.

Colquhoun M. K. and Morley A. 1943. Vertical zonation in woodland bird communities. J.Anim.Ecol. 12:75–81.

Cornwallis R. K. 1961. Four invasions of Waxwings during 1956–60. Brit.Birds 54:1–30.

Crowell K. 1961. The effects of reduced competition in birds. Proc.Nat. Acad.Sci. 47:240–243.

Crowell K. L. 1962. Reduced interspecific competition among the birds of Bermuda. Ecology 43:75–88.

Crowell K. L. 1968. Competition between two West Indian flycatchers *Elaenia*. Auk 85:265–286

REFERENCES

Cullen J. M., Guiton P. E., Horridge G. A. and Peirson J. 1952. Birds on Palma and Gomera (Canary Islands). Ibis 94:68–84.

Curio E. 1959. Beobachtungen am Halbringschnäpper, *Ficedula semitorquata*, im mazedonischen Brutgebeit. J.Orn. 100:176–209.

Curio E. and Kramer P. 1964. Vom Mangrovefinken (*Cactospiza heliobates* Snodgrass and Heller). Zeits.Tierpsychol. 21:223–234.

Danford C. G. 1878. A contribution to the ornithology of Asia Minor. Ibis 1878:1–35, esp. pp. 10–12.

Dawson B. V. and Foss B. M. 1965. Observational learning in Budgerigars. Anim.Behav. 13:470–474.

DeBenedictis P. A. 1966. The bill-brace feeding behavior of the Galapagos finch *Geospiza conirostris*. Condor 68:206–208.

Deignan H. G. 1938. A new nuthatch from Yunnan. Smithsonian Misc. Coll. 97 no. 9.

Deignan H. G. 1945. The Birds of Northern Thailand. U.S.Nat.Mus.Bull. 186.

Delacour J. and Jabouille P. 1931. Les Oiseaux de l'Indochine Française. (Paris). Vol. 4.

Delacour J. and Mayr E. 1946. Birds of the Philippines. (New York).

Dementiev G. P. and Gladkov N. A. 1968. Birds of the Soviet Union. (Moscow). Vol. 6. (transl. A. Birron and Z. S. Cole, Jerusalem).

de Schauensee R. M. 1934. Zoological results of the third De Schauensee Siamese expedition, Part II—Birds from Siam and the Southern Shan States. Proc.Acad.Nat.Sci. Philadelphia 86:165–280.

Dixon K. L. 1961. Habitat distribution and niche relationships in North American species of *Parus*. pp. 179–216 in Vertebrate Speciation ed. F. W. Blair. (Austin, Texas).

Dorward D. F. 1962. Comparative biology of the White Booby and Brown Booby *Sula* spp. at Ascension. Ibis 103b:174–220.

Durango S. 1950. (The influence of climate on distribution and breeding-success of the Red-backed Shrike.) Fauna och Flora 45:49–78.

Durango S. 1954. (The habitats of the Red-backed and Woodchat Shrikes *(Lanius collurio* and *L.senator).*) Fauna och Flora 54:1–16.

Ehlert W. 1964. Zur Ökologie und Biologie der Ernährung einiger Limikolen-Arten. J. Orn. 105:1–53.

Eisentraut M. 1963. Die Wirbeltiere des Kamerungebirges. (Hamburg and Berlin).

Eisentraut M. 1968. Beitrag zur Vogelfauna von Fernando Poo und Westkamerun. Bonn.zool.Beitr. 19:49–68.

Elgood J. H. and Sibley F. C. 1964. The tropical forest edge avifauna of Ibadan, Nigeria. Ibis 106:221–248.

Elton C. S. 1927. Animal Ecology. (London).

Elton C. S. 1946. Competition and the structure of animal communities. J.Anim.Ecol. 15:54–68.

REFERENCES

Ern H. 1966. Zur Ökologie und Verbreitung des Blaukehlchens, *Luscinia svecica*, in Spanien. J.Orn. 107:310-314.

Fabricius E. 1951. Zur Ethologie jungen Anatiden. Acta Zool.Fenn. 68:1-175.

Falla R.A., Sibson R.B. and Turbott E.G. 1966. A Field Guide to the Birds of New Zealand. (London).

Ferguson-Lees I.J. 1968. Serins breeding in southern England. Brit. Birds, 61:87-88.

Ferry C. and Deschaintre A. 1966. *Hippolais icterina* et *polyglotta* dans leur zone de sympatrie. Abstr. XIV Cong.Int.Orn. pp. 57-58. (Oxford).

Ficken M.S. and Ficken R.W. 1967. Age-specific differences in the breeding behavior and ecology of the American Redstart. Wilson Bull. 79:188-199.

Fisher J. and Lockley R.M. 1954. Sea-Birds. (London). p. 284.

Friedmann H. 1948. The Parasitic Cuckoos of Africa. Washington Adad.Sci.Monog. 1.

Friedmann H. 1955. The Honeyguides. U.S.Nat.Mus.Bull. 208.

Friedmann H. 1960. The Parasitic Weaverbirds. U.S.Nat.Mus.Bull. 233.

Friedmann H. 1964. Evolutionary trends in the avian genus *Clamator*. Smithsonian Misc.Coll. 146 no. 4.

Friedmann H. 1967a. Evolutionary terms for parasitic species. Syst. Zool. 16:175.

Friedmann J. 1967b. Alloxenia in three sympatric African species of *Cuculus*. Proc.U.S.Nat.Mus. 124:no. 3633.

Friedmann H. 1968. The Evolutionary History of the Avian Genus *Chrysococcyx*. U.S.Nat.Mus.Bull. 265.

Gaston A.J. 1968. The birds of the Ala Dagh mountains, southern Turkey. Ibis 110:17-26.

Gause G.F. 1934. The Struggle for Existence. (Baltimore).

Gause G.F. 1939. In discussion of Park T. Analytical population studies in relation to general ecology. Am.Midl.Nat. 21:255.

Gibb J. 1954. Feeding ecology of tits, with notes on Treecreeper and Goldcrest. Ibis 96:513-543.

Gifford E.W. 1919. Field notes on the land birds of the Galapagos Islands and Cocos Island, Costa Rica. Proc.Calif.Acad.Sci. Ser. 4 Vol. 2 pt. 2:189-258.

Gilliard E.T. 1969. Birds of Paradise and Bower Birds. (London).

Glas P. 1960. Factors governing density in the Chaffinch *(Fringilla coelebs)* in different types of wood. In L.Tinbergen, The Dynamics of Insect and Bird Populations in Pine Woods. (Leiden). Arch.neerl. Zool. 13:466-472.

Glenister A.G. 1951. The Birds of the Malay Peninsula, Singapore and Penang. (London). p. 194.

REFERENCES

Glutz U.N. von Blotzheim. 1962. Die Brutvögel der Schweiz. (Aarau).

Godfrey W.E. 1966. The Birds of Canada. (Ottawa). esp. pp. 277–281.

Gosse P.H. 1847. The Birds of Jamaica. (London).

Gosse P.H. 1851. A Naturalist's Sojourn in Jamaica. (London).

Grant P.R. 1968. Bill size, body size and the ecological adaptations of bird species to the competitive situations on islands. Syst.Zool. 17:319–333.

Grant P.R. 1969. Colonization of islands by ecologically dissimilar species of birds. Canad.J.Zool. 47:41–43.

Greenway J.C. 1967a. Family Sittidae. Check-List of Birds of the World vol. 12 (ed. R.A. Paynter). pp. 125–149. (Cambridge, Mass.).

Greenway J.C. 1967b. Extinct and Vanishing Birds of the World. 2nd rev. ed. (New York).

Grinnell J. 1904. The origin and distribution of the Chestnut-backed Chickadee. Auk 21:364–382.

Grinnell J. 1917a. The niche-relationships of the California Thrasher. Auk. 34:427–433.

Grinnell J. 1917b. Field tests of theories concerning distributional control. Amer.Nat. 51:115–128.

Grinnell J. 1924. Geography and evolution. Ecology 5:225–229.

Grinnell J. 1928. Presence and absence of animals. Reprinted in *Joseph Grinnell's Philosophy of Nature* (1943). (Berkeley).

Grinnell J. 1943. Philosophy of Nature. (Berkeley and Los Angeles, California).

Grinnell J. and Miller A.H. 1944. The Distribution of the Birds of California. (Berkeley, California).

Hachisuka M. and Udagwa T. 1951. Contribution to the ornithology of Formosa. Pt. 2. Q.J.Taiwan Mus. 4:1–180, esp. pp. 23–25.

Haftorn S. 1956. Contribution to the food biology of tits especially about storing of surplus food. Pt. IV. Kgl.Norske Vidensk.Selsk. Skrifter. 1956 no. 4.

Hagen Y. 1952. Birds of Tristan da Cunha. Results of the Norwegian Scientific Expedition to Tristan da Cunha 1937–8. no. 20. (Oslo).

Hall B.P. 1960. The ecology and taxonomy of some Angola birds. Bull. Brit.Mus.Nat.Hist. 6:369–453.

Hamilton T.H. 1958. Adaptive variation in the genus *Vireo*. Wilson Bull. 70:307–346.

Hamilton T.H. 1962. Species relationships and adaptations for sympatry in the avian genus *Vireo*. Condor 64:40–68.

Hardin G. 1960. The competitive exclusion principle. Science 131:1292–1297.

Hardy J.W. 1967. Evolutionary and ecological relationships between three species of blackbirds *(Icteridae)* in central Mexico. Evolution 21:196–197.

REFERENCES

Hardy J.W. and Dickermann R.W. 1965. Relationships between two forms of the Red-winged Blackbird in Mexico. Living Bird 4:107–129.

Harris M.P. 1963. Some aspects of the biology of Larus gulls. Ph.D. thesis, Univ.Coll.Swansea.

Harris M.P. 1969. The biology of storm petrels in the Galapagos islands. Proc.Calif.Acad.Sci. (4) 37:95–166.

Harrison C.J.O. 1966. The validity of some genera of larks (A.landidae) Ibis 108:573–583.

Harrison J.M. 1933. A contribution to the ornithology of Bulgaria. Ibis 1933:494–521, 589–611.

Hartley P.H.T. 1949. The biology of the Mourning Chat in winter quarters. Ibis 91:393–413.

Hemmingsen A.M. 1951. Observations on birds in North Eastern China, especially the migration at Pei-tai-Ho beach. Spolia Zool.Mus. Hauniensis 11 (pp.227).

Hildén O. 1964. Ecology of duck populations in the island group of Valassaaret, Gulf of Bothnia. Ann.Zool.Fenn. 1:153–279.

Hildén O. 1965. Habitat selection in birds. Ann.Zool.Fenn. 2:53–75.

Hinde R.A. 1959. Behaviour and speciation in birds and lower vertebrates. Biol.Rev. 34:85–128.

Höglund N.H. 1964. Über die Ernährung des Habichts (Accipiter gentilis Lin.) in Schweden. Viltrevy 2:271–328.

Holyoak D. 1968. A comparative study of the food of some British Corvidae. Bird Study 15:147–153.

Holmes R.T. and Pitelka F.A. 1968. Food overlap among coexisting sandpipers on northern Alaskan tundra. Syst.Zool. 17:305–318.

Howell A.H. 1924. Birds of Alabama. (Montgomery, Alabama).

Howell A.H. 1932. Florida Bird Life. (New York).

Hutchinson G.E. 1959. Homage to Santa Rosalia, or Why are there so many kinds of animals? Amer.Nat. 93:145–159.

Huxley J.S. 1942. Evolution. The Modern Synthesis. (London). p. 280.

Huxley J.S. 1950. Natural history in Iceland. Discovery March 1950 73–78.

Jespersen P. 1944. (Immigration and distribution of the Crested Tit, Parus cristatus L., in Denmark.) Dansk Orn.For.Tids. 38:1–13.

Jewett S.G., Taylor W.P., Shaw W.T. and Aldrich J.W. 1953. Birds of Washington State. (Seattle).

Johnson N.K. 1963. Biosystematics of sibling species of flycatchers in the Empidonax hammondii-oberholseri-wrightii complex. Univ.Calif. Publ.Zool. 66:79–238.

Johnson N.K. 1966. Bill size and the question of competition in allopatric and sympatric populations of Dusky and Gray Flycatchers. Syst.Zool. 15:70–87.

REFERENCES

Johnston D.W. 1969. The thrushes of Grand Cayman Island, B.W.I. Condor 71:120-128.

Jourdain F.C.R. 1911-12. Notes on the ornithology of Corsica. Ibis 1911:189-208, 437-458; 1912:63-82, 314-332.

Kartaschew N.N. 1960. Die Alkenvögel des Nordatlantiks. Neue Brehm-Bücherie. (Wittenberg Lutherstadt).

Kear J. 1962. Food selection in finches with special reference to interspecific differences. Proc.Zool.Soc.Lond. 138:163-204.

Keast, A. 1968. Competitive interactions and the evolution of ecological niches as illustrated by the Australian honeyeater genus *Melithreptus* (Meliphagidae). Evolution 22:762-784.

Kepler C.B. and Kepler A.K. (1969). Luquillo and Guanica Forests: preliminary comparison of bird species diversity and density. in Odum E.T. A Tropical Rain Forest. (Read in MS.).

Koenig O. 1952. Ökologie und Verhalten der Vögel des Neusiedlersee-Schilfgürtels. J.Orn. 93:207-289.

Klopfer R.H. 1957. An experiment on empathic learning in ducks. Amer.Nat. 91:61-63.

Klopfer P.H. 1959. Social interactions in discrimination learning with special reference to feeding behaviour in birds. Behaviour 14:282-299.

Klopfer P.H. 1963. Behavioral aspects of habitat selection: the role of early experience. Wilson Bull. 75:15-22.

Klopfer P.H. and MacArthur R.H. 1961. On the causes of tropical species diversity: niche overlap. Amer.Nat. 95:223-226.

Kluyver H.N. and Tinbergen L. 1953. Territory and the regulation of density in titmice. Arch.neerl.Zool. 10:265-289.

Kozlova E.V. 1933. The birds of South-West Transbaikalia, Northern Mongolia and Central Gobi. Pt. V. Ibis 1933 pp. 303-304.

Kruuk H. 1967. Competition for food between vultures in East Africa. Ardea 55:171-193.

Lack D. 1933. Habitat selection in birds. J.Anim.Ecol. 2:239-262.

Lack D. 1934. Habitat distribution in certain Icelandic birds. J.Anim. Ecol. 3:81-90.

Lack D. 1937. The psychological factor in bird distribution. Brit.Birds 31:130-136.

Lack D. 1940. Habitat selection and speciation in birds. Brit.Birds 34:80-84.

Lack D. 1942-3. The breeding birds of Orkney. Ibis 1942:461-484, 1943:1-27.

Lack D. 1944. Ecological aspects of species-formation in passerine birds. Ibis 1944:260-286.

Lack D. 1945a. The ecology of closely related species with special reference to Cormorant *(Phalacrocorax carbo)* and Shag *(P.aristotelis)*. J.Anim.Ecol. 14:12-16.

REFERENCES

Lack D. 1945b. The Galapagos finches *(Geospizinae)*. Occ.Pap.Calif. Acad.Sci. 21.

Lack D. 1946. Competition for food by birds of prey. J.Anim.Ecol. 15:123-129.

Lack D. 1947. Darwin's Finches. (Cambridge).

Lack D. 1949. The significance of ecological isolation. Genetics, Palaeontology and Evolution ed. G. L. Jepsen *et al.* (Princeton). pp. 299-308.

Lack D. 1954. The Natural Regulation of Animal Numbers. (Oxford).

Lack D. 1955. The species of *Apus.* Ibis 98:34-62.

Lack D. 1966. Population Studies of Birds. (Oxford).

Lack D. 1968. Ecological Adaptations for Breeding in Birds. (London).

Lack D. 1969. The numbers of bird species on islands. Bird Study 16:193-209.

Lack D. and Southern H. N. 1949. Birds on Tenerife. Ibis 91:607-626.

Lanyon W. E. 1956a. Territory in the meadowlarks, genus *Sturnella.* Ibis 98:485-489.

Lanyon W. E. 1956b. Ecological aspects of the sympatric distribution of meadowlarks in the north-central states. Ecology 37:98-108.

Lanyon W. E. 1957. The comparative biology of the meadowlarks *(Sturnella)* in Wisconsin. Pub.Nuttall Orn.Club no. 1. (Cambridge, Mass.).

Lanyon W. E. 1967. Revision and probable evolution of the *Myiarchus* flycatchers of the West Indies. Bull.Amer.Mus.Nat.Hist. 136(6): 329-370.

La Touche J. D. D. 1925-30. A Handbook of the Birds of Eastern China. Vol. 1. p. 29. (London).

Lawrence L. de K. 1958. On regional movements and body weight of Black-capped Chickadees in winter. Auk 75:415-443.

Lockie J. D. 1956. The food and feeding behaviour of the Jackdaw, Rook and Carrion Crow. J.Anim.Ecol. 25:421-428.

Löhrl H. 1959. Zur Frage des Zeitpunktes einer Prägung auf die Heimatregion beim Halsbandschnäpper *(Ficedula albicollis).* J.Orn. 132-140.

Löhrl H. 1960-1. Vergleichende Studien über Brutbiologie und Verhalten der Kleiber *Sitta whiteheadi* Sharpe und *Sitta canadensis* L. J.Orn. 101:245-264; 102:111-132.

Löhrl H. 1965. Zur Vogelwelt der griechischen Insel Lesbos (Mytilene). Vogelwelt 86:105-112.

Löhrl H. 1966a. Experiments on the nesting ecology and ethology of the Great Tit, *Parus major,* and related species. Abstr.XIV Congr.Int. Orn. (Oxford). pp. 81-82.

Löhrl H. 1966b. Zur Biologie der Trauermeise *(Parus lugubris).* J.Orn. 107:167-186.

Löhrl H. and Thielcke G. 1969. Zur Brutbiologie und Systematik einiger Waldvögel Afghanistans. Bonn.zool.Beitr. 20:85-98.

365

REFERENCES

Ludlow F. 1944. The birds of South-eastern Tibet. Ibis 1944:43–86, 176–208, 348–389.

MacArthur R.H. 1958. Population ecology of some warblers of northeastern coniferous forests. Ecology 39:599–619.

MacArthur R.H. 1964. Environmental factors affecting bird species diversity. Amer.Nat. 98:387–397.

MacArthur R.H. 1965. Patterns of species diversity. Biol.Rev. 40:510–533.

MacArthur R.H. 1969. Patterns of communities in the tropics. Biol.J. Linn.Soc. 1:19–30. Speciation in Tropical Environments ed. R.H. Lowe-McConnell.

MacArthur, R.H. and Levins R. 1964. Competition, habitat selection, and character displacement in a patchy environment. Proc.Nat. Acad.Sci. 51:1207–1210.

MacArthur R.H. and Levins R. 1967. The limiting similarity, convergence, and divergence of coexisting species. Amer.Nat. 101:377–385.

MacArthur R.H. and Pianka E.R. 1966. On optimal use of a patchy environment. Amer.Nat. 100:603–609.

MacArthur R.H., Recher H. and Cody M. 1966. On the relation between habitat selection and species diversity. Amer.Nat. 100:319–332.

MacArthur R.H. and Wilson E.O. 1967. The Theory of Island Biogeography. (Princeton).

Madsen F.J. 1957. On the food habits of some fish-eating birds in Denmark. Danish Rev.Game Bird. 3:19–83.

Makatsch W. 1950. Die Vogelwelt Macedoniens. (Leipzig).

Marshall J.T. 1957. Birds of pine-oak woodland in southern Arizona and adjacent Mexico. Pacific Coast Avif. 32. (Berkeley, Cal.).

Mastrovic A. 1942. Die Vögel des Küstenlandes Kroatiens. (Zagreb).

Mathews G.M. 1928. The Birds of Norfolk and Lord Howe Islands. (London).

Mayaud N. 1958. La Gorge-Bleue à Miroir Luscinia svecica, en Europe. Alauda 26:290–301.

Mayr E. 1926. Die Ausbreitung des Girlitz (Serinus canaria serinus L.). J.Orn. 74:571–671.

Mayr E. 1931. Birds collected during the Whitney South Sea expedition. XVI. Notes on Fantails of the genus Rhipidura. Amer.Mus.Novit. 502.

Mayr E. 1933. Die Vogelwelt Polynesiens. Mitt.Zool.Mus.Berlin 19:306–323.

Mayr E. 1940. The origin and history of the bird fauna of Polynesia. Proc.6th Pacific Sci.Congr. 4:197–216.

Mayr E. 1942. Systematics and the Origin of Species. (New York).

Mayr E. 1945. Birds of the Southwest Pacific. (New York).

Mayr E. 1963. Animal Species and Evolution. (Cambridge, Mass.).

REFERENCES

Mayr E. 1964. Inferences concerning the Tertiary American bird faunas. Proc.Nat.Acad.Sci. 51:280–288.

Mayr E. 1965. What is a fauna? Zool.Jarhb.Syst. 92:473–486.

Mayr E. 1967. Family Zosteropidae. Indo-Australian Taxa. Check-List of Birds of the World Vol. 12 (ed. R.A.Paynter) pp. 289–326. (Cambridge, Mass.).

Mayr E. and Phelps W.H. 1967. The origin of the bird fauna of the South Venezuelan highlands. Bull.Amer.Mus.Nat.Hist. 136:269–328.

Mees G.F. 1957, 1961, 1969. A systematic review of the Indo-Australian Zosteropidae. Pts. 1,2,3. Zool.Verhandl. nos. 35,50,102. (Leiden).

Meiklejohn R.F. 1934, 1935. Notes on Rüppell's Warbler *Sylvia rüppelli* (Temminck). Ibis 1934:301–305; 1935:432–435.

Merikallio E. 1951. Der Einfluss der letzten Wärmeperiode (1930–49) auf die Vogelfauna Nordfinnlands. Proc.Int.Orn.Cong. 10:484–493.

Miller R.S. 1968. Conditions of competition between Redwings and Yellow-headed Blackbirds. J.Anim.Ecol. 37:43–62.

Milon Ph. 1951. Notes sur l'avifaunne actuelle de l'ile de la Réunion. Terre et Vie 98:129–178.

Moreau R.E. 1948. Ecological isolation in a rich tropical avifauna. J.Anim.Ecol. 17:113–126.

Moreau R.E. 1957. Variation in the western Zosteropidae. Bull.Brit. Mus.(Nat.Hist.) Zool. 4:309–433.

Moreau R.E. 1958. Some aspects of the Musophagidae. Ibis 100:67–112, 238–270.

Moreau R.E. 1966. The Bird Faunas of Africa and its Islands. (New York, London).

Moreau R.E. 1967. Family Zosteropidae. African and Indian Ocean Taxa. Check-List of Birds of the World (ed. R.A.Paynter). Vol. 12 pp. 326–337.

Moreau R.E. *in prep.* The Palaearctic-African Bird Migration System. (Academic Press).

Morse D.H. 1968. A quantitative study of foraging of male and female spruce-woods warblers. Ecology 49:779–784.

Munro J.A. and Cowan I.McT. 1947. A review of the bird fauna of British Columbia. B.C.Prov.Mus.Dept.Ed. special Publ. 2 (Victoria, B.C.).

Murton R.K. 1968. Darwin's theory and the five British pigeons. Animals 10:400–404.

Murton R.K. and Clarke S.P. 1968. Breeding biology of Rock Doves. Brit.Birds 61:429–448.

Murton R.K. and Isaacson A.J. 1964. The feeding habits of the Wood Pigeon *Columba palumbus*, Stock Dove *C.oenas* and Turtle Dove *Streptopelia turtur*. Ibis 106:174–188.

REFERENCES

Murton R. K. and Westwood N. J. 1966. The foods of the Rock Dove and
. Feral Pigeon. Bird Study 13:130-146.

Nelson B. 1968. Galapagos. Island of Birds. (London).

Newton I. 1964. Bud-eating by Bullfinches in relation to the natural
food-supply. J.appl.Ecol. 1:265-279.

Newton I. 1967a. The adaptive radiation and feeding ecology of some
British finches. Ibis 109:33-98.

Newton I. 1967b. Attacks on fruit buds by Redpolls Carduelis flammea.
Ibis 109:440-441.

Newton I. 1967c. The feeding ecology of the Bullfinch (Pyrrhula
pyrrhula L.) in southern England. J.Anim.Ecol. 36:721-744.

Nicolai J. 1964. Der Brutparasitismus der Viduinae als ethologisches
Problem. Zeits. Tierpsychol. 21:129-204.

Niethammer G. 1937, 1938, 1942. Handbuch der Deutschen Vogelkunde.
Vols.1-3 (Leipzig).

Niethammer G. 1943. Beiträge zu Kenntnis der Brutvögel des Pelo-
ponnes. J.Orn. 91:167-238.

Niethammer G. 1957. Zur Vogelwelt der Sierra Nevada. Bonn.zool.Beitr.
8:230-247

Norris R. A. 1958. Comparative biosystematics and life history of the
nuthatches, Sitta pygmaea and Sitta pusilla. Univ.Calif.Publ.Zool.
56:119-300

Norton-Griffiths M. 1967. Some ecological aspects of the feeding
behaviour of the Oystercatcher Haematopus ostralegus on the
Edible Mussel Mytilus edulis. Ibis 109:412-24.

Norton-Griffiths M. 1968. The feeding behaviour of the Oystercatcher
Haematopus ostralegus. D.Phil. Thesis. (Oxford Univ.).

Olney P.J.S. 1963. The autumn and winter feeding biology of certain
sympatric ducks. Int.Union Game Biologists. Trans.6th Conf.
309-320. (London).

Orians G. H. 1961. The ecology of Blackbird (Agelaius) social systems.
Ecol.Monog. 31:285-312.

Orians G. H. and Willson M. F. 1964. Interspecific territories of birds.
Ecology 45:736-745.

Paludan K. 1938. Zur Ornis des Zagrossgebietes, W.-Iran. J.Orn.
86:562-638.

Parker S. A. and Harrison C. J. O. 1963. The validity of the genus
Lusciniola Gray. Bull.Brit.Orn.Club. 83:65-69.

Parslow J. L. F. 1967-68. Changes in status among breeding birds in
Britain and Ireland. Brit.Birds 60:2-47, 97-123, 177-202, 261-285,
396-404, 493-508; 61:49-64, 241-255.

Pearson T. H. 1968. The feeding biology of sea-bird species breeding
on the Farne Islands, Northumberland. J.Anim.Ecol. 37:521-552.

REFERENCES

Perkins R.C.L. 1903. Fauna Hawaiiensis. Vol. 1. pt. IV. Vertebrata. pp. 365–466. (Cambridge).

Perrins C.M. 1965. Population fluctuations and clutch-size in the Great Tit, *Parus major* L. J.Anim.Ecol. 34:601–647.

Peus F. 1954. Zur Kenntnis der Brutvögel Griechenlands. Bonn.zool.-Beitr. Sond. 1:1–50.

Pitelka F.A. 1950. Geographic variation and the species problem in the shore-bird genus *Limnodromus*. Univ.Calif.Publ.Zool. 50:1–108.

Pitelka F.A., Tomich P.Q. and Treichel G.W. 1955. Ecological relations of jaegers and owls as lemming predators near Barrow, Alaska. Ecol.Monog. 25:85–117.

Pleske T. 1912. Zur Lösung der Frage, ob *Cyanistes pleskei* Cab.eine selbständige Art darstelt, oder für einen Bastard von *Cyanistes caeruleus* (Linn.) und *Cyanistes cyanus* (Pall.) angesprochen werden muss. J.Orn. 60:96–109.

Praed C.M. and Grant C.H.B. 1955. Birds of Eastern and North Eastern Africa. (London). Vol. 2.

Rand A.L. and Gilliard E.T. 1967. Handbook of New Guinea Birds. (London). pp. 586–592.

Rand A.L. and Rabor D.S. 1960. Birds of the Philippine Islands: Siquijor, Mount Malindang, Bohol, and Samar. Fieldiana Zool. 35:223–441.

Reiser O. 1894, 1905, 1939. Materialen zu einer Ornis Balcanica (Wien). III Griechenland und die griechischen Inseln, II Bulgarien, I Bosnien und Herzegowina nebst Teilen von Serbien und Dalmatien.

Rensch B. 1928. Grenzfälle von Rasse und Art. J.Orn. 76:222–231.

Rensch B. 1929. Das Prinzip geographischer Rassenkreise und das Problem der Artbildung. (Berlin).

Rensch B. 1931. Die Vogelwelt von Lombok, Sumbawa und Flores. Mitt.Zool.Mus.Berlin. 17:451–637.

Rensch B. 1933. Zoologische Systematik und Artsbildungs Problem. Verh.Deutsch.Zool.Ges. 1933:19–83.

Reynard G.B. and Wetmore A. 1962. The rediscovery of the Puerto Rican Whip-poor-will. Living Bird 1:51–60.

Rice J.N. 1969. The decline of the Peregrine population in Pennsylvania. Peregrine Falcon Populations *ed.* J.J. Hickey. (Madison, Wisconsin). pp. 155–163.

Ridgway R. 1902, 1904, 1911. The Birds of North and Middle America. Pts.2,3,5. U.S.Nat.Mus.Bull. 50.

Riley J.H. 1938. Birds from Siam and the Malay Peninsula in the United States National Museum. U.S.Nat.Mus.Bull. 172.

Ripley S.D. 1959. Character displacement in Indian nuthatches *(Sitta)*. Postilla 42.

REFERENCES

Ripley S.D. 1961. A Synopsis of the Birds of India and Pakistan. (Bombay).

Rising J.D. 1968. A multivariate assessment of interbreeding between the chickadees, *Parus atricapillus* and *P.carolinensis*. Syst.Zool. 17:160–169.

Robertson W.B. 1962. Observations on the birds of St. John, Virgin Islands. Auk 79:44–76.

Robson G.C. and Richards O.W. 1936. The Variation of Animals in Nature. (London).

Root R.B. 1964. Ecological interactions of the Chestnut-backed Chickadee following a range expansion. Condor 66:229–238.

Rothschild W. 1893–1900. The Avifauna of Laysan and the Neighbouring Islands. (London).

Rountree F.R.G., Guérin R., Pelte S. and Vinson J. 1952. Catalogue of the birds of Mauritius. Mauritius Inst.Bull. 3:155–217 (esp. pp. 193–194).

Royama T. 1963. Cuckoo hosts in Japan. Bird Study 10:201–202.

Rucner D. 1956 (1954). Die Vögel im Nationalpark Plitvicaseen. Ein Beitrag zur Kenntnis Ornithofauna von Lika. Larus 8:56–64.

Rucner D. 1967. Über die Verbreitung der *Hippolais*-Arten im Küstenlande, Jugoslawiens. J.Orn. 108:71–75.

Safriel U. 1967. Population and food study of the Oystercatcher. D.Phil. thesis. Oxford Univ.

Salomonsen F. 1932. Description of three new Guillemots *(Uria aalge)*. Ibis 1932: 128–132.

Salomonsen F. 1955. The evolutionary significance of bird-migration. Dan.Biol.Medd. 22:1–62.

Salomonsen F. 1962. The mountain bird fauna of Palawan, Philippine Islands. Dansk. Orn.For.Tidsk. 56:129–134.

Schäfer E. 1938. Ornithologische Ergebnisse zweier Forschungsreisen nach Tibet. J.Orn. 86 Sond. pp. 349.

Schäfer E. and de Schauensee R.M. 1938. Zoological results of the second Dolan expedition to western China and eastern Tibet, 1934–6. Part II Birds. Proc.Acad.Nat.Sci. Philadelphia 90:185–260.

Scharnke H. and Wolf A. 1938. Beiträge zur Kenntnis der Vogelwelt Bulgarisch-Mazedoniens. J.Orn. 86:309–327.

Schoener T.W. 1965. The evolution of bill size differences among sympatric congeneric species of birds. Evolution 19:189–213.

Schoener T.W. 1968. Sizes of feeding territories among birds. Ecology 49:123–141.

Schoener T.W. 1969. Models of optimal size for solitary predators. Amer.Nat. 103:277–313.

Sclater P.L. 1858. On the general geographical distribution of the members of the Class Aves. J.Proc.Linn.Soc. (Zool.). 2:130–145.

REFERENCES

Serventy D.L. 1967. Aspects of the population ecology of the Short-tailed Shearwater *Puffinus tenuirostris*. Proc.Int.Orn.Cong. 14:165–190.

Serventy D.L. and Whittell H.M. 1962. Birds of Western Australia. (Perth).

Selander R.K. 1966. Sexual dimorphism and differential niche utilization in birds. Condor 68:113–151.

Selander R.K. and Giller D.R. 1959. Interspecific relations of woodpeckers in Texas. Wilson Bull. 71:107–124.

Selander R.K. and Giller D.R. 1961. Analysis of sympatry of Great-tailed and Boat-tailed Grackles. Condor 63:29–86.

Selander R.K. and Giller D.R. 1963. Species limits in the woodpecker genus *Centurus* (Aves). Bull.Amer.Mus.Nat.Hist. 124:213–274.

Sibley C.G. 1950. Species formation in the Red-eyed Towhees of Mexico. Univ.Calif.Publ.Zool. 50:109–194.

Sibley C.G. 1954. Hybridization in the Red-eyed Towhee of Mexico. Evolution 8:252–290.

Sibley C.G. 1958. Hybridization in some Columbian tanagers, avian genus *Ramphocelus*. Proc.Amer.Phil.Soc. 102:448–453.

Sibley C.G. 1961. Hybridization and isolating mechanisms. Vertebrate Speciation *ed.* W.F. Blair. pp. 69–81. (Austin, Texas).

Sibley C.G. and Short L.L. 1959. Hybridization in the buntings *(Passerina)* of the Great Plains. Auk 76:443–463.

Sibley C.G. and Short L.L. 1964. Hybridization in the orioles of the Great Plains. Condor 66:130–150.

Sibley C.G. and Sibley F.C. 1964. Hybridization in the Red-eyed Towhees of Mexico: the populations of the southeastern plateau region. Auk 81:479–504.

Simmons K.E.L. 1951. Interspecific territorialism. Ibis 93:407–413.

Sims R.W. 1955. The morphology of the head of the Hawfinch *(Coccothraustes coccothraustes)*. Bull.Brit.Mus.Nat.Hist. 2:369–393.

Smith S.M. 1967. An ecological study of winter flocks of Black-capped and Chestnut-backed Chickadees. Wilson Bull. 79:200–207.

Smythies B.E. 1953. The Birds of Burma. (Edinburgh).

Smythies B.E. 1960. The Birds of Borneo. (Edinburgh).

Snow D.W. 1949. (Selection of feeding niches among birds in a Swedish wood.) Vår Fågelvärld 8:156–169.

Snow D.W. 1950. The birds of Sao Tomé and Principe in the Gulf of Guinea. Ibis 92:579–595.

Snow D.W. 1952a. The winter avifauna of arctic Lapland. Ibis 94:133–143.

Snow D.W. 1952b. A contribution to the ornithology of North-West Africa. Ibis 94:473–498.

REFERENCES

Snow D. W. 1953. Systematics and Comparative Ecology of the Genus *Parus* in the Palaearctic Region. (D.Phil. thesis, Oxford Univ.).

Snow D. W. 1954. The habitats of Eurasian tits (*Parus* spp.). Ibis 96:565–585.

Snow D. W. 1955. Geographical variation of the Coal Tit *Parus ater* L. Ardea 43:195–226.

Snow D. W. 1962. A field study of the Black and White Manakin *Manacus manacus*, in Trinidad. Zoologica 47:65–104.

Snow D. W. 1967. Family Paridae. Check-List of Birds of the World, vol. 12 (*ed.* R. A. Paynter) pp. 70–124. (Cambridge, Mass.).

Snow D. W. *in prep.* Evolutionary aspects of fruit-eating in birds. Evolution

Snow B. K. and Snow D. W. *in prep.* The feeding ecology of tanagers and honeycreepers in Trinidad. Auk.

Stanford J. K. and Mayr E. 1941. The Vernay-Cutting expedition to Northern Burma. Pt.II. Ibis 1941 pp. 56–57.

Stanford J. K. and Ticehurst C. B. 1938. On the birds of Northern Burma Pt. I. Ibis 1938 pp. 87–88.

Steere J. B. 1894. On the distribution of genera and species of non-migratory land-birds in the Philippines. Ibis 1894:411–420.

Steinbacher J. 1960. Zum Brutvogelleben in Sardinien. Vogelwelt 81:73–90.

Stenhouse D. 1962. A new habit of the Redpoll *Carduelis flammea* in New Zealand. Ibis 104:250–252.

Stonehouse B. 1962. Ascension Island and the British Ornithologists' Union Centenary Expedition 1957–59. Ibis 103B:107–123.

Storer R. W. 1955. Weight, wing area and skeletal proportions in three accipiters. Proc.Int.Orn.Cong. 11:287–290.

Storer R. W. 1966. Sexual dimorphism and food habits in three North American accipiters. Auk 83:423–436.

Stresemann E. 1920. Avifauna Macedonica. (München).

Stresemann E. 1939. *Zosterops siamensis* Blyth—einer gelbbaüchige Rasse von *Zosterops palpebrosa*. J.Orn. 87:156–164.

Stresemann E. 1940. DieVögel von Celebes. Pt. 2. J.Orn. 87:60–68.

Stresemann E. 1943. Ueberblick über die Vögel Kretas und den Vogelzug in der Aegaeis. J.Orn. 91:448–514.

Stresemann E. 1950. Interspecific competition in chats. Ibis 92:148.

Stresemann E. and Heinrich G. 1940. Die Vögel des Mount Victoria. Mitt.Zool.Mus. Berlin 24:151–264.

Stresemann E. and Portenko L. A. 1960. Atlas der Verbreitung Palaearktischer Vögel. (Berlin).

Sturman W. A. 1968. Description and analysis of breeding habits of the chickadees *Parus atricapillus* and *P.rufescens*. Ecology 49:418–431.

REFERENCES

Sunkel W. 1928. Bedeutung optischer Eindrücke der Vögel für die Wahl ihres Aufenthaltortes. Zeits.Wiss.Zool. 132:171–175.

Suomalainen H. 1936. Der Grüne Laubsänger, *Phylloscopus nitidus viridanus* Blyth, in Finnland, nebst einigen Hauptzügen seiner Ausbreitungsgeschichte. Orn.Fenn. 13:89–124.

Svärdson G. 1949. Competition and habitat selection in birds. Oikos 1:157–174.

Swarth H.S. 1931. The avifauna of the Galapagos Islands. Occ.Pap. Calif.Acad.Sci. 18.

Tanner H.T. 1952. Black-capped and Carolina Chickadees in the southern Appalachian mountains. Auk 69:407–424.

Tischler F. 1942. *Muscicapa albicollis* Temm. im Urwalde von Bialowies. Orn.Monatsber. 50:125–127.

Turner R.A. 1964. Social feeding in birds. Behaviour 24:1–46.

Tyrväinen H. 1969. The breeding biology of the Redwing (*Turdus iliacus* L.). Ann. Zool.Fenn. 6:1–46.

Udvardy M.F.D. 1959. Notes on the ecological concepts of habitat, and niche. Ecology 40:725–728.

Van Tyne J. 1951. The distribution of the Kirtland Warbler *(Dendroica kirtlandii)* Proc.Int.Orn.Cong. 10:537–544.

Vaurie C. 1950. Notes on some Asiatic nuthatches and creepers. Amer. Mus.Novit. No. 1472.

Vaurie C. 1951. Adaptive differences between two sympatric species of nuthatches *(Sitta).* Proc.Int.Orn.Cong. 10:163–166.

Vaurie C. 1957. Systematic notes on palearctic birds No. 29. The Subfamilies Tichodrominae and Sittinae. Amer.Mus.Novit. no. 1854.

Vaurie C. 1959, 1965. The Birds of the Palearctic Fauna. Vols. 1, 2. (London).

Voipio P. 1956. The biological zonation of Finland as reflected in zootaxonomy. Ann.Zool.Soc.Zool.Bot.Fenn.Vanamo 18:1–36.

Volsøe H. 1955. The breeding birds of the Canary Islands. II. Origin and History of the Canarian avifauna. Vidensk.Medd.Dansk Naturh.Foren. 117:117–178.

Voous K.H. 1960. Atlas of European Birds. (Amsterdam, London).

Voous K.H. and Van Marle J.G. 1953. The distributional history of the Nuthatch *Sitta europaea* L. Ardea 41 extra no. p. 68.

Ward P. 1965. Feeding ecology of the Black-faced Dioch *Quelea quelea* in Nigeria. Ibis 107:173–214.

Ward P. 1968. Origin of the avifauna of urban and suburban Singapore. Ibis 110:239–255.

Watson G.E. 1961. Aegean bird notes including two breeding records new to Europe. J.Orn. 102:301–307.

Watson G.E. 1962a. Three sibling species of *Alectoris* partridge. Ibis 104:353–367.

REFERENCES

Watson G. E. 1962b. Sympatry in palaearctic *Alectoris* partridges. Evolution 16:11-19.

Watson G. 1964. Ecology and Evolution of passerine Birds on the Islands of the Aegean Sea. Ph.D. thesis (Yale University).

Wetmore A. 1927. The Birds of Porto Rico and the Virgin Islands. Sci. Survey of Porto Rico and the Virgin Islands. 9 pts. 3, 4. pp. 245-598.

Wetmore A. and Lincoln F. C. 1933. Additional Notes on the Birds of Haiti and the Dominican Republic. Proc.U.S. Nat.Mus. 82 art. 25 pp. 68.

Wetmore A. and Swales B. H. 1931. The Birds of Haiti and the Dominican Republic. U.S.Nat.Mus.Bull. 155 (Washington, D.C.).

Wettstein O. V. 1938. Die Vogelwelt der Ägäis. J.Orn. 86:9-52.

Whistler H. 1941. Popular Handbook of Indian Birds. (London and Edinburgh).

Willgohs J. F. 1951. Bidrag til trostenes forplantningsbiologi. Univ. Bergen Årbok 1951 no. 2.

Williams C. B. 1947. The generic relations of species in small animal communities. J.Anim.Ecol. 16:11-18. (also J.Anim.Ecol 20:246-253).

Williamson K. 1969. Habitat preferences of the Wren on English farmland. Bird Study 16:53-59.

Willis E. O. 1966a. Competitive exclusion and birds at fruiting trees in western Colombia. Auk 83:479-480.

Willis E. O. 1966b. Interspecific competition and the foraging behavior of Plain-brown Woodcreepers. Ecology 47:667-672.

Willson M. F. 1969. Avian niche size and morphological variation. Amer.Nat. 103:531-542.

Witherby H. F., Jourdain F. C. R., Ticehurst N. F., and Tucker B. W. 1938-41. The Handbook of British Birds vols. 1-5. (London).

Zahavi A. 1957. The breeding birds of the Huleh swamp and lake (Northern Israel). Ibis 99:600-607.

Zimmerman E. C. 1948. Insects of Hawaii vol. 1 Introduction. (Honolulu).

Zink G. 1963. Populationsuntersuchungen am Weissen Storch *(Ciconia ciconia)* in S.W.-Deutschland. Proc.Int.Orn.Cong. 13:812-818 (extended in Lack 1968 p. 218).

Zusi R. L. 1969. Ecology and adaptations of the Trembler on the island of Dominica. Living Bird. 8:137-164.

Index of Birds and Other Animals

All species are indexed in alphabetical sequence of their generic names and, within each genus, of their specific names, but the latter are not listed independently. Family and sub-family names have also been indexed. The vernacular names are also listed alphabetically, but where there are two words in the name, the group name is placed first.

Accentor   305
Accipiter   133–4, 237, 310, 349
   badius   163
   brevipes   310
   cooperii   133, 237
   gentilis   118, 133, 237, 310
   gundlachi   237, 349
   nisus   134, 310
   striatus   133, 237, 349
Accipitridae   118
Acridotheres fuscus   163, 164
   tristis   163, 164
Acrocephalus   100, 149, 194, 290–1,
      302, 323, 324
   arundinaceus   290, 302
   dumetorum   290, 302
   melanopogon   290, 302
   paludicola   290, 302
   palustris   100, 248, 290, 302
   schoenobaenus   100, 248, 290, 302
   scirpaceus   100, 248, 290, 302
Aegithalos caudatus   326
Aegithina   162
Aegypius monachus   152
Aethia cristatella   136
   pusilla   136
Agelaius humeralis   349
   phoeniceus   127, 128, 257, 349
   xanthomus   349
Ailuroedius   158, 330–1
   buccoides   330
   crassirostris   331
   melanotis   330, 331
Alauda arborea   254, 304
   arvensis   4, 254, 304
   eristata   305
   thekia   305
Alca torda   122, 321
Alcidae   321–2
Alcids   137

Alectoris barbara   120, 311
   chukar   120, 121, 311
   graeca   120, 311
   rufa   120, 311
Alethe fülleborni   325
   montana   325
Allenia   346
Alseonax cinereus   324
   minimus   324
Amauresthes fringilloides   324
   (see Lonchura)
Amaurornis   162
Amazona   234, 350, 353
   agilis   350
   arausiaca   350
   collaria   350
   guildingii   350
   imperialis   350
   ventralis   350
   vittata   350
Amblyornis   158, 331
   flavifrons   331
   inornatus   331
   macgregoriae   331
   subalaris   331
Ammodytes tobianus   322
Anaplectes melanotis   324, 326
   (see Ploceus)
Anas.   16, 122, 318
   angustirostris   318
   acuta   118
   clypeata   258, 318
   crecca   318
   penelope   318
   platyrhynchos   318
   strepera   318
Anatidae   318–19
Andropadus   322, 324
   masukuensis   325
   milangensis   325

Andropadus (*continued*)
  tephrolaema 325
  vivens 325
Anomalospiza imberbis 150
Anous 170, 263
  stolidus 168–9, 333
  tenuirostris 168, 170, 333
Anser albifrons 319
  anser 319
  brachyrhynchus 319
  erythropus 319
  fabalis 319
Ant-thrush 267
Ants, army 267
Anthracothorax 228–232, 351
  dominicus 231, 351, 253
  mango 228, *229*, 231, 351, 353
  prevostii 231, 351
  viridis 231, 351, 353
Anthreptes 162, 323
  collaris 325
  neglectus 325
  orientalis 326
  pallidigaster 325
  reichenowi 325
Anthus 30, 96, 162, 287
  campestris 287, 301
  cervinus 96, 287, 301
  pratensis 10, 96, 254, *254*, 287, 301
  spinoletta petrosus 96, 114, 255, 287, 301
  spinoletta spinoletta 96, 114, 287, 301
  trivialis 10, 96, 254, *254*, 287, 301
Antichromus minutus 324
  (see *Tschagra*)
Apalis 323
  caniceps 325
  flavida 326
  melanocephala 326
  moreaui 326
  murina 325
Aphids 27
Aplonis 162, 354
Apodidae 118, 313
Apus 12, 121, 162, 313
  affinis 164
  apus 169, 313
  caffer 313
  melba 313
  pallidus 313
Aquila chysaetos 310
  clanga 310

Aquila (*continued*)
  heliaca 310
  pomarina 310
Aratinga chloroptera 350
  euops 350
  nana 350
  pertinax 350
Ardea cinerea 314
  purpure 314
Arizelocichla masukuensis 324
  milanjensis 324
  tephrolaema 324
  (see *Andropadus*)
Artisornis metopias 325
Asio otus 313
  flammeus 313, 350
  stygius 350
Astrapia 158, 330
  mayeri 330
  nigra 330
  rothschildi 330
  splendidissima 330
  stephaniae 330
Astrapias 330
Auk 122, 321–2
  little 124, 137, 321
Auklet 126–7, 136–7
  crested 136–7
  least 136
  parakeet 136, 137
Aythya ferina 318
  fuligula 258, 318
  marila 318
  nyroca 318

Babbler 148, 151
Baeolophus 14, 49–52, 278
Batis 322
  minor 326
  mixta 325
  molitor 326
Bee-eater 149
Beetles, cerambycid 335
Belone 333
Benthodesmus 333
Blenniidae 168, 170
Birds of Paradise 157, 158, 329–31
  Sickle-billed 330
  Six-wired 330
Blackbird 256, 261, 305, 306, 327
  Redwinged 127, 128, 129, 257, 349

Blackbird (*continued*)
  Tawny-shouldered 349
  Yellowheaded 128, 229
  Yellow-shouldered 349
Blackcap 104, 105, 294, 295, 296,
    297, 327
Bluethroat 106, 298
Bombycilla garrulus 86, 249, 251
Booby 166, 183
  Bluefaced 172
  Brown 166, 333
  White 166, 172, 333
Bower bird 158, 330-1
  Fawn-breasted 331
  Gardener 331
  Great Grey 331
  Regent 331
  Spotted 331
Bradornis 322
  griseus 326
  microrhynchus 326
Bradypterus 323, 324
  cinnamomeus 144, 325
  mariae 144, 325
Brambling 77, 78, 79, 81, 83, 89,
    90, 91, 115, 244, 245,
    284, 327
Branta bernicla 319
  leucopsis 319
Bubo bubo 313
Buccanodon 323
  leucotis 326
  olivaceum 325
Bucephala clangula 318
  islandica 318
Bulbul 138, 139, 148, 151, 163, 325
Bullfinch 75, 77, 78, 79, 80, 81, 82,
    83, 84, 85, 86, 176, 256, 284,
    327
  Cuban 226
  Greater Antillean 349
  Lesser Antillean 349
  West Indian 226, 228, 349
Bunting 113, 149, 307-8, 310
  Blackheaded 307
  Cinereous 308
  Cirl 114, 307
  Corn 307
  Cretzschmar's 307
  Indigo 127
  Little 308
  Reed 259, 308
  Rock 307

Bunting (*continued*)
  Rustic 308
  Yellowbreasted 307, 308
Bustard 118, 312
  Great 263, 312
  Little 312
Buteo buteo 311
  jamaicensis 237, 349
  lagopus 311
  lineatus 237
  platypterus 237, 349
  ridgwayi 349
  rufinus 311
Buzzard 237, 349
  Common 311
  Long-legged 311
  Rough-legged 311

Cactospiza 187
Calandrella brachydactyla 305
  rufescens 305
Calamoecetor leptorhyncha 324-5
Calamonastes simplex 324
Calanus finmarchicus 136
Calidris 122, 134-6, 316, 317
  alba 136, 316
  alpina 134, 135, 316
  bairdii 135
  canutus 136, 316
  ferruginea 136
  fuscicollis 136
  maritima 316
  mauri 136
  melanotos 134, 135
  minutus 316
  pusilla 135
  ruficollis 136
  temminckii 316
Callichelidon cyaneoviridis 226
Calliphlox evelynae 228-231
Camarhynchus 176, 187-91, 192,
    193, 199
  heliobates 187, 188, 192
  pallidus 187, 188, 189, 190, *191*,
    192
  parvulus 187, 188, 189, 192
  pauper 187, 189, 190, 192
  psittacula 187, 188, 189, 190,
    191, 192
Camaroptera 323, 324
  brevicaudata 325

Campethera taeniolaema 144
Canary 284
Capella 315
Capercaillie 263
Capitonidae 323, 324, 325
Caprimulgidae 118, 313
Caprimulgus 162, 313, 351
  cayennensis 237, 351
  cubanensis 351
  europaeus 313
  noctitherus 351, 353
  ruficollis 313
  rufus 351
Carduelinae 83, 283
Carduelis 78–89, 115, 283, 284, 285
  cannabina 78, 80, 82, 89, 266,
    283, 285
  carduelis 78, 80, 81, 283, 285
  chloris 78, 80, 81, 82, 266, 283,
    285
  flammea 74, 78, 81, 82, 89, 256,
    283, 284, 285, 327
  flavirostris 87, 283, 285
  hornemanni 87, 283, 284, 285
  spinus 78, 80, 81, 283
Carpodacus 85, 283
  erythrina 86, 283
Cassidix major 128
  mexicanus 128
Catbirds 330–1
Caterpillars 21, 24, 83, 97, 195,
    289, 333
Catharacta skua
Catharospeza bishopi 226
Cattle 288
Centrolophus 333
Centropus 162
Centurus 234, 237, 238, 351
  radiolatus 351, 353
  striatus 351
  superciliaris 351
Cepphus grylle 124, 321
Cercotrichas 323, 324
Cercococcyx 148
Certhia 114, 253, 306–7
  brachydactyla 306, 307, 327
  familiaris 306, 307
Certhidea olivacea 175–7
Cettia 149
Chaetoptila 194
Chaetura brachyura 237, 251
  cinereiventris 237, 351
  martinica 351

Chaffinch 4, 9, 77, 78, 79, 80, 81,
    83, 84, 90, 91, 115, 244, 245,
    257, 266, 284, 327
  Blue 9
Chalcomitra 324
Charadriidae 315
Charadrius alexandrinus 315
  dubius 315
  hiaticula 245, 315
Chasiempsis 194
Chats 4, 108, 110, 149, 300, 304,
    325
  desert 299–300
  palm 227
Chickadee, Black-capped 50, 51,
    52–5, 56, 57, 58, 277
  Boreal 50, 55, 56, 58, 277
  Brown-capped 50, 55, 56, 277–8
  Carolina 50, 51, 52, 53, 54, 56,
    57, 58, 59, 277
  Chestnut-backed 50, 51, 55, 56,
    57, 58, 278
  Hudsonian 56, 58, 277
  Mexican 50, 53, 55, 277
  Mountain 50, 51, 52, 53, 54, 55,
    56, 57, 58, 277
Chlamydera 158, 331
  cerviniventris 331
  lauterbachi 331
  maculata 331
  nuchalis 331
Chlidonias hybrida 319
  leucoptera 319
  niger 320
Chlorocharis 215
  emiliae 340
Chloropeta natalensis 326
Chloropetella holochlorus 325, 326
Chlorophanes 156
Chlorophoneus nigrifrons 324
  sulphureopectus 324
  (see *Malaconotus*)
Chlorostilbon maugaeus 231, 351,
    353
  ricordii 228, 229, 231, 351
  swainsonii 231, 351
Chondrohierax uncinatus 349
  wilsonii 349
Choristoneura fumiferana 131
Chough, Alpine 310
  (Redbilled) 310
Chrysococcyx caprius 148
  cupreus 148

Chrysococcyx (continued)
  flavogularis 148
  klaas 148
Cichladusa 323
Ciconia ciconia 257, 314
  nigra 314
Ciconiiformes 314
Cinclocerthia ruficauda 238, 239
Cinnyris 324
Circus aeruginosus 311
  cyaneus 311
  macrourus 311
  pygargus 311
Ciridops anna 335
Cisticola 150, 323
  erythrops 145
Citril 89, 284
Clamator glandarius 148, 149, 150
  jacobinus 148, 151
  levaillantii 148, 151
Climacteris 161
Clupeus sprattus 322
Clytorhynchus 226, 355
Coccopygia melanotis 324
Coccothraustes coccothraustes 78, 81,
      82, 85, 284, 327
Coccyzus americanus 350, 353
  minor 350, 353
Cockles 264
Coereba 157
  flaveola 226, 237, 352
Colaptes 127
Coliuspasser albonotatus 324
  ardens 324
  (see *Euplectes*)
Collocalia 162
  francica 164
Columba 121, 162, 312, 349
  caribaea 349
  inornata 349
  leucocephala 349
  livia 312
  oenas 312, 327
  palumbus 256, 312, 327
  squamosa 349
Columbidae 118, 312, 353
Contopus caribaeus 352
Coot 319
  Crested 319
Copepod 136
Copsychus saularis 164
Coracina 354
Cormorant 8, 124, 318

Cormorant (*continued*)
  Pygmy 318
Corvids 149, 150, 264
Corvus 162, 194, 237, 309, 346, 354
  corax 309
  cornix 115, 116, 309
  corone 115, 116, 309
  frugilegus 309
  jamaicensis 346, 352
  leucognaphalus 346, 352
  macrorhynchos 163
  monedula 309
  nasicus 345
  palmarum 346
  splendens 163, 165
Corythaixoides 145, 147, 328
  concolor 328
  leucogaster 328
  personata 328
Cossypha 323
  heuglini 144
  natalensis 325
  semirufa 144
Cotinga 157
Cractes infaustus 327
Crake, Baillon's 314
  Little 314
  Spotted 314
Crane 252
Crinifer 145, 328
  piscator 328
  zonurus 328
Crossbill 86-7, 246
  Common 78, 79, 81, 82, 83, 86, 87
  Parrot 86, 284, 327
  Two-barred or White-winged 86,
      87, 284, 327
Crow 113, 124, 148, 163, 194, 237,
      309, 310, 346, 354
  Carrion 115, 116, 309
  Cuban 346
  Hooded 115, 116, 309
  House 165
  Palm 346
  Whitenecked 346
Crustacea 170, 317, 321
  gammarid 136
  hyperiid 136, 137
  mysid 136
Cryptospiza reichenowi 325
Crypsirhina temia 163
Cuckoo 148, 149, 150, 151, 327,
      350, 353

Cuckoo (*continued*)
  Bay-breasted 350
  Chestnut-bellied 350
  Emerald 148
  Lizard 234, 350
  Mangrove 234, 350
  Yellow-billed 350
Cuckoo-shrike 354
Cuculus canorus 148, 149, 327
  clamosus 148
  fugax 149
  poliocephalus 149
  saturatus 149
  solitarius 148
Curculionidae 197
Curlew 122, 245, 260, 315
  Bristle-thighed 249
Cyanerpes 156
  cyaneus 226, 237
Cyanistes 14, 19, 272
Cyanomitra 324
Cyanophaia bicolor 228–232
Cyanoptila 149
Cyclorhynchus psittacula 136
Cygnus columbianus 319
  cygnus 319
  olor 319
Cypseloides niger 315
Cypsilurus 333
Cypsiurus parvus 162, 164

Dabchick 317
Dacnis 156
Dafila 16
Danichthys 333
'Darzee' 164
Decapterus 333
Delichon urbica 286
Dendrocopos 162, 314
  leucotos 313, 327
  major 256, 257, 314, 327
  medius 314, 327
  minor 314, 327
  syriacus 313, 327
Dendroica 129–132, 174, 226, 227,
  237, 348
  adelaidae 225, 348, 352
  castanea 129, 130, 131
  coronata 129, 130, 131,
  dominica 348
  fusca 129, 130, 131
  kirtlandii 249, 250

Dendroica (*continued*)
  magnolia 131
  petechia 348, 352
  pharetra 225, 348, 352
  pinus 348
  pityophila 348
  plumbea 348
  tigrina 129, 130, 131
  virens 129, 130, 131
  vitellina 348
Dendrocincla fuliginosa 267
Dendropicus obsoletus 144
Diaphus 333
Dicaeum 161, 162, 165, 354
Diphyllodes magnificus 330
  respublica 330
Diver (Loon) Black-throated 317
  Great Northern 317
  Red-throated 317
  White-billed 317
Dove, Bridled Quail 350
  Collared 256, 312
  Crested 350
  Grey-headed 350
  Ground 237
  Key-West 350
  Mourning 350
  Rock 312
  Ruddy Quail 350
  Stock 312, 327
  Turtle 312
  Violet-eared 350
  White-winged 350
  Zenaida .350
Dowitcher, Longbilled 263
Dragonflies 311, 319
Drepanididae 334–5, 336
Drepanis 197, 335, 336
  funerea 335, 336
  pacifica 335, 336
Drepanorhynchus 324
Drepanornis albertisii 330
  brouijnii 330
Dryoscopus 322
  cubla 325
  martius 327
  pringlii 326
Duck, Dabbling 14, 122, *123*
  Ferruginous 318
  Sawbill 122
  Tufted 258, 318
Dulidae 227
Dunlin 316, 317

Dunnock   104–5, 305, 327

Eagle   118
  Bonelli's   311
  Booted   311
  Golden   310
  Imperial   310
  Lesser-Spotted   310
  Spotted   310
Earthworm   264
Edilosoma   354
Egret, Little   314
Egretta alba   314
  garzetta   314
Eider, Common   258, 319
  King   319
Elaenia, Greater Antillean   346
  Yellow-bellied   346
  Yellow-crowned   346
Elaenia fallax   346, 352
  flavogaster   227, 346
  martinica   227, 346, 352
Emberiza   149, 307–8
  aureola   307, 308
  caesia   307, 308
  calandra   307, 308
  cia   307, 308
  cineracea   308
  cirlus   114, 307, 308
  citrinella   307, 308
  hortulana   307, 308
  melanocephala   307, 308
  pusilla   308
  rustica   308
  schoeniclus   259, 308
Empidonax oberholseri   127
  hammondi   127
  wrightii   127
Engraulis   333
Epimachus   158, 330
  fastosus   330
  meyeri   330
Eremomela   323
  griseoflava   326
  scotops   326
Erithacus rubecula   327
Erythropygia, leucophrys   324
  leucoptera   324
Erythrura   355
Estrilda   323, 324
  astrild   150
Estrildidae   139, 323, 324, 325

Estrildines   138, 139, 141, 150, 325
Eulampis jugularis   228–232
Euneornis campestris   352
Euphausiids   136
Euphonia, Blue-hooded   226
  Jamaican   226
Euplectes   323, 324
Eurilla virens   324
Eurocephalus rüppellii   326
Exocoetidae   168, 169
Exocoetus   333

Falco   118, 119, 311
  biarmicus   311
  cherrug   311
  columbarius   311
  eleonorae   311
  naumanni   311
  peregrinus   258, 311
  rusticolus   311
  tinnunculus   311
  subbuteo   311
  vespertinus   311
Falcon   118, 241
  Eleonora's   311
  Gyr   311
  Lanner   311
  Peregrine   118, *119*, 258
  Redfooted   311
  Saker   311
Ficedula   289–90
Fieldfare   306, 327
Finches   74–92, 93, 94, 108, 113,
    115, 125, 198, 199, 265, 283–5
  Cocos   176
  Darwin's   3, 6, 9, 68, 74, 174–193,
    194, 197, 199, 200, 202, 215,
    220, 222, 227, 268
Firecrest   305, 327
Fish   169, 171, 173, 311, 318, 319,
    321, 322, 333
  flying   169, 170, 171
Fistularia   333
Flycatcher   19, 108, 109, 110, 115,
    139, 143, 149, 194, 226, 227,
    243, 289, 302, 325, 354, 355
  Crested   224, 225, 345
  Dusky   127
  Dusky-capped   345
  Fantail   165
  Garden   165
  Grey   127

Flycatcher (*continued*)
  Paradise 143, 144
  Pied 97, 289, 290, 327
  Redbreasted 97, 100, 290, 327
  Rufous-tailed 345
  Rusty-tailed 345
  Spotted 97, 100, 289, 290, 327
  Tyrant 174, 175
Flicker 127
Flowerpecker 161, 165, 354
Fox 264
Fratercula arctica 122, 123, 321
Fregata aquila 333
Frigate bird 333
Fringilla 89–91, 284, 315
  coelebs 4, 9, 77, 90, 115, 244,
    266, 284, 285, 327
  montifringilla 77, 78, 81, 90, 115,
    244, 284, 285, 327
  teydea 9
  fringillidae 74–92, 283–4
Fulica atra 319
  cristata 319
Fulicinae 319

Gadwall 318
Galerida 304, 305
Galliformes 118, 311–12
Gallinago gallinago 315
  media 315
Gamebirds 118, 311–12
Garganey 256, 318
Garrulus glandarius 326
Gavia adamsii 317
  arctica 317
  immer 317
  stellata 317
Gaviidae 317
Geese 252
Gempylidae 168, 170
Geokichla gurneyi 324
Geopelia striata 164
Geospiza 176, 177–87, 189, 190,
    191, 192, 199
  conirostris 182, 183, 184, *185*,
    187, 192
  difficilis 177, 178, *179*, 182, 183,
    184, 185, *185*, 192
  fortis 177, 179, 180, *181*, 182,
    183, 192
  fuliginosa 177–185, 192
  magnirostris 177–185, 187, 192

Geospiza (*continued*)
  scandens 177, 178, *179*, 182, 183,
    185, 192
Geospizinae 3, 174
Geotrygon 237, 350
  caniceps 350
  chrysia 350
  montana 350
  mystacea 350
  versicolor 350
Glaucidium passerinum 313
Glaucis hirsuta 228–232
Gnatcatcher, Cuban 347
  Blue-Grey 347
Godwit, Bar-tailed 315
  Black-tailed 315
Goldcrest 114, 117, 305, 327
Goldeneye 318
  Barrow's 318
Goldfinch 78, 80, 81, 82, 83, 84, 88,
    283
Goosander 318
Goose, Barnacle 319
  Bean 319
  Brent 319
  Greylag 319
  Lesser Whitefront 319
  Pinkfoot 319
  Whitefront 319
Goshawk 118, 133, 134, 237, 310
Grackle 263, 348
  Boat-tailed 128
  Great-tailed 128
  Greater Antillean 348
  Lesser Antillean 348
Granatina granatina 150
  ianthinogaster 150
Grasshoppers 311
Grassquit, Black-faced 349
  Cuban 349
  Yellow-faced 349
Grebe 122, 317
  Blacknecked 317
  Great Crested 317
  Horned (Slavonian) 317
  Rednecked 317
Greenfinch 78, 80, 81, 82, 84, 88,
    266, 283
Greenshank 14, *15*, 315, 316
Griffon 151, 152, 153
  White-backed 151
Grosbeak, Pine 85, 284, 327
Grosbeak, Scarlet 86, 283

Grouse, Red 312
Guillemot, Black 124, 321
    Brünnich's 10, 122, 245, 321
    Common 10, 122, 245, 321, 322
Gull 252, 320-1
    Audouin's 320
    Herring 260, 320
    Lesser Blackback 320
    Little 320
Gygis alba 168, 170, 333
Gypaetus barbatus 154
Gyps africanus 151
    fulvus 152
    rüppellii 151

Haematopus ostralegus 264
Halcyon 162
    chloris 164
    smyrnensis 164
Halobates 170
Harrier 311
    Marsh 311
    Montagu's 311
    Pallid 311
Hawfinch 78, 79, 80, 81, 82, 83, 85,
    284, 327
Hawk 118, 133-4, 163, 241, 310,
    349
    Broadwinged 133, 237, 249
    Cooper's 133
    Redshouldered 237
    Redtailed 237, 349
    Sharpshinned 237, 349
Heleia 215, 218
    crassirostris 340
    muelleri 340
Heliolais erythroptera 324
Hemignathus 196, 197, 199, 334, 336
    lucidus 196, 334, 336
    obscurus 196, 334, 336
    procerus 196, 334, 336
    wilsoni 196, 197, 334, 336
Hemiprocne 162
    longipennis 164
Hemiramphus 333
Heron, Great White 314
    Grey 314
    Purple 314
Heteralocha acutirostris 239, 262,
Hieraaetus fasciatus 311
    pennatus 311
Himatione sanguinea 197, 335
Hippolais 14, 101-3, 291-2, 302

Hippolais (continued)
    icterina 101, 102, 103, 291, 302,
        327
    olivetorum 101, 102, 103, 292,
        302, 327
    pallida 101, 102, 103, 292, 302
    polyglotta 101, 102, 103, 292,
        302, 327
Hirundichthyes 333
Hirundines 94, 95, 121, 124, 258,
    311
Hirundinidae 94, 313
Hirundo 94, 162, 286, 301
    daurica 286, 301, 313
    (Ptyonoprogne) rupestris 286, 301
    rustica 169, 286, 301, 313
Hobby 118, 119, 311
Holocentrus 333
Honeycreeper 156, 157, 226, 237
Honey-eater 155, 174, 354, 355
Honeyguide 149, 150
Hoopoe 149
Huia 239, 262
Hummingbird 228-34, 241, 243
    Bee 351
    Emerald 351
    Mango 351
Hyaloteuthis 333
Hyetornis pluvialis 350, 353
    rufigularis 350
Hyliota australis 325
Hypargos 323, 324
    niveoguttatus 325
    nitidulus 325
Hypochera chalybeata 150
Hypocryptadius 344

Icterus bonana 348
    dominicensis 348, 352
    galbula 127
    icterus 348
    laudabilis 348
    oberi 348
    leucopteryx 348, 353
Indicator indicator 149, 150
    minor 149
    variegatus 149

Jackdaw 309
Jaeger, (see Skua) 320
Jay 326
    Siberian 327
Jynx torquilla 327

Kalochelidon euchrysaea 225
Kestrel 311
  Lesser 311
Kingbird, Giant 345
  Grey 345
  Loggerhead 345
  Tropical 345
Kingfisher 164, 246
Kinglet 305
Kite 163, 165, 349
  Black 311
  Cuban 249
  Hook-billed 249
  Red 311
Kittiwake 124
Knot 136, 316, 317

Lagonosticta 323
  senegala 150
Lagopus lagopus 312
  mutus 312
  scoticus 312
Lalage 162
Lammergeier 154
Lampanyctus 333
Lamprotornis ornatus 335
Laniarius 322
  ferrugineus 325
  fülleborni 325
Laniidae 139, 322, 324, 325
Lanius 97, 108, 109, 149, 288–9,
    301, 322
  cabanisi 326
  collurio 97, 99, 116, 256, 288,
    301
  excubitor 97, 98, 301
  minor 97, 98, 288, 301
  nubicus 97, 289, 301
  senator 97, 99, 116, 288, 289, 301
Lapwing 257
Larus argentatus 260, 320
  audouinii 320
  fuscus 320
  minutus 320
Lark 116, 260, 304–5, 311
  Crested 305
  Lesser Short-toed 305
  Short-toed 305
  Thekla 305
Larvae, acridean 136
  chironomid 135
  dipterous 188
  ostomid 188

Larvae (*continued*)
  phalaenid 188
  spruce budworm 131
  tipulid 135
Lemming 7, 264, 265, 320
Leptocoma 162
Limnodromus scolopaceus 263
Limosa lapponica 315
  limosa 315
Limpet 264
Linnet 78, 80, 81, 82, 84, 86, 87,
    89, 91, 266, 283
Locustella fluviatilis 103, 292–3, 302
  luscinioides 292–3, 302
  naevia 103, 292–3, 302
Lonchura 323, 324
Loon, see Diver
Lophophanes 10, 19, 272, 273
Lophozosterops 214, 215, 218
  dohertyi 340
  goodfellowi 342
  javanica 339, 340
  pinaiae 341
  squamiceps 341
  superciliaris 340
Loxia 86–7, 246, 284, 285
  curvirostra 78, 81, 82, 86, 284,
    285, 327
  leucoptera 86, 284, 285, 327
  pityopsittaca 86
Loxigilla 226, 349
  noctis 228, 349
  portoricensis 228, 349, 353
  violacea 349, 353
Loxipasser anoxanthus 353
Loxops 194, 196, 199, 334, 336
  coccinea 194, 334, 336
  maculata 194, 334, 336
  parva 195, 334, 336
  sagittirostris 195, 334, 336
  virens 194, 195, 334, 336
Lullula 304
Luscinia 106, 149, 253, 298, 303
  luscinia 9, 107, 116, 256, 298,
    303, 327
  megarhynchos 9, 107, 116, 298,
    303, 327
  svecica 106, 298, 303
Lusciniola 290
  (see *Acrocephalus*)
Lybius 323, 324
  melanopterus 326
  torquatus 326

Machlolophus 271, 272
Madanga 213, 215
  ruficollis 341
Magpie Robin 164
Malacocincla 162
Malaconotus 322, 324
  nigrifrons 325
  quadricolor 325
  sulphureopectus 326
Mallard 16, *123*, 318
Mammals 133, 154, 288, 309,
  311
Mandingoa nitidula 324
  (see *Hypargos*)
Manucode 329
Manucodia 158, 329
  ater 329
  chalybatus 329
  comrii 329
  jobiensis 329
Mareca 16
Margarops fuscatus 346, 352
  fuscus 346
Martin 109, 118, *119*, 174, 286
  Crag 94, *95*, 286
  House 94, *95*, 286
  Sand 94, *95*, 286
Mayrornis 354
Meadowlark, Eastern 127
  Western 127
Megalaema haemacephala 163
Megazosterops 208, 209, 214, 215,
  344
  palauensis 208, *210*
  (*Rukia*) *palauensis* 344
Melanerpes 234, 237, 238, 351
  (*Centurus*) *aurifrons* 128, 239, 240
  (*Centurus*) *carolinus* 128
  herminieri 351
  portoricensis 351, 353
  striatus 239, 240
  (*Centurus*) *striatus* 239
Melanitta fusca 318
  nigra 318
Melastomaceae 155
Meliphagid 158, 194
Melithreptus 158-61
  affinis *159*, 160, 161
  albogularis 159, 160
  brevirostris 159, 160
  gularis *159*, 160
  lunatus 159, 160
  validirostris *159*, 160, 161

Mellisuga helenae 229-232, 351
  minima 229-232, 351, 353
Melopyrrha nigra 226
Merganser, American, see Goosander
  Redbreasted 319
Mergus 122, 319
  albellus 319
  merganser 319
  serrator 319
Merlin 311
Merops 162
Mesopicos griseocephalus 144
Mice 264
Microligea montana 348
  palustris 348
Milvus migrans 163, 311
  milvus 311
Mimocichla plumbea 347, 352
  ravida 347
Mimus gilvus 346
  gundlachii 346, 352
  magnirostris 346
  polyglottos 346, 352
Mockingbird 174, 346
  Bahaman 346
  Northern 346
  Saint Andrew's 346
  Tropical 346
Moho 194
Monarcha 354
Monticola saxatilis 106, 298, 304
  solitarius 106, 298, 304
Motacilla 7, 96, 288, 301
  alba 98, 288, 301
  cinerea 96, 288, 301
  flava 288, 301
Munia 162
  maja 164
  punctulata 164
Muscicapa 97, 109, 115, 289-90,
  302, 322, 324
  (*Ficedula*) *albicollis* 252, 289, 302,
  327
  cinerea 325
  (F.) *hypoleuca* 289, 302
  minima 325
  (F.) *parva* 97, 290, 302, 327
  (F.) *semitorquata* 115, 289, 290, 302
  striata 97, 289, 290, 302, 327
Muscicapinae 139, 322, 324, 325
Musophaga 145, 328
  rossae 328
  violacea 328

Musophagidae 138, 139, 145-7,
    323, 324, 325, 328-9
Mussel 264, 265
Muttonbird 252
Myadestes elisabeth 347
    genibarbis 347, 352
Myiagra 355
Myiarchus 174, 224, 225, 227, 345
    antillarum 224, 225, 345
    barbirostris 345
    nugator 224, 225, 345
    oberi 224, 225, 345
    sagrae 224, 225, 345
    stolidus 224, 225, 345, 352
    tuberculifer 345, 352
    tyrranulus 345
    validus 225, 345, 352
Myiopagis cotta 346, 352
Mynah 163, 164
Myzomela 345, 355

Necrosyrtes monachus 152
Nectarinia 220, 323, 324, 355
    bifasciatus 326
    hunteri 326
    mediocris 325
    olivacea 326
    senegalensis 326
    veroxii 326
Nectariniidae 138, 139, 323, 324,
    325
Neocossyphus rufus 325
Neophron monachus 152
Neositta 161
Nesomimus 174
Nesospar nigerrimus 353
Nesospingus speculiferus 352
Nesospiza acunhae 198
    wilkinsi 198
Nettion 16
Nicator chloris 325
Nightingale 9, 106, 107, 110, 116,
    253, 298, 303, 327
    Thrush 107, 256, 298, 327
Nightjar 118, 237, 313, 351
    Red-necked 313
Nilaus minor 326
Noddy, Black 166, *167*, 168, 170,
    171, 172, 333
    Blue-grey 166, *167*, 168, 170, 172
    Brown 166, *167*, 168, 169, 170,
    171, 172, 333

Nomeus 333
Nucifraga caryocatactes 86, 246, 327
Numenius arquata 122, 245, 260,
    315
    phaeopus 122, 245, 315
    tahitiensis 249
Nutcracker 86, 246, 264, 327
Nuthatch 12, 63-73, 93, 94, 126,
    254, 280-2
    Australian 161
    Beautiful 72, 280
    Blue 73
    Brown-headed 66
    Chestnut-bellied 68, 280
    Chinese 72, 281
    Chin Hills 281
    Corsican *64*, 65, 327
    European 63 *64*, 65, 280, 327
    Giant 72, 281
    Lilac 73
    Naga Hills 69, 281
    Pygmy 66
    Redbreasted 65, 66
    Rock 66-8, 72
        Eastern 9, 66, 67, 281
        Western 9, 63, 66, 67
    Velvetfronted 68, 280
    Whitebreasted 65, 66
    White-cheeked 71, 280
    White-tailed 69, 280
    Yunnan 71, 280

Oceanites gracilis 172
Oceanodroma castro 172
    tethys 172
Oculocincta 216, 217
    squamifrons 340
Oenanthe 106, 108, 111, 299-300,
    304
    deserti 299, 300
    hispanica 106, 115, 299, 304
    isabellina 299
    leucomela 299, 304
    leucura 299, 304
    oenanthe 106, 115, 299, 304
Ophioblennius 333
Oriole 127, 165, 348
    Black-cowled 348
    Golden 327
Oriolus 162
    chinensis 163, 165
    oriolus 327

INDEX OF BIRDS AND OTHER ANIMALS

Orthorhyncus cristatus 228–232, 353
Orthotmus 162, 255
  sericeus 164
  sutorius 163, 164
Ortolan 307
Otis tarda 263, 312
  terax 312
Otididae 118, 312
Otus 162
Ousel, Ring 114, 255, 306, 327
Owl 117, 118, 163, 264, 313, 327, 353
  Barn 269
  Eagle 313
  Eared 350
  Great Grey 313
  Pygmy 313
  Short-eared 313, 350
  Stygian 350
  Tawny 246, 313 264
  Ural 313
  Woodland 234, 264
Oxyporamphus 333
Oystercatcher 264, 265

Pachycephala 266, 354
Pachycoccyx audeberti 148
Padda 162
  oryzivora 165
Palmeria dolei 335
Paradigalla 158, 329
  brevicauda 329
  carunculata 329
Paradigalla 329
Paradisaea 157, 158, 330
  apoda 157, 330
  gulielmi 157, 330
  minor 157, 330
  raggiana 157, 330
  rubra 157, 330
  rudolphi 157, 330
Parakeet 350
Pardaliparus 271, 272
Parisoma plumbeum 325
Parotia carolae 330
  lawesi 330
  sefilata 330
  wahnesi 330
Parrot 163, 234, 349, 350, 353
  Blackbilled 350
  Yellowbilled 350
Parrot-finch 355

Partridge, Barbary 120, 311
  Chukor 120, 121, 311
  Redlegged 120, 311
  Rock 120, 311
Parulidae 59
Parus 4, 18–37, 38–62, 115, 139, 244, 254, 271–8
  afer 45, 46, 274, 275, 276
  afer afer 45, 46, 274, 275
  afer cinerascens 45, 46, 274, 275, 276
  afer thruppi 45, 46, 274, 275, 276
  albiventris 45, 46, 47, 48, 274, 275, 276
  amabilis 271
  ater 19, 20, 21, 22, 24, 28, 34, 36, 43, 45, 50, 55, 263, 271, 272, 273, 327
  ater aemodius 42, 43, 45, 272
  ater phaeonotus 34
  ater rufipectus 42, 272
  atricapillus 50, 52–5, 277, 278
  atricristatus 49, 50, 276, 278
  bicolor 49, 50, 277, 278, 279
  caeruleus 10, 19, 20, 21, 22, 24, 31, 34, 35, 36, 50, 187, 263, 266, 327
  caeruleus teneriffae 34
  carolinensis 50, 52, 53, 277, 278, 279
  cinctus 19, 20, 21, 30, 36, 50, 53, 55, 273, 277–8, 327
  cristatus 19, 20, 21, 28, 36, 50, 327
  cyanus 19, 20, 21, 30, 31, 36, 42, 272, 273, 327
  davidi 41, 42, 53, 272, 273
  dichrous 42, 43, 44, 272, 273
  elegans 271
  fasciiventer 45, 46, 274, 275, 276
  fringillinus 47, 48, 275, 276
  funereus 48, 275, 276
  gambeli 50, 52, 53, 277, 278
  griseiventris 45, 46, 47, 274, 275, 276
  holsti 271
  hudsonicus 50, 53, 55, 277, 278
  inornatus 49, 50, 277, 278, 279
  leucomelas 45, 46, 47, 274, 275, 276
  leuconotus 46, 47, 274, 275, 276
  lugubris 19, 20, 21, 31, 32, 36, 53, 327

INDEX OF BIRDS AND OTHER ANIMALS

Parus (continued)
  major 11, 19, 20, 21, 22, 24, 36,
    42, 43, 44, 50, 266, 272, 273,
    275, 326
  major major 40
  major bokharensis 40, 273
  major minor 40, 273
  major cinereus 40, 273
  melanolophus 41, 42, 43, *43*, 44,
    272, 273
  montanus 19, 20, 21, 22, 28, 35,
    36, 42, 50, 53, 272, 273, 326
  montanus affinis 42
  montanus songarus 42, 272
  monticolus 41, 42, 43, 44, 271,
    272, 273
  niger 45, 46, 47, 274, 275, 276
  nuchalis 40, 273
  palustris 19, 20, 21, 22, 24, 36,
    42, 44, 50, 53, 272, 327
  palustris hypermelaena (or
    hypermelas) 272, 273
  rubidiventris 42, 43, 44, 45, 272,
    273
  rubidiventris rubidiventris 42
  rubidiventris rufonuchalis 273
  rufescens 50, 53, 55, 278
  rufiventris 47, 48, 275, 276
  sclateri 50, 53, 277, 278
  semilarvatus 271, 272
  spilonotus 40, 41, 42, 272, 273
  superciliosus 41, 42, 53, 272,
    273
  (Sittiparus) varius 39, 40, 271,
    272, 273
  venustulus 40, 41, 42, 272, 273
  wollweberi 50, 52, 277, 278, 279
  xanthogenys 40, 41, 42, 272, 273
Passer domesticus 18, 114, 163, 308
  flaveolus 163
  hispaniolensis 10, 114, 255, 308
  iagoensis 255
  italiae 308
  montanus 309
Passerina 127
Pastor, Rose-coloured (or Starling)
  309
Peeps, (Stints) 126–37, 316–17
Pelecaniformes 318
Pelecanus crispus 318
  onocrotalus 318
Pelican, Dalmatian 318
  White 318

Periparus 14, 19, 43, 271, 272
Petrel, Phoenix 171
  Storm 172
Petronia 149
  petronia 10, 255
Phaenicophilus 224, 348
  palmarum 348
  poliocephalus 348
Phalarope, Grey (Red) 317
  Rednecked 317
Phaenostictus 267
Phaeornis 194, 198, 199
  obscurus 198
  palmeri 198
Phaethon aethereus 333
  lepturus 333
  rubricaudus 171
Phalacrocorax
  aristotelis 8, 124, 318
  carbo 8, 124, 318
  pygmaeus 318
Phalaropus fulicarius 317
  lobatus 317
Phoenicurus 108, 300, 304
  ochuros 256, 300, 304
  phoenicurus 300, 304, 327
Phyllastrephus 322
  cerviniventris 326
  fischeri 326
  flavostriatus 325, 326
  strepitans 326
  terrestris 326
Phylloscopus 14, 101, 103–4, 109,
  116, 149, 200, 293–4, 302, 354
  bonelli 104, 293, 302, 327
  borealis 293, 302, 327
  collybita 103, 261, 293, 302, 327
  sibilatrix 104, 256, 293, 302, 327
  trochilus 4, 103, 261, 293, 302,
    327
  trochiloides 293, 302, 327
Picidae 118, 314
Picoides tridactylus 327
Picus canus 314, 327
  viridis 314, 327
Pigeon 118, 119, 121, 234, 327,
  249, 353
  Plain 349
  Red-necked 349
  Ring-tailed 349
  White-crowned 349
Pinicola enucleator 85, 284, 327
Pinaroloxias inornata 176, *176*

388

Pintail   16, *123*, 318
Pipilo erythrophthalmus   126
   ocai   126
Pipit   96, 116, 260, 287, 301, 311
   Alpine   96, 104, 109, 255, 287
   Meadow   10, 96, 109, 254, 287,
      315
   Redthroated   96, 287
   Rock   96, 109, 114, 255, 287, 315
   Tawny   287
   Tree   10, 96, 109, 254, *254*, 287
Platypsarus niger   353
Platyspiza crassirostris   176, *176*
Platysteira peltata   326
Plautus alle   124, 137, 321
Ploceidae   139, 323, 324, 325
Ploceines   141
Ploceus   323, 324, 355
   bicolor   325
   insignis   144
   intermedius   326
   ocularis   144
Plover, Golden   315
   Grey   315
   Kentish   315
   Little Ringed   315
   Ringed   245, 315
Pluvialis apricaria   315
   squatarola   315
Pochard   318
Podiceps   122, 317
   auritus   317
   cristatus   317
   griseigena   317
   nigricollis   317
   ruficollis   317
Podicipitidae   317
Poecile   19, 41, 49, 53, 272
Pogoniulus   149, 323, 324
   bilineatus   144, 324, 325
   leucomystax   144, 325
   pusillus   144, 326
   simplex   324, 325
Pogonocichla stellata   325
Polioptila caerulea   347
   lembeyi
Porzana parva   314
   porzana   314
   pusilla   314
Prawns   318
Prinia   323, 324
Prionopidae   139, 322, 324, 325
Prionops poliocephala   326

Prionops (continued)
   retzii   326
   scopifrons   325
Procellariiformes   320
Procelsterna cerulea   168, 170
Prodotiscus insignis   149
   regulus   149
Progne   174
Prunella   4, 305
   collaris   305
   modularis   104–5, 305, 327
Pseudonestor xanthophrys   335
Psittacula krameri   163
Psittirostra   197, 199, 335, 336
   bailleui   335, 336
   cantans   335, 336
   flaviceps   335, 336
   kona   335, 336
   palmeri   335, 336
   psittacea   335, 336
Ptarmigan (Rock)   312
   Willow   312
Puffin   122, 124, 321, 322
Puffinus (Procellaria) diomedea   320
   nativitatis   171
   puffinus   320
   tenuirostris   252
Pyrocephalus   175
Pyrrhocorax graculus   310
   pyrrhocorax   309
Pyrrhula pyrrhula   75, 78, 81, 85,
      176, 256, 284, 327
Pyrrhuphonia jamaica   226, 352
Pytilia afra   150
   hypogrammica   150
   melba   150

Quelea   323
Quiscalus   263, 348
   lugubris   348
   niger   348

Rails   314
Rallidae   314
Ramphocelus   127, 332
   carbo   156, 332
Ranzania   333
Raven   309
Razorbill   122, 124, 321, 322
Redpoll (Lesser)   74, 77, 78, 79, 80,
      81, 82, 83, 87, 88, 256, 283,
      327
   Arctic   87, 89, 91, 283, 284

Redpoll (*continued*)
  Mealy 91
Redshank 14, *15*, 316
  Spotted 14, *15*, 315, 316
Redstart 108, 300, 304
  American 257
  Black 256
  Common 300, 327
Redwing 306, 327, 349
Regulus 114, 305
  ignicapillus 305, 327
  regulus 306, 327
Rhipidura 162, 354
  javanica 165
Rifle Bird 329
Riparia riparia 286
Rissa tridactyla 124
Robin 327
Rodents 311
Rook 309
Rosefinch, Scarlet (or Grosbeak) 86,
  283
Rukia 208, 209, 214, 215, 344
  longirostra 210, 344
  oleaginea 344
  ruki 344

Sand-eels 322
Sanderling 136, 316, 317
Sandgrouse 118, 312
  Black-bellied 312
  Pintailed 312
Sandpiper 122, 137, 315-7
  Baird's 135
  Common 14, *15*, 316
  Curlew 136
  Dunlin 134, 135, 316, 317
  Green 316
  Marsh 316
  Pectoral 134, 135
  Purple 316, 317
  Redbacked 135
  Rufous-necked 136
  Semipalmated 135
  Western 136
  White-rumped 136
  Wood 14, *15*, 316
Sathrocercus cinnamomeus 324
  mariae 324
Saurothera merlini 350
  longirostris 350
  vetula 350, 353
  vieilloti 350

Saxicola 4, 108, 110, 300, 304
  rubetra 300, 304
  torquata 300, 304
Scaup 318
Scolopacidae 315-17
Scomberesox 333
Scombridae 168, 169
Scoter, Common 318
  Velvet 318
Seicercus ruficapilla 325
Selar 333
Sericotes hotosericeus 229-232, 353
Sericulus 158, 331
  aurens 331
  bakeri 331
  chrysocephalus 331
Serin 75, 76, 89, 256, 284, 327
Serinus 85, 89, 284, 285
  canaria 284
  citrinella 89, 284, 285
  serinus 75, 76, 89, 256, 285, 327
Setophaga ruticilla 257
Shag 8, 124, 318
Shearwater 171
  Manx 320
  North Atlantic 320
Shelduck 319
  Ruddy 319
Sheppardia 148
  sharpei 325
Shoveller 16, *123*, 258, 318
Shrike 97, 108, 109, 110, 116, 125,
  139, 148, 149, 288, 301, 325
  Great Grey 97, 98, 288, 289
  Helmet 139, 148
  Lesser Grey 97, 98, 288, 289
  Masked 97, 289
  Redbacked 97, 99, 116, 246, 247,
    256, 288, 289
  Woodchat 97, 99, 116, 288, 289
Sicklebill, Hawaiian 174-99,
    esp. 193-7, 200, 215, 220,
    227, 239, 262, 334-5
Sigmodus retzii 324
  scopifrons 324
Siskin 78, 79, 80, 81, 82, 83, 87, 88,
    89, 283, 327
Sitta 63-73, 254, 279-83
  azurea 73
  canadensis 65, 279, 280
  carolinensis 65, 279, 280
  castanea 68, 69, 70, 71, 280, 281,
    282, 283

Sitta (*continued*)
  castanea cashmirensis   70, 281, 283
  castanea tonkinensis   70, 283
  europaea   63, 64, 65, 72, 279, 280, 281, 282, 327
  formosa   72, 280, 282
  frontalis   68, 71, 72, 73, 280, 282, 283
  himalayensis   69, 70, 71, 72, 280, 282, 283
  krüperi   65, 279, 280
  leucopsis   69, 71, 72, 280, 282
  magna   72, 281, 282, 283
  nagaensis   69, 70, 71, 72, 282, 283, 291
  neumayer   9, 63, 65, 66, 67, 279
  pusilla   66, 279, 280
  pygmaea   66, 279, 280
  solangiae   73
  tephronota   9, 66, 67, 72, 281, 282
  victoriae   69, 71, 281, 282
  villosa   72, 281, 282
  whiteheadi   64, 65, 279, 280, 327
  yunnanensis   69, 71, 72, 281, 282
Sittiparus   40, 271
Sittinae   12
Skua (or Jaeger)   264, 320
  Arctic   320
  Great   320
  Long-tailed   320
  Pomarine
Skylark   4, 254, 304
Smew   319
Snails   316
Snipe, Common   315
  Great   315
Solitaire, Cuban   347
  Rufous-throated   347
Sparrow, Hedge (Dunnock)   4, 305
  House   18, 144, 163, 308, 309
  Java   165
  Rock   10, 149, 255
  Spanish   10, 114, 255, 308, 309
  Tree   309
Sparrowhawk   134, 310
  Levant   310
Spatula   16
Sprats   322
Speirops   202, 215
  brunnea   337
  leucophaea   337
  lugubris   337
  melanocephala   337

Spermestes cucullatus   324
  nigriceps   324
Spermophaga ruficapilla   325
Spindalis zena   352
Spizella passerina   258
Spreo bicolor   150
Squid   168, 169, 171
Starling   116, 148, 150, 163, 253, 309, 354, 355
  Rose-coloured   309
  Spotless   116, 309
Steganura obtusa   150
  orientalis   150
  paradisaea   150
  togoensis   150
Stercorariidae   320
Stercorarius longicaudus   320
  parasiticus   320
  pomarinus   320
Sterna albifrons   320
  caspia   320
  dougallii   321
  fuscata   168, 169, 333
  hirundo   321
  lunata   166
  paradisaea   321
  sandvicensis   321
Sternoptyx   333
Stint, Little   316
  Temminck's   316–17
Stoat   264
Stonechat   300
Stork, Black   314
  White   257, 308, 314
Streptopelia chinensis   162, 164,
  decaocto   256, 312
  turtur   312
Streptoprocne collaris   351
Strigidae   118, 313, 353
Strix aluco   264, 313
  nebulosa   313
  uralensis   313
Sturnella magna   127
  neglecta   127
Sturnus nigricollis   163
  roseus   309
  unicolor   116, 253, 309
  vulgaris   116, 253, 309
Suaheliornis kretschmeri   325
Sula   183
  dactylatra   166, 172, 333
  leucogaster   166, 333
  nebouxii   172

Sula (continued)
  sula 172
Sunbird 138, 139, 148, 220, 239,
  325, 355
Swallow 94, 109, 226, 286, 301
  Barn 94, 95, 169, 286, 313
  Bahama 226
  Golden 225
  Red-rumped 94, 95, 286, 313
Swan 252
  Bewick's 319
  Mute 319
  Whooper 319
Swift 12, 118, 121, 124, 234, 237,
  259, 265, 313
  Alpine 313
  Black 351
  Chimney 351
  Collared 351
  Common 169, 313
  Grey-rumped 351
  Lesser Antillean 351
  Pallid 313
  Short-tailed 351
  White-rumped 313
Sylvia 14, 101, 104–5, 116, 294–7,
  303
  atricapilla 104, 294, 303, 327
  borin 104, 294, 303, 327
  cantillans 105, 295, 303
  communis 294, 303
  conspicillata 295, 303
  curruca 104, 295, 303, 327
  hortensis 295, 303, 327
  melanocephala 105, 295, 303
  nisoria 295, 303
  rüppelli 115, 295, 303
  sarda 105, 295, 303
  undata 105, 247, 295, 303
Sylvietta 323
  brachyura 326
Sylviinae 139, 323, 324, 325

Tachyphonus 156, 332
  luctuosa 332
  rufus 332
Tadorna casarca 319
  tadorna 319
Tailorbird 163, 164, 254
Tanager 127, 155, 156, 264, 332
  Black-crowned 348
  Grey-crowned 348

Tanager (*continued*)
  Palm 224, 348
Tanagra 155, 156, 332
  guttata 332
  gyrola 332
  mexicana 332
  musica 226, 352
  trinitatis 156, 332
  violacea 156, 332
Tauraco 145–7, 323, 328–9
  bannermani 147, 328
  corythaix 147, 328
  c. corythaix 147
  c. fischeri 147
  c. livingstonii 147
  c. persa 147
  c. schutti 147
  erythrolophus 147, 328
  fischeri 324, 325, 329
  hartlaubi 147, 324, 325, 328, 329
  johnstoni 147, 328, 329
  leucolophus 147, 328
  leucotis 147, 328
  livingstonii 329
  macrorhynchus 147, 328, 329
  persa 329
  porphyreolophus 147, 328, 329
  ruspolii 147, 328
  schütti 328
Tchitraea 322
  perspicillata 325
  quadricola 324
Teal 16, *123*, 318
  Marbled 318
Tephrozosterops 215, 218
  stalkeri 341
Teretistris 224, 348
  fernandinae 348
  fornsi 348
Termites 7, 264
Tern 124, 166, 170, 171, 252,
  319, 320–1
  Artic 321
  Black 320
  Caspian 320
  Common 321
  Crested 166
  Fairy 166, *167*, 168, 170, 172,
  333
  Greybacked 166
  Little 320
  Noddy 263
  Roseate 231

Tern (*continued*)
  Sandwich 321
  Sooty 166, *167*, 168, 169, 171
    172, 333
  Whiskered 319
  Whitewinged Black 319
Terpsiphone 355
  viridis 144
Tetraenura fischeri 150
  regia 150
Tetrao urogallus 263
Thalasseus bergii 166
Thrasher, Pearly-eyed 346
  Scaly-breasted 346
Thraupidae 156
Thaupis palmarum 156, 332
  virens 332
Thrush 4, 138, 139, 144, 194, 198,
    199, 227, 255, 305-6, 316,
    325, 346-7
  Bare-eyed 347
  Cocoa 347
  Forest 148
  Grand Cayman 347
  La Selle 347
  Mistle 260, 306, 326
  Red-legged 347
  Rock 298-9, 304, 106, 110
    Blue 106, 110, 298, 299
  Song 261, 306, 326
  White-eyed 346
Thysanoessa 136
Tiaris bicolor 349
  canora 349
  olivacea 349
Tinkerbird 149
Tit 4, 14, 18-27, 38-62, 83, 93, 94,
    108, 109, 110, 113, 115, 116,
    125, 126, 128, 129, 244, 246,
    247, 250, 254, 257, 259,
    271-9
  Azure 19, 20, 21, 30, 31, 39, 272,
    327
  Black 45, 272, 274
  Black-crested 49, 50, 52, 54, 58,
    60, 276
  Black-spotted yellow 272
  Blue 10, 19, 20, 21, 22, 24, 25,
    26, 27, 31, 32, 33, *34*, 35, 36,
    39, 40, 50, *51*, 52, 58, 59, 60,
    187, 263, 266, 267, 327
  Bridled 50, *51*, 52, 55, 58, 60, 276
  Brown-crested 272

Tits (*continued*)
  Canarian Blue *34*
  Cinnamon-breasted 275
  Coal 19, 20, 21, 22, 24, 25, 26,
    27, 28, 29, 30, 32, 33, 34, *34*,
    36, 39, 43, 45, 50, 51, 55, 57,
    59, 263, 271, 272, 273, 327
  Crested 19, 20, 21, 27, 28, *29*,
    30, 32, 33, 35, 39, 40, 50, *51*,
    53, 57, 327
  Crested Black 272
  Dusky 275
  Elegant 271
  Formosa Yellow 271
  Great 11, 19, 20, 21, *22*, *23*, 24,
    25, 26, 27, 28, 30, 31, 32, 33,
    39, 40, *40*, 41, 44, 50, *51*, 52,
    58, 59, 60, 257, 266, 267, 272,
    326
  Green-backed 41, 272, 371
  Grey 45, 274
  Long Tailed 326
  Marsh 19, 20, 21, *22*, 24, 26, 29,
    31, 33, 41, 44, 50, *51*, 53, 58,
    59, 272
  Pére David's 272
  Persian Coal *34*
  Plain 49, 52, 54, 56, 277
  Red-throated 275
  Siberian, (or Lapp) 19, 20, 21,
    30, 39, 50, 55, 56, 277, 327
  Sombre, 19, 20, 21, 31, 33, 39, 327
  Southern Black 274
  Striped-breasted 274
  Tufted 49-52, 54, 55, 59, 276-7
  Varied 39, 40, 271
  White-breasted 274
  White-breasted Black 274
  White-browed 272
  White-fronted 271
  Willow 19; 20, 21, 22, 25, 26, 27,
    28, *28*, *29*, 30, 32, 33, 35, 36,
    39, 41, 50, *51*, 53, 56, 57, 59,
    272, 326
  Yellow-bellied 272
  Yellow-cheeked 272
Todidae 234
Todus angustirostris 351
  mexicanus 351, 353
  multicolor 351
  sublatus 351
  todus 351, 353
Tody, Broadbilled 351

Tody (*continued*)
  Narrowbilled 351
Torgos tracheliotus 152
Tortoise 154
Towhee, Mexican Collared 126, 127
  Red-eyed 126, 127
Trachyphonus 323
Tree-creeper 114, 116, 117, 133,
    253, 306, 327, 306–7
  Australian 161
  Short-toed 306, 327
Tree-pie 163
Trembler 238, 239
Treron 162
Tricholaema melanocephalum 324
Trigonoceps occipitalis 152, 154
Tringa 14, *15*, 122, 136, 315–16
  erythropus 315
  glareola 316
  hypoleucos 316
  nebularia 315
  ochropus 316
  stagnatilis 316
  totanus 316
Trochilus polytmus 228–232, 353
Trochocercus 322
  albonotatus 143, 144, 325
  bivittatus 325
Troglodytes troglodytes 149, 257,
    326
Tropicbird 166
  Redbilled 333
  Redtailed 171
  Yellowbilled 333
Troupial 348
Tschagra 322, 324
Tuna 169, 170
Turaco 138, 139, 145–7, 158, 324,
    325, 328–9
Turdinae 138, 139, 323, 324, 325
Turdus 227, 323, 324, 335–6, 346–7
  aurantius 346, 352
  fumigatus 347
  gurneyi 325
  iliacus 316
  jamaicensis 346, 352
  merula 256, 261, 305, 327
  musicus 327
  nudigensis 347
  olivaceus 325
  philomelos 261, 306, 326
  pilaris 306, 327
  poliocephalus 255

Turdus (*continued*)
  swalesi 347
  torquatus 114, 255, 306, 327
  viscivorus 260, 306, 326
Turnix 162, 345
Twite 87, 88, 91, 283
Tyrannus caudifasciatus 345, 352
  cubensis 345
  melancholicus 345
Tyto alba 163, 269

Uraeginthus bengalensis 324
  cyanocephalus 324
Uria 321–2
  aalge 10, 122, 245, 321
  lomvia 10, 122, 245, 321
Urobrachya axillaris 324

Vanellus vanellus 257
Vermivora bachmanii 249
Vestiaria coccinea 335
Vidua 323
  macroura 150
Viduinae 139, 323
Viduines 141, 150, 151
Vireo 132, 237, 347
  altiloquus 227, 347, 352
  caribaeus 347
  crassirostris 347
  flavifrons 132
  gilvus 132
  griseus 132, 347
  gundlachii 347
  latimeri 225, 347, 352
  modestus 132, 225, 347, 352
  nanus 347
  olivaceus 132
  osburni 132, 228, 347, 352
  solitarius 132
Vireo 59, 60, 129–32, 137, 227,
    236, 237
  Black-whiskered 227, 347
  Blue Mountain 132, 347
  Jamaican White-eyed 132
  Red-eyed 132
  Thick-billed 228
  White-eyed 132, 225, 227, 228, 347
  Yellow-throated 132
  Yucatan 347
Vireonidae 59
Vireosylvia 132

Viridibucco leucomystax 324
  simplex 324
Voles 118, *119*
Vulture 151–4
  Bearded 154
  Black 152, *153*
  Egyptian 152, *153*
  Griffon 151, 152, *153*
  Hooded 152
  Lappet-faced 152, 164
  White-backed 151
  White-headed 152, 154

Wagtail 96–7, 288, 301
  Grey 96, 288
  White 96, 288
  Yellow 288
Water-strider (marine) 170
Warblers 14, 59, 101, 104–5, 108,
    110, 116, 129–32, 139, 144,
    149, 150, 224, 226, 227, 237,
    241, 247, 248
  acrocephaline 100, 116, 149, 194,
    248, 260, 269, 290–1
  grass 145
  ground 348
  leaf 14, 101, 103–4, 116, 149,
    200, 293–4, 302, 354
  parulid 61, 129, 137, 174, 175,
    177
  Aquatic 290, 291
  Arctic 293, 294, 327
  Arrow-headed 348
  Bachman's 249
  Barred 295, 297
  Bay-breasted 129, 130, 131
  Bonelli's 104, 109, 293, 294, 327
  Blackburnian 129, 130, 131
  Black-throated Green 129, 130,
    131
  Blyth's Reed 290, 291
  Cape May 129, 130, 131
  Dartford 105, 247, 295, 296, 297
  Garden 14, *15*, 104, 105, 294,
    295, 297, 327
  Grasshopper 103, 292–3, 302
  Great Reed 290, 291
  Green-tailed 348
  Greenish 293, 294, 327
  Icterine 14, *15*, 101, 102, 103,
    291, 292, 327
  Kirtland's 249, 250

Warbler (*continued*)
  Marmora's 105, 115, 295, 297
  Marsh 100, 248, 290, 291
  Melodious 101, 102, 103, 291,
    292, 327
  Moustached 290, 291
  Myrtle 129, 130, 131
  Olivaceous 101, 102, 103, 292
  Olive-capped 348
  Olive-tree 101, 102, 103, 291,
    292, 327
  Orphean 295, 296, 297, 327
  Oriente 348
  Pine 348
  Reed 100, 248, 290, 291, 302
  River 103, 292, 293
  Rüppell's 115, 295, 296, 297
  Sardinian 105, 295, 296, 297
  Savi's 292, 293
  Sedge 100, 248, 290, 291
  Spectacled 295, 296, 297
  Subalpine 105, 295, 296, 297
  Whistling 226
  White-winged 348
  Willow 4, 14, *15*, 103, 104, 109,
    261, 293, 294, 327
  Wood 104, 256, 293, 294, 327
  Yellow 348
  Yellow-headed 348
  Yellow-throated 348
Waterfowl 318–19
Waxwing 86, 249, 251
Weaverbird, Ploceine 138, 139, 141,
    148, 150, 325
  Viduine 149
Weevil 197, 334
Wheatear 106–7, 111, 299–300, 304
  Black 299
  Black-eared 106, 115, 299, 300
  Common 106, 115, 299, 300
  Isabelline 299, 300
  Pied 299
Whimbrel 122, 245, 315
Whinchat 300
White-eye 12, 68, 149, 194,
    200–221, 222, 227, 244, 262,
    263, 336–44, 354, 355
Whitethroat 294, 295, 297
  Lesser 104, 295, 297, 327
Whydah 139
Wigeon 16, *123*, 318
Woodchat 97, 99, 116, 288, 289
Woodcreeper 267

Woodfordia   215
  lacertosa   343
  superciliosa   343
Woodlark   254, 304
Woodpecker   117, 118, 120, 121,
     144, 149, 234, 351, 353
  Black   327
  Golden-fronted   128, 239, 240
  Great Spotted   120, 256, 314, 327
  Green   314, 327
  Grey   314, 327
  Hispaniola   239, 240
  Jamaican   239
  Lesser (Little) Spotted   120, 314,
     327
  Middle Spotted   314, 327
  Red-bellied   128
  Syrian   314, 327
  Three-toed   327
  White-backed   314, 327
Worm, nereid   317
Wren   149, 257, 326
Wryneck   327

Xanthocephalus xanthocephalus   128

Yellowhammer   307
Yungipicus   144

Zenaida asiatica   350
  auriculata   350
  aurita   350
  macroura   350
Zosteropidae   12, 200–21, 336–44
Zosterops   162, 194, 200–21, 223,
     224, 237, 242, 244, 254, 255
  abyssinica   337
  albogularis   343
  anomala   340
  atricapilla   339, 340
  atriceps   341
  atrifrons   340, 341, 342, 344
  atrifrons hypoxantha   344
  borbonica   337, 338
  buruensis   341
  ceylonensis   339
  chloris   213, 338, 339, 340, 341
  cinerea   208, 209, *210*, 344
  consobrinorum   340
  conspicillata   209, *210*, 344
  erythropleura   338
  everetti   203, 338, 340, 342

Zosterops (continued)
  explorator   343
  ficedulina   337
  flava   343
  flavifrons   343
  fuscicapilla   341, 342
  grayi   341
  griseotincta   342, 344
  griseovirescens   337
  (atrifrons) hypoxantha   342
  inornata   343
  japonica   338, 339, 340, 341, 342
  kuehni   341
  (rendovae) kulambangrae   342, 343
  lateralis 208, 213, 216, 217, 338 342
  lutea   338
  luteirostris   342
  luteirostris splendida   344
  madaraspatana   337
  mayottensis   337
  (atrifrons) meeki   342
  metcalfii   342, 343
  minuta   343
  modesta   337
  montana   214, 339, 340, 341, 342
  mouroniensis   337
  murphyi   343
  mysorensis   341
  natalis   339
  nigrorum   342
  novaeguineae   341
  olivacea   337, 338
  pallida   337
  palpebrosa   203, 338, 339, 340
  poliogastra   202, 337
  rendovae   342, 344
  rendovae kulambangrae   344
  renelliana   344
  renelliana (griseotincta)   343
  salvadorii   339
  samoensis   342
  sanctaecrucis   343
  senegalensis   337
  (luteirostris) splendida   342
  stresemanni   342
  strenua   343
  tenuirostris   343
  ugiensis   342, 343, 344
  uropygialis   341
  vellalavella   342
  virens   337
  wallacei   339, 340
  xanthochroa   343

General Index

Abingdon I. 175, 180, 181, 189
Abruzzi Mts. 96
Abyssinia 46, 47, 274, 286, 298,
    300, 337
acknowledgements 16–17
adaptation of subspecies 33–6, 87
adaptations to habitat 10, 21, 33–6,
    53, 254
adaptations to range 10, 244–5
adaptive radiation 1, 174, 194, 227,
    267–8
Aegean Sea 299
  islands 33, 106
afforestation, effects of 3–4
Afghanistan 66, 280
Africa 11, 18, 45–9, 61, 94, 97, 100,
    104, 106, 111, 138–54, 202,
    203, 244, 246, 247, 250, 269,
    276, 336–8
  east 102, 107, 143–5, 154, 269,
    286, 288, 291, 297, 298, 329
  islands 202
  mountains 143
  north 9, 38, 100, 103, 107, 289,
    291, 294, 299, 300, 306
  northeast 105
  northwest 35, 87
  southern 94, 100, 102, 274–5,
    287, 291, 294, 297
  tropical 179, 200, 247, 280, 286,
    287, 291, 292, 293, 298, 350
  west 103, 202, 275, 288, 290,
    291, 292, 297, 329
Alaska 55, 134, 136, 277, 278,
    321
Albemarle I. (Galapagos) 175, 180,
    181, 182, 186, 189, 190
Alberta 278
Aldabra I. 337

Algeria 33
Allen's rule 244
Alor I. 204, 205
Alps 33, 35, 89, 283, 284, 291, 293
Ambon I. 204, 205, 341
America Central 87, 222, 338
  North 18, 21, 49–62, 65, 66,
    126–37, 244, 253, 269,
    276–80, 316, 317, 347, 348
  South 157, 222, 237, 269, 344
Andaman Is. 339
Angola 145, 274, 275
Anjouan Is. 203, 337
Annam 69, 73, 281
Annobon I. 202, 203, 216, 217, 337,
    355
Antilles, Greater 222, 234, 238, 344,
    345, 346, 348, 349, 350, 351
Antilles, Lesser 222, 226, 228, 230,
    232, 233, 238, 239, 344, 345,
    346, 347, 348, 349, 350, 351
Appenines, Mts. 33
Arabia 291
Arfak Mts. 341
Arizona 52, 277, 280
Aru I. 157, 204, 205, 329, 330, 331,
    341
Aruba I. 344
Ascension I. 166, 169, 333
Asia 12, 18, 21, 38–62, (esp. 41–5,
    61), 63, 104, 107, 203–5, 293,
    299, 305, 317, 338–9
  central 49, 72, 248
  mountains of 42, 61, 68–73,
    272–4, 280–3
  eastern 200, 309
  mountains of 44, 53, 59, 106
  northern 280
  southern 200

Asia (*continued*)
  southeast 72, 73, 309, 338
    mountains of 68–73, 280–3
  western 38
Asia Minor 66, 308
Assam 72, 280
Atlantic Ocean 137, 166, 245, 277, 298
  North 319
  coast 88
Atlas Mts. 298, 306
Austin (Texas) 128
Australia 158–61, 173, 200, 201, 203–5, 207, 212, 213, 216, 220, 252, 329, 338
Austria 100, 248, 269
avifaunas 1, 269–70

Bagga I. 342
Bahama Is. 223, 222–6, 230, 232, 238, 249, 250, 345, 346, 347, 348, 349, 350, 351
Baja peninsula 239, 271
Balabac I. 271
Balearic Is. 87, 297
Bali I. 204, 205, 340
Balkans 19, 20, 31, 32, 66, 288, 289, 292, 311
Baltic Sea 75, 96, 283, 287, 290, 293, 315, 318
Baluchistan 43, 66
Banda Sea 214, 216
Banks I. 207, 343
Barbuda I. 348
Barents Sea 166
Barrington I. 175
Barro Colorado 142, 143
Barrow 134, 136
Batan I. 342
Bear I. 321
Bergmann's rule 244
Bialowies Forest 117, 290
Bindloe I. 175, 180, 181, 185, 189, 190, 191
birds of prey, see raptors
birthplace, return to 252
Bismarck Archipelago 207, 342
Black Sea 319, 321
Bonaire I. 344
Borneo 203, 205–6, 215, 216, 217, 340, 354
Bothnia, Gulf of 318

Botswana 289, 292, 298
Bougainville I. 206, 207, 343
British Columbia 54, 55, 56, 65, 66, 277
British Isles 283, 305, 306, 307, 312
brood parasites 147–51
Burma 41, 68, 69, 70, 71, 72, 163, 164, 272, 280, 281, 291, 294, 338
Buru I. 204, 206, 214, 215, 216, 218, 219, 341

Caicos Is. 346
California 54, 56, 65, 193, 278
Cameroon, Mt. 202, 214, 215, 337
Cameroons 274, 292
Canada 55, 65, 277
Canadian zone, North America, 56, 58, 65
Canary Islands 10, 34, 35, 104, 109, 255, 308
Cantabrian Mts. 33
Cape Province 274, 289
Cape Verde Is. 255
Cape York 329, 331, 338
Caribbean Sea 346
Caroline Is. 204, 205, 208, 209, 210, 214, 215, 216, 237, 344
Caucasus Mts. 33, 39, 66, 269
Cayman Is. 223, 346, 347, 348
Cebu I. 342
Celebes I. 204, 205–6, 213, 340
Ceram I. 204, 205, 206, 215, 216, 218, 219, 341
Ceylon I. 291, 339
character displacement 10, 43, 63, 66–8, 70, 159–60, 180–5, 189–91, 195, 215, 228, 349
Chad, Lake 289, 297
Charles I. 175, 180, 181, 188, 189
Chatham I. 175, 177, 180, 181, 186, 189, 190, 191
China 40, 41, 69, 71, 72, 272, 280, 281, 338
Choiseul I. 342
Cholo Mt. 144, 145
Christmas I. (Indian Ocean) 204, 205, 216, 217, 339
Christmas I. (Pacific Ocean) 166, 167, 168, 170, 171
classification 12–16
climate, adaptations to 244

climate, influence of   91, 122, 246–7
Cocos I.   176
Colombia   127, 155
Comoro Is.   216, 337
Congo   202, 274, 275, 286, 287, 290, 300
convergent evolution   50, 51, 63, 182–5
Corsica   65, 89, 279, 284, 297, 305, 307
Coto Donana   305
Crete   106, 298
Crossman Is.   175, 181, 182
Cuba   222, 223, 224, 226, 230, 232, 237, 239, 241, 250, 344, 345, 346, 347, 348, 349, 350, 351
Culpepper I.   175, 182, 183, 184, 185
Curacao I.   344
Cyprus   87

Dalmatian Coast   292
Daphne I.   175, 181, 182
Darfur   299
Denmark   32, 39, 75, 76, 315, 317
D'Entrecasteaux Is.   157, 207, 329, 330, 342
District of Columbia (U.S.A.)   58
Doi Hua Mot Mts.   70
Dominica I.   223, 230, 233, 234, 347, 348, 350
Dominican Republic (see Hispaniola)
Duncan I.   175, 180

ecological niche   3
Ecuador   143, 174
Egypt   152, 287
Elburz Mts.   33, 34
England   3, 4, 20, 22, 24, 25, 26, 33, 35, 52, 74, 75, 77–80, 81, 82, 83–5, 120, 122, 124, 169, 244, 256, 257, 261, 291, 295, 307, 318, 321, 322
Engano I.   204
Eritrea   46, 274, 294, 297
Europe   7, 12, 14, 18, 36, 37, 44, 49, 51, 53, 57, 58, 61, 63–65, 74–92, 93–111, 112–25, 132, 134, 137, 141, 142, 143, 149, 179, 246, 247, 248, 250, 253, 279–80, 284, 285, 286–300, 301–22, 326–7

Europe (*continued*)
  central   19, 104, 114, 255, 284, 288, 289, 293–4, 295, 298, 300, 305, 306
  mountains   85, 287, 305
  eastern   9, 31, 75, 288, 289, 290, 291, 298, 299, 306, 309, 316
  middle latitudes   103, 104, 106, 109, 209, 269, 287, 292, 293, 294, 298, 305, 306, 307, 309
  northern   19, 96, 104, 109, 114, 258, 284, 287, 290, 293, 295, 300, 304, 306, 307, 309, 316, 317, 320, 321
  south   19, 27, 75, 76, 89, 106, 152, 284, 286, 287, 288, 289, 290, 292, 295, 298, 299, 300, 304, 305, 306, 307, 313, 314, 320
  west   85, 105, 248, 293–4, 295, 298, 314, 317

Farne Is.   124, 166
Fenno-Scandia   19, 91, 103, 115, 122, 244, 245, 249, 256, 288, 291, 306, 316, 319
Fernando Po I.   202, 214, 216, 219, 337, 355
Fiji Is.   207, 343, 354–5
Flores I.   204, 205, 216, 218, 339–40
Florida   222, 223
Florida I.   342
Fokien   69, 281
Forest of Dean   21, 22, 24
Formosa   41, 271–2
France   35, 103, 105, 106, 291, 292, 295, 296, 298, 299
fruit-eating birds   138, 145–7, 155–8, 243, 249, 251, 264, 328–32

Galapagos archipelago   3, 172, 174–99, 220, 235, 262, 355
Ganonga I.   206, 342
Gause's principle   3, 7
genus, nature of   7, 13–16
geographical range, adaptations to, see under adaptations
  changes in   54, 75, 76, 89–91, 122, 163–4, 244–5
Germany   76, 257, 260, 289, 298, 315, 317

Gibraltar   120, 311
Giluwe, Mt.   330
Gizo I.   206, 342
Gonave I.   348
Goodenough I.   206, 207, 342
Gotland   32, 289
Gran Canaria I.   120, 255
Grand Cayman I.   226, 345, 347, 348
Grand Comoro I.   202, 215, 216, 337
Great Basin   54, 127
Great Britain   19, 40, 74, 76, 87, 88,
      91, 96, 251, 283, 288, 320
Great Kei   206, 216, 341
Greater Antilles   222, 234, 238, 344,
      345, 346, 349, 350, 351
Greece   285, 299, 308, 311
   islands   297
Grenada I.   222, 223, 227, 230, 231,
      237, 345, 346, 347, 349, 351
Grenadine Is.   346
Guadalcanal, I.   207, 342
Guadeloupe, I.   223, 238, 348, 350,
      351
Guan I.   208
Guinea, Gulf of   202, 216, 337, 355
Guinea islands   215, 241, 270
Gulf Coast   277

habitat, differences between summer
      and winter   93, 100
   selection   4–5, 258–61
   variations within same species
      32–6, 39, 44, 46, 47, 71, 96,
      97, 114, 143–5, 182, 203, 225,
      227, 228, 255–8
Hagen, Mt.   330
Haifan Mts.   41
Haiti (see Hispaniola)   344, 348, 349
Halmahera I.   204, 205
Hanang, Mt.   144, 145
Hanshan Mts.   41
Hawaiian Islands   174–99, esp.
      193–9, 200, 215, 220, 227, 235,
      239, 262, 334–5, 336, 355
Himalayan Mts.   41, 43, 45, 69, 70,
      71, 72, 87, 272, 280, 281, 298
Hispaniola I. (Haiti and Dominican
      Republic)   222, 223, 224,
      226, 227, 228, 230, 232, 233,
      234, 239, 240, 241, 250, 344,
      345, 346, 347, 348, 349, 350,
      351, 352
historical survey   2–9, 11

Holland   76, 257, 315
Honduras   223
Hood I.   175, 182, 183, 184, 185
Hudsonian Zone   56
hummingbirds   228–33, 351
Huon Peninsula   157, 330
Huron, Lake   250
hybrids   126–7, 308

Ibadan   161, 162, 165, 256
Iberia   19, 288, 292, 295
Iceland   96, 316, 318, 319, 321
Idaho   277
Illinois   54
Inaccessible I.   198
Indefatigable I.   175, 180, 181, 182,
      187, 189, 190
India   40, 68, 69, 100, 148, 149, 151,
      161, 164, 201, 203, 212, 220,
      273, 280, 290, 291, 294, 338,
Indian Ocean   200, 202, 216, 217,
      220, 337
Indo-Australia   201, 212, 220,
      338–44
Indochina   69, 87, 204, 205, 280,
      281, 294
Indonesia   200, 212, 215, 280, 294
interspecific territories   127–9, 253
invasion birds   86
Iowa   277
Iran   31, 32, 66, 68, 281
Ireland   32, 33, 283, 309
island birds   160–1, 174–242, 243–4
Italy   19, 288, 290, 292, 295, 308,
      309

Jamaica   132, 222, 223, 224, 225,
      226, 227, 228, 230, 232, 233,
      234, 235, 344, 345, 347, 348,
      349, 350, 351, 352
James I.   175, 178, 180, 181, 185,
      186, 187, 190, 191
Japan   39, 40, 149, 272, 273, 338
Japen I.   157, 204, 205, 329, 330,
      341
Java I.   73, 203, 204, 205, 213, 339
Jervis I.   75, 181
Jugoslavia   31, 117

Kangaroo I.   159, 160
Kansas   54, 277

Kansu   69, 72, 272, 280, 281
Kashmir   69, 70, 71, 280, 281
Kauai I.   194, 195, 196, 198, 334
Kei Is.   204, 205, 341 (see also
   Great Kei, Little Kei)
Kentucky   54
Kenya   48, 275, 287, 289, 292, 293,
   297, 298, 337
Kokonor Mts.   72, 272, 281
Korea   39, 273
Krakatau I.   213
Kulambangra I.   206, 207, 343
Kunlun Mts.   41, 272
Kurile Is.   39
Kusaie I.   208, 209, 215, 216, 217,
   344

Laccadive Is.   339
Lanai I.   198, 334
Laos   69, 70, 72, 280
Lapland   30, 315
Laysan I.   197, 335
Lesser Antilles   222, 226, 228, 230,
   232, 233, 238, 239, 344, 345,
   346, 347, 348, 349, 350, 351
Lesser Sunda Is.   215
Lifu I.   207, 215, 343
Little Kei   206, 216, 341
Lombok I.   204, 205, 340
London   264
Lord Howe I.   207, 208, 215, 216,
   217, 219, 343
Louiade Arch.   207, 216, 329, 342
Loyalty Is.   207, 343
Luzon I.   204, 205, 342, 354

Macedonia   32, 308
Madagascar, I.   203, 217, 337
Maderia I.   120, 255
Maghreb   104, 106, 109, 294, 296,
   297, 306, 308
Mahé I.   203, 337
Majorca I.   33, 305
Malawi   46, 144, 145, 274, 290, 297,
   300
Malay Archipelago   204, 205
Malay Peninsula   165, 203, 338
Malaya   73, 161, 162, 164, 165, 204,
   205
man-modified habitats   74-80,
   83-5, 91, 126-7, 161-5, 256

Mariana Is.   204, 205, 208, 344
Marianne I.   203, 337
Marley wood   20, 21, 22, 24, 59
Marquesas Is.   193
Martinique I.   223, 230, 232, 233,
   237, 345, 348, 350, 351
Massachusetts   277
Maui I.   334
Mauretania   299
Mauritius I.   202, 215, 216, 218, 219,
   338
Mayotte I.   203, 337
Mbulu Mts.   144, 145
Mediterranean   105, 110, 116, 248,
   306, 307, 320
   basin   294, 297
   islands   105, 115, 295, 296, 297
      309, 311, 284, 287, 293
   littoral   19
Melanesia   200, 212
Mexico   49, 52, 65, 66, 127, 128,
   222, 223, 276, 277
Mexico, Gulf of   128, 277
Michigan, Lake   250
Micronesia   200
migratory birds   93-111, 249-51,
   286-304
Mindanao I.   204, 205, 342, 354
Mindoro I.   342
Misol I.   157, 330, 331
Moco, Mt.   145
Moheli I.   337
Molokai I.   198, 335
Molucca Is.   205-6, 341
   Northern   341
Mongolia   39
Montana   278
Montserrat I.   348
Morocco   33, 108
Mozambique   274, 275
Mytilene I.   65

Narborough I.   175, 186
Natuna I.   340
Nebraska   277
Negros I.   342
Nepal   43
nest-sites, competition for   26-7,
   94-5, 122, 124, 258
Neusiedlersee   100
New Caledonia I.   207, 343
New England   143

New Georgia I.   206, 342
New Guinea   158, 204, 205–6, 208,
    329, 330, 331, 341, 342
New Hebrides Is.   207, 343
New Mexico   277
New Zealand   75, 114, 213, 239, 307
Nicobar Is.   339
Nigeria   161, 274, 286, 287, 297,
    299
Nightingale I.   198
Nissan I.   342
nomenclature   16
Norfolk I.   207, 208, 213, 215, 216,
    217, 218, 219, 343
North Carolina   54
North Sea   315
Norway   27–8, 30, 87, 283

Oahu I. 334
Oaxaca   127
Old Providence I.   231, 351
Oregon   54, 277
origin of species   1, 5, 6, 243, 253,
    267–8
Orkney Is.   88
Oxford   18, 77, 78, 83, 91, 257

Pacific Ocean   166, 206, 220
Pacific coast   66, 277
Pacific islands   206, 207–10, 338–44

    North   137
    Western   205, 207
Palau I.   204, 208, 209, 215, 216,
    219, 344
Palawan I.   204, 271, 342
Panama   142, 236
Pantapui   255, 270
Pemba I.   337
Pennsylvania   277
Persia   9, 11, 20, 33, 67
Philippine Is.   41, 203, 204, 206–7,
    208, 271–2, 280, 342, 344, 354
Pines, Isle of   223
Plitvicke Forest   117
Poland   117, 290
Ponape I.   208, 209, 215, 216, 219,
    344
Portugal   34, 35
Principe I.   202, 219, 220, 270, 337,
    355
proximate factors   5

Puerto Rico   222, 223, 224, 226,
    228, 230, 232, 233, 235, 237,
    238, 239, 241, 344, 345, 346,
    347, 348, 349, 350, 351, 352
Punjab   69
Pyrenees Mts.   33

Queensland   160, 329

range, adaptations to   10, 244–5
    changes in, see geographical range
raptors   118–20, 133–4, 151–4,
    310–11, 349
recognition of habitat   4–5, 258–61
Red Sea   298
Rendova I.   206, 342
Rennell I.   207, 215, 216, 217, 219,
    255, 343
Reunion I.   1, 202, 215, 216, 218,
    337
Rhodesia   45, 274, 275, 292
Rocky Mts.   52, 56, 277
Rossel I.   207, 216, 342
Rudolph, Lake   328
Ruwenzori Mts.   329

Sahara Desert   106, 111, 293, 299,
    (trans-Saharan migrants
    93–111)
St Andrew I. 223, 231, 346, 347, 348,
    351
St John I.   235, 352
St Kitts I.   288, 349
St Lawrence I.   136
St Lucia   223, 225, 237, 345, 348,
    350
St Thomas I.   350
St Vincent I.   223, 226, 227, 237,
    345, 346, 347, 350, 351
Salawati I.   330
Samar I.   342
Samoa Is.   207, 216, 343
San Cristobal I.   342
San Juan Is.   56
Santa Cruz I.   207, 215, 216, 218,
    219, 343
Sao Tome I.   202, 216, 217, 220,
    239, 337, 355
Sardinia I.   33, 89, 120, 284, 297,
    308, 309, 311

Scandinavia   20, 33, 35, 91, 261,
    293, 305, 313, 315
Schouten Is.   341
Scotland   291
sea birds   122, 124, 136-7, 166-72,
    320-2, 333
semi-species   38
Senegal   286, 294, 297, 299
Serengeti   151-4
Seymour I.   175, 180
Seychelle Is.   216, 337
sex differences   81, 82, 239-41,
    262-3
Siberia   30
Sierra Leone   298
Sikkim   72, 280
Singapore I.   161-5, 173, 214, 246,
    254, 255, 256
Socotra I.   337
Soepiori I.   204, 205, 206, 341
Solomon Islands   206-7, 213, 255,
    342-3, 354
Somalia   45, 274, 275, 300, 337
South Africa   45, 46, 47, 150, 286,
    290, 328
South Island, New Zealand   213
Southwest Africa   289, 292, 297
Spain   35, 89, 101, 105, 106, 284,
    292, 294, 295, 296, 298, 299,
    305, 306, 307, 308, 309, 313,
    319
speciation   1, 5, 6, 243, 253, 267-8
species diversity   1, 142-3, 234-9,
    248, 268-9
Spitsbergen   316
suboptimal habitats   256-8
subspecific adaptations   33-6, 87
Sudan   274, 286, 297, 400
Sumatra I.   73, 203, 204, 205, 213,
    339
Sumba I.   204, 205, 340
Sumbawa I.   204, 205, 216, 218,
    339-40
Sunda Is.   205
superabundant foods   7, 155, 264
superspecies   13, 201, 213
Sweden   20, 32, 34, 35, 75, 76, 246,
    256
Switzerland   25, 63, 305
Szechwan   69, 71, 72, 203, 272, 280,
    281, 338

Tagula I.   206, 207, 342

Taiwan I.   354
Tanimbar Is.   204, 205
Tanzania   138, 139, 144, 145, 151,
    152, 154, 274, 287, 290, 292,
    297, 298, 299, 322-6, 337
Tasmania   158, 159, 160, 161
Tenerife I.   9
Tennessee   54
territories, interspecific   127-9, 253
Tetipara I.   206, 342
Texas   49, 128, 276, 277
Thailand   68, 69, 70, 72, 281
Tian Shan Mts.   41, 66, 72, 272,
    280, 281
Tibet   41, 69, 71, 272, 280, 281
Timor I. (Mal. Arch)   204, 205, 340
Tobago I.   223, 344
Transvaal   292, 293, 297, 298
Tonga archipelago   226, 243
Tonkin   69, 70, 71, 73, 280
Tower I.   175, 182, 183, 184, 185, 187
Transcaspia   34
Transition Zone (N. America)   56, 65
Tristan da Cunha I.   198, 199
Trinidad I.   155, 156, 237, 332, 344,
    351
Trobriand Is.   329
Truk I.   208, 209, 216, 217, 218,
    219, 344
Tschad   299
Tunisia   33, 34
Turkey   31, 65, 298

Urup I.   39
U.S.A.   49, 52, 54, 55, 60, 65, 66,
    128, 129, 142, 259, 269, 277
U.S.S.R.   19, 20, 30, 31, 293, 313,
    315, 320, 321, 322
Asiatic   293
ultimate factors   5
Usambara   138-54, esp. 138-42,
    143, 235, 236, 322-6
Ussuria   338

Vellavella I.   206, 342
Venezuela   222, 223, 255, 344
Vermont   142, 143
Victoria, Mt.   39, 44, 69, 71, 272,
    281, 282
Virgin Is.   223, 230, 231, 232, 233,
    235, 345, 346, 350, 352

Virginia   54
Vosges, Mts.   96

Waigeu I.   157, 330
Wales   322
Wash, The   89
Washington State   55, 56
Wenman I.   175, 182, 183, 184, 185
West Indies   87, 222–42, 243,
      344–53, 354, 355
West Papuan Archipelago   329, 330,
      331

Wetar I.   204, 205
Yap I.   204, 205, 208, 209, 216, 218,
      219, 344

Ysabel I.   342
Yucatan   347
Yukon   55, 277
Yunnan   39, 41, 43, 44, 69, 70, 71,
      203, 272, 280, 281, 338

Zagross Mts.   33
Zambia   46, 287, 290, 292